AF691645

DIGITAL COMMUNITIES IN A NETWORKED SOCIETY

e-Commerce, e-Business and e-Government

IFIP – The International Federation for Information Processing

IFIP was founded in 1960 under the auspices of UNESCO, following the First World Computer Congress held in Paris the previous year. An umbrella organization for societies working in information processing, IFIP's aim is two-fold: to support information processing within its member countries and to encourage technology transfer to developing nations. As its mission statement clearly states,

> *IFIP's mission is to be the leading, truly international, apolitical organization which encourages and assists in the development, exploitation and application of information technology for the benefit of all people.*

IFIP is a non-profitmaking organization, run almost solely by 2500 volunteers. It operates through a number of technical committees, which organize events and publications. IFIP's events range from an international congress to local seminars, but the most important are:

- The IFIP World Computer Congress, held every second year;
- Open conferences;
- Working conferences.

The flagship event is the IFIP World Computer Congress, at which both invited and contributed papers are presented. Contributed papers are rigorously refereed and the rejection rate is high.

As with the Congress, participation in the open conferences is open to all and papers may be invited or submitted. Again, submitted papers are stringently refereed.

The working conferences are structured differently. They are usually run by a working group and attendance is small and by invitation only. Their purpose is to create an atmosphere conducive to innovation and development. Refereeing is less rigorous and papers are subjected to extensive group discussion.

Publications arising from IFIP events vary. The papers presented at the IFIP World Computer Congress and at open conferences are published as conference proceedings, while the results of the working conferences are often published as collections of selected and edited papers.

Any national society whose primary activity is in information may apply to become a full member of IFIP, although full membership is restricted to one society per country. Full members are entitled to vote at the annual General Assembly, National societies preferring a less committed involvement may apply for associate or corresponding membership. Associate members enjoy the same benefits as full members, but without voting rights. Corresponding members are not represented in IFIP bodies. Affiliated membership is open to non-national societies, and individual and honorary membership schemes are also offered.

DIGITAL COMMUNITIES IN A NETWORKED SOCIETY

e-Commerce, e-Business and e-Government

The Third IFIP Conference on e-Commerce, e-Business and e-Government (I3E 2003) September 21–24, 2003, São Paulo, Brazil

Edited by

Manuel J. Mendes
UNISANTOS, Universidade Católica de Santos
Brazil

Reima Suomi
Turku School of Economics and Business Administration
Finland

Carlos Passos
CenPRA, Centro de Pesquisas Renato Archer
Brazil

KLUWER ACADEMIC PUBLISHERS
BOSTON / DORDRECHT / LONDON

Distributors for North, Central and South America:
Kluwer Academic Publishers
101 Philip Drive
Assinippi Park
Norwell, Massachusetts 02061 USA
Telephone (781) 871-6600
Fax (781) 681-9045
E-Mail <kluwer@wkap.com>

Distributors for all other countries:
Kluwer Academic Publishers Group
Post Office Box 322
3300 AH Dordrecht, THE NETHERLANDS
Telephone 31 78 6576 000
Fax 31 78 6576 254
E-Mail <services@wkap.nl>

Electronic Services <http://www.wkap.nl>

Library of Congress Cataloging-in-Publication Data

A C.I.P. Catalogue record for this book is available from the Library of Congress.

Digital Communities in a Networked Society:
e-Commerce, e-Business and e-Government
Edited by Manuel J. Mendes, Reima Suomi and Carlos Passos
ISBN 1-4020-7795-5 eBook ISBN 1-4020-7907-9

Copyright © 2004 by International Federation for Information Processing.

All rights reserved. No part of this work may be reproduced, stored in a retrieval system, or transmitted in any form or by any means, electronic, mechanical, photo-copying, microfilming, recording, or otherwise, without written permission from the Publisher (Kluwer Academic Publishers, 101 Philip Drive, Assinippi Park, Norwell, Massachusetts 02061), with the exception of any material supplied specifically for the purpose of being entered and executed on a computer system, for exclusive use by the purchaser of the work.

Printed on acid-free paper.

TABLE OF CONTENTS

SECTION 4: AUCTIONS AND E-PAYMENT

SECTION 5: FUTURE ASPECTS OF COMMUNICATION

SECTION 6: INTERNET AND THE WEB

SECTION 7: ADVANCED PLATFORMS AND GRID COMPUTING

I3E International Program Committee

J. Adán Coello, Brazil
C. Arias Mendez, Chile
D. Avison, France
M. Bichler, Germany
W.Cellary, Poland
N. Cerpa, Chile
Y. Chen, USA
J. Cooper, Australia
R. Debreceny, Singapore
A. Economides, Greece
M. Funabashi, Japan
W.Golden, Irland
R. Grimm, Germany
F. Kamoun, Tunisia
C. Kirner, Brazil
A. Iyengar, USA
W. Lamersdorf, Germany
R. Lee, Holland
C. Linnhoff-Popien, Germany
T. Magedanz, Germany
M. Mendes, Brazil (General Chair)
A. Molina, England
J. Monteiro, Portugal
V. Ouzounis, Belgium
C. Passos, Brazil (Progr Co-Chair)
H. Pohl, Germany
R. Rabelo, Brazil
K. Rannenberg, Germany
B. Roberts, England
H. Rudin, Switzerland
B. Schmidt, Switzerland
M. Singh, Australia
J. Soto Mejia, Colombia
K. Stanoevska, Switzerland
C. Steinfield, United States
L. A. M. Strous, Holland
R. Suomi, Finland (Progr.co-chair)
P. Swatman, Germany
S. Teufel, Switzerland
R. Traunmueller, Austria
A. Tsalgatidou,, Greece
V. Tschammer, Germany
V. Tuunainen, Finland
H. Werthner, Italy
J. Wielki, Poland
H. Zimmermann, Switzerland
Y. Zhang, Australia

I3E Main Supporters and Sponsors

Centro de Pesquisas Renato Archer

International Federation for Information Processing

State Government of S.Paulo

Caixa Econômica Federal

Fundação de Estudos e Projetos

Fundação de Apoio à Pesquisa no Estado de S.Paulo

FOREWORD

Supporting a Revolution with Information

A Citizen-Centered State

Information is one of the basic resources of the society of the new millennium and therefore, a common asset in the realm of the government, the private initiatives, or the individual. Its generation, distribution and use should take place in a two-direction channel of easy access in order to be used productively by all.

Thus, the government should not only inform the population about the services it offers but also supply guidance about the use of the services offered.

That is the reason why the State has been changing its structure, so that it can increase efficiency and lower costs for the citizens. As a result, the use of information technology by the government has been a tool to facilitate such process.

The information transfer by electronic means has made the government to adopt a new style of administration, in other words, the e-government. E-government means a commitment with the use of information technology for the society. It will make possible the continuous improvement of the actions of the State focusing on the efficiency of the internal administration and establishing a system of information management to arrange internal processes and to speed up decision-making at all levels of the government. It also allows the establishment of an information network integrating the State Public Administration and the municipal and federal areas, and the Executive area to the Legislative and Judiciary areas, and facilitating democratic access to information by the citizens through its suitability to the socio-cultural reality of the majority of the population.

E-Government

E-government's aim is to place the government within the reach of all citizens increasing transparency and citizen's participation. Thus, the development of electronic government should promote universal access to government's services, integrate administrative systems, networks, and databases, and make such information available to the citizens via Internet.

In the last decade, the rendering of public services in Brazil has been changing substantially. Some aspects are the indicators of an increasing concern about the quality of the services rendered to the population: the proliferation of Customer Attendance Service in some state companies and the Ombudsman Systems in the majority of the public departments.

The recent practices of private companies that establish a relationship with their clients (CRM) have influenced the implementation of these channels of communication. Either effective or not, they have offered to the population the possibility of establishing an interaction with the public administration. On the other hand, with the advance of the democratic process in the country, the civil society has demanded, among many of their requirements, more transparency, speed and efficiency in the public administration.

Thus, we can observe, in all areas of the government, an increasing concern about projects of bureaucracy reduction (such as, the implementation of Programs of Bureaucracy Reduction of Federal and State Governments) and about initiatives that aim to shape the public services as the resulting products of administrative activities.

Following the same direction, the public departments in the last decade have been concerned about building indicators of attendance, implementing mechanisms of assessing productivity and quality, elaborating specific laws of protection of the user's rights.

In this context, during the 1990's, the creation of Citizen's Services Centers (e.g. Poupatempo) in almost all the Brazilian states (nowadays 23 out of 27 existing States), which gather several agencies carrying out services from any area of the government in a unique space, has created a great advance on the answering of the demands of the civil society: initiatives which have contributed to improve significantly the image of the public service in Brazil.

Before these initiatives, the public services were considered archaic places, where reign the image of bureaucracy, lack of information and explanations, bleak workplaces and services rendered with no respect and dignity to the citizen. Today, the Citizen's Services Centers have been transformed in paradigms of efficiency, effectiveness and respect to the citizens' rights not only for the public administration but also for the private sector.

However, the facility introduced by these Centers, contributing to the performance of hundreds of services in a single space, do not resolve the problems of the citizens presented by the specification of public services. Even when carried out in one single space, the citizens are required to present several times the same personal data and documents to the rendering of several services in these Centers. In its relationship with the government, the citizen assumes several conditions: as a driver, a worker, a family supporter, someone with criminal records, a taxpayer, a customer (of gas, electricity, etc). In other words, the rendering of each of these services depends on the database belonging to the different sectors of the public administration.

These sectarian databases, some of them built more than three decades ago, cannot respond to the new demands placed by the civil society that, as mentioned above, require a new type of relationship with the State. That means the need of a new structure of databases and information, which has the ability of incorporating these new demands and functionalities.

The significant public resources applied in the legacy, the difficult rescue of memory of transactional rules (not systematized or scarce documentation, absence of the assigned database programmers, etc) and the complexity of requirements and the used logic require the decision of how to solve the use and updating of the legacy simultaneously aiming the new demands by the current administrators of these systems.

On the other hand, we should consider that 90% of the public services rendered in Brazil are still in the presential mode. The rendering of services by electronic means, also do not solve the mentioned fragmentation of the citizen in the several categories in which he/she is required to be submitted, according to the service carried out. On the contrary, the public sites reflect the division in sectors and similarly to the presential mode, "force" the user to surf in several pages and to register several times the same demands to the rendering of the several services.

The great challenge presented in Brazil is the possibility of the construction of virtual citizen's services centers, where it will be possible, by the integration of the legacy systems, the access to public services and information without the obligatory repeated certifications and where it will be possible to establish a new form of relationship between State and Citizen unlike the current fragmented one.

Conclusions

We can highlight some decisive factors in the implementation and success of initiatives for the use of Information and Communication Technology and that have been transforming and revolutionizing the State Government:

- Unconditional support and incentive by the governor;
- General policies: not many and flexible - prioritizing connectivity and its activities; the intensive use of existing resources; the obligatory participation by the administrators and of whom produces the information or service; the use of Intranet and Internet to speed up the exchange of information and to eliminate administrative divisions and excessive hierarchies;

- Partnerships with suppliers under the guidance of the government;
- Priority on action and not on excessive planning, a willingness to learn from mistakes;
- The use of the existing legacy systems as much as possible to create new and better services;
- Flexibility to change;
- From the singular and anarchical spirit of the Internet, to stimulate and support the development of public servants´ and agencies' ideals and projects;
- Absolute priority for digital inclusion programs.

However, all the efforts for the use of Information and Communication Technology in the building of the e-Government will not be successful if the government does not prioritize the universality of access to electronic means to the entire population, especially to the poor classes. Only by this way, it will achieve its main objective, that is, the implementation of the Electronic Democracy - e-Democracy – which allows the effective integration and participation of all the SP State's 37 million citizens.

We are pleased to welcome IFIP I3E in our State, we recognize how important are the topics discussed and wish all a good reading of this book!

Roberto Meizi Agune
S. Paulo State Government
Dezember 2003

PREFACE

In the last years we have observed a accelerating evolution in the computerization of the society. This evolution, or should we call it a revolution, is dominantly driven by the Internet, and documented in several ways:

- The Information and Communication Technologies (ICT) bring, year per year, novelties: new processing architectures, new software methodologies, new systems and products, new communication networks. Distributed Processing Architectures spread in the Internet (e.g. Enterprise Distributed System, Distributed Object Computing, Grid Computing). Due to the proliferation of Platform and Middleware, some old software development approaches mature (e.g. MDA - Model Driven Architecture). In the field of Knowledge, the last years saw an interesting development of Metadata Techniques (e.g. based on MOF-OMG). Otherwise, representation of Knowledge and Semantic Processing, introduced in the past by the AI Community saw a strong push with the proposals of Semantic Web. And, without any question, the new communication technologies, bringing mobility, ubiquity and personalization, will change the ways in which individuals and public organizations perform their activities.
- The application fields of those technologies are expanding constantly transferring high benefits for the users, human beings (clients, consumers, citizens) and organizations (SME's and big enterprises, public administration in the spheres of federal, state and local governments activities). Not only do the technologies cause profound modifications in the enterprise structures , but also give new tools to the quest for new organizational forms that bring more productivity and the chance of survival in the new global world of commerce, business and government. In Electronic Business, enterprises build production networks and proceed to expressive reorganization of their internal activities. And in Electronic Government, still in its infancy, practically all nations in the world -rich or poor - search the way to use ICT, to reach efficiency, and to eliminate old problems such as corruption. It is not yet possible to foresee the impacts for the citizen but, by sure, the old democracy is being reshaped.

We assembled, in this book, several contributions towards our title of

"Digital Communities in a Networked Society"

with base themselves on the papers, contributions and ideas discussed during the 3rd IFIP Conference I3E eCommerce, eBusiness and eGovernment, which took place in September in Guarujá, SP, Brazil.

Conference proceedings were distributed, containing 52 papers selected by the International Program Committee. The present book is a posterior effort, where 16 papers have been selected by the IPC and 25 other papers were proposed but subjected to major revisions. From them, 14 have been selected for the book. Besides that, five of our distinguished Keynote Speakers submitted papers. And finally 9 new papers have been submitted after the conference, and 4 of them have been selected to this volume, by the Editors. The book was organized in 9 sections comprising 33 chapters.

We want to express our words of gratitude to all of those that somehow contributed to the success of the Conference and helped compiling this book. First of all, to the hundreds of authors that spent precious time, bringing their ideas and work to paper. We regret that so much of them could not find their place in the book and, in fact, we were obliged to disconsider very good contributions, because of evident space restrictions. Then, we are very grateful for the members (and their co-workers) of the International Program Committee, for their evaluations, suggestions and discussions. A very special word of gratitude goes to the members of the Local Organizing Committees that were not only indefatigable but also able to introduce the kind and warm human Brazilian way of handling things around the conference.

Finally we have to thank the entities that participated as Organizers, Supporters and Sponsors. We traversed a difficult economical situation in Brazil during the organization of the event. The hard conflicts in other parts of the world had a profound influence during the year of 2003. Nevertheless, many organizations could support us with special means and some were able to grant generous financial support. From these we thank, specially, different entities of the State Government S.Paulo (e.g. Secretaria da Casa Civil, Imprensa Oficial), the federal Bank Caixa Econômica Federal, the federal agency FINEP and the SP research state agency FAPESP. This book would be impossible without their direct support.

The Editors

Manuel de Jesus Mendes, Cenpra /Unisantos, Brazil
Reima Suomi, Turku School of Economy, Finland
Carlos Passos, CenPRA, Brazil

Dezember 2003

SECTION 1
e-Government

Chapter 1

E-GOVERNMENT – A ROADMAP FOR PROGRESS

Roland Traunmüller, Maria Wimmer

Abstract: E-government can transform and improve the entire scope of administrative action and the political processes. So e-government is both, vision of a future government and the reality we have to live with today. Sketching a roadmap may give us indications where we are heading. To begin with, e-government is not an objective per se; more it has to be seen as means in organizing public governance for better serving citizens and enterprises. This makes service provision essential. Reflecting the viewpoints of individual citizens (or of companies) is an obligation. When looking from outside, portals and forms of service delivery become key success factors. Moving ahead implies having an integrated view, clear strategies and concepts that are both innovative and feasible. Two guiding visions will have strong impacts on developments. First, a holistic approach is necessary to create work-processes and work-situations, as they are highly knowledge-intensive and rely on close forms of interaction between individual persons and IT. Next, knowledge enhanced government is a leading idea and management of legal/administrative domain knowledge becomes a decisive driver in governance. Designing for governmental applications touches several vital issues: transferring concepts and systems from the private to the public sector; making use of standards; safeguarding trust and security; enhancing usability. These lines have to be blended with an adequate management for change.

Key words: E-government, roadmap for e-government, knowledge enhanced government, holistic view

1. ADVANCEMENTS TOWARDS E-GOVERNMENT

1.1 A Comprehensive View of Modernization

Both, e-government and e-commerce are largely driven by the hopes and perspectives which the new wave of technology has prompted and the desire to renew national economies is another important driver of developments. E-government can transform and improve the entire scope of administrative action and the political processes. So e-government is both, vision of a future Government and the reality we have to live in today. Sketching a roadmap may give us indications where we are heading.

E-government is more than a new wave of administrative modernization. It means a permanent e-transformation opening up entirely new ways for public governance:

- Electronic Government concerns the whole scope of administrative action and the connected political processes. IT as an enabling force will enhance effectiveness, quality and efficiency of public action as well as its legitimacy.
- Thus legislature, executive and judiciary should be called to mind.
- The task of sustaining democratic deliberations (e-democracy and e-voting) becomes important.
- As governance is a vast concept one may discriminate different spheres: so according to the Speyerer definitions [4], [8] an inner sphere aims at novel organization and a thorough rethinking of the machinery of Government; an outer realm considers the changing roles of the state as well as a new balancing of public and private activities.

1.2 Systemic Features Distinguishing the Governmental Realm

When looking from afar striking correspondences appear between e-government and e-commerce. Both involve reengineering and integrating flows of information, of money and of goods, and they exhibit a trend toward spatially distributed organization. And both started with customer interface problems but soon came to dig deeper in an effort to overhaul the businesses in question completely. Not to forget that both are only successful if there exists a vision and a novel business model as matrix for shaping reality.

There is an expression coined by Wallace Sayre "public and private management are fundamentally alike in all unimportant respects". Therefore, no wonder that on closer inspection more differences appear: the specific tasks of government, the role of law and negotiation, the special significance of knowledge (see next paragraphs). Accordingly, feasibility needs attention for each individual case when transferring concepts and systems. Often even minor distinctions may exert essential influence on design. Reproducing concepts and systems from the commercial domain has to be done with thoughtfulness and sensitivity.

The ways in which branches of Government work are manifold. Often, they differ from what can be found in the private sector. The variety and diversity of policy fields and of forms of action in state, politics and administration is high. Legal and political preconditions vary and the context and situational factors are influential. Thus a mere replication of commercial concepts and systems will not suffice. Moreover, systems have to cope with distinctiveness of the governmental realm. Here some demarcations are shortly outlined [11]:

- An extraordinarily complex goal structure distinguishes the public sector from private business.
- Legal norms are a standard vehicle of communication; yet they have to be supplemented by legal interpretation, negotiation and consensus building.
- Equality before the law calls for social inclusion; e-identity is needed in nearly all administrative transactions.
- Legal norms give particular meaning to administrative structures posing several limitations on process reengineering (protecting privacy, safeguarding legality etc.)
- Public Administration mostly works via a complex tissue of cooperation involving quite many acting entities (which is rather contrary to the private sector).

This contribution concerns the topic of e-government in general and as view on the state of the art. As basic reference, the reader is referred to some recent collective volumes and conference proceedings [1], [2], [7], [4], [8], [10]. Chapter 2 describes e-government as the novel paradigm. Following, chapter 3 details some routes to pass through: portals, processes, cooperation and knowledge. Finally, chapter 4 sketches a plan for moving ahead.

2. A ROADMAP FOR SUCCESS

The alarm bells ring as take-up of e-services remains low. As a result, defining strategies for e-government is an urgent task. Since e-government is a new paradigm, strategies will be distinct from previous ones.

New Public Management [5] that dominated the last decade had brought considerable change to many branches of public administration. Now e-government has emerged as a paradigm that builds on NPM, however goes far beyond. Especially e-government deals directly with the administrative processes themselves. To say it with other words: NPM focuses primarily on better ways of managing processes; in e-government, the processes themselves are reengineered. Changing paradigms means changing strategies and criteria - a new roadmap for achieving success is needed.

Recognizing the way to success needs above all a point of view that offers global perspectives. From such a vista a roadmap for success can be sketched:

- Considerations have to start with taking a holistic approach. This means integrating several aspects: users, technology, organization, law, knowledge, culture, society and politics.
- Next the whole machinery of Government comes under scrutiny: providing administrative services, running work processes, and modes of cooperative work have to be defined in a new way.
- In addition, future Government will be knowledge enhanced and innovative solutions have to mirror that fact.
- All these redesign efforts - public services, processes, cooperation and knowledge management – will lead up to rethinking the institutional structures of government.
- For such changes a sound engineering approach is essential. This is a broad claim so let us mention just some key requests: building a secure and reliable infrastructure, developing standards, adequate interface design.
- Competent change management and improving the innovative capacity of the public sector is a must.

3. ROUTES TO PASS THROUGH: PORTALS, PROCESS, COOPERATION, KNOWLEDGE

3.1 Portals Open the Way to Service Provision

Portals for delivering services to business, individual citizens and communities reflect a view from outside. Portals are of prime concern, however as Reinermann stated already years ago (IFIP World Conference 1998) "This is only the tip of the iceberg". Hence design has to aim at the entire scope of administrative action. So in designing electronic service delivery one has to regard processes from two sides: from the standpoint of the citizen and from the view of the producer of the service. There are typically five stages (with some parallel to commercial services) which have to be looked at. Seen from the citizen point of view these are: information, intention, contracting, settlement, aftercare.

Low user take-up of e-Services has become a main problem. It shows that resistance to change includes many stakeholders and one has to answer the question: What has gone wrong with e-government projects? Hence low uptake is a key issue and in the language of the users the culprit has a name: measly usability. In terms of user-friendliness many existing portals are far off from being satisfactory. Many examinations and assessments have revealed deficiencies. Long is the list of shortcomings: a general lack in targeting the audience; an inadequate and inconsistent design lacking of comments and adequate examples; a sloppiness in maintenance showing unreliable and outdated pieces of information. It is a distressing picture that comes out from in-depth analyses of typical interaction processes: users cannot cope with the logic of administrative thinking, other users do not comprehend the administrative jargon and some other clients who pilot helplessly through the jungle of information.

3.2 Redefining Governmental Processes

Online One-stop Government means that external service structures are adequately mapped to the internal process structures of public authorities [11]. Therefore, the addressee's perspectives have to be complemented by a restructuring of the business processes. Process design has to break new ground by taking into account several aspects:

- Different locations of service production and delivery
- Organizational front office / back office connection
- Combining processes according to life situations
- Including distinct processes from strict workflows to collaborative decision-making

Process reorganization in the public sector may often have to stop short of established structures; but finally they will lead to rethinking the institutional structures of Government. In many respects the legal framework of these processes has to be changed. Also new institutions may emerge which fit the new ways of producing and delivering public services.

A further point is that design has to consider the very different ways of administrative processes. For each of them, IT support will rather be different:

- Recurrent and well-structured processes
- Processing of cases: individualized decision-making
- Negotiation processes and consensus finding
- Weakly structured processes in the field of policy-making

Process structure is not the only perspective when discussing the changes. Two complementary perspectives are of equal importance: cooperation and knowledge. This leads to the next two sections.

3.3 Strengthening a Broad Cooperation View

The cooperation view is of special importance to those activities that are related to higher order administrative work. They include e.g. negotiation, consensus finding, planning and policy formulation. Especially for the higher ranks of bureaucrats such mode of work becomes prevalent. However, not only intra-governmental activities need extensive cooperation, when communicating with citizens such modes of work occur as well. Examples are plentiful: negotiating with citizens, giving advice in complex questions, mediation – they all have to be seen as cooperative settings.

So, what has to be sustained is cooperation in the broad. Support of computer-mediated cooperation in a comprehensive sense means sophisticated tools, multiple media for these contacts become a must. To give a flavor of the capabilities, some illustrations are added:

- Meeting as well as related activities take hold of a substantial part of administrative work. Many occurring activities are cooperative in nature and claim for IT-support.
- First, the meeting activity per se may be performed via video techniques – so economizing on travel costs and time.
- Next, many activities associated with meetings can be largely improved by tools using multimedia. Examples are plentiful: clarifying procedural questions; scheduling of meetings and implied sub-activities; supporting the agenda setting and spotting experts, supporting brainstorming sessions, structuring issues etc.
- For the illustration of advanced systems using multimedia, we regard a future scenario "citizen advice for solving complex questions". A citizen may go to mediating persons at the counter of public one-stop service shops. The mediators will use the system with its diverse repositories. In case the issue is too complex it is possible to invoke further expertise from distant experts via a multimedia link between the service outlet and back-offices: dialogue becomes trialogue.
- As the accessed expert himself may use knowledge repositories, finally, human and machine expertise become intensely interwoven. So this example leads to the next issue: knowledge enhanced government.

3.4 Knowledge Enhanced Government

In a novel concept of governance the role of knowledge becomes dominant. Building a modern administration with novel patterns of cooperation is tantamount to changing the distribution of knowledge. Redistribution of knowledge has to be designed and orchestrated carefully. Managing knowledge becomes a major responsibility for officials. All these facts point to the concept "knowledge enhanced Government".

Prospects for knowledge management in Government are remarkable from the point of demand: nearly all administrative tasks are informational in nature, decision making is a public official's daily bread, and for any agency its particular domain knowledge is an asset of key importance. Such a new direction will engender considerable progress:

- The focus of attention is shifted away from a discussion of structures and processes towards issues of content. It reaches the very heart of administrative work: making decisions.
- In some aspect, a regained focus on decision-making will help to propagate comprehensive systems thinking.
- Eventually, a better management of knowledge will lead to forms of "smart government". Knowledge derived from previous action or gained through policy evaluation will be fed back to policymaking in an effort to better target policies.
- Management of legal and administrative domain knowledge is a critical factor in governance. In addition, a deeper understanding of the connections between processes and knowledge will improve design. In the public agencies of the future, human and software expertise will become intensely interwoven – knowledge enhancement at its best.

4. A PLAN FOR MOVING AHEAD

4.1 Building on a Sound Engineering Approach

A sound engineering approach is indispensable to bring about an IT-induced modernization of public administration and public governance. At the bottom level this means a suitable IT infrastructure for unimpeded communication and cooperation meeting high demands on availability and security as well. At the application level objectives are smooth cooperation, high usability and a design integrating all these before mentioned aspects: citizen service, process reorganization, cooperation and knowledge enhancement.

4.2 Interoperability and Standards

If one compares the public and the commercial domain one can see both, communalities as well as differences. The former ones occur at the technical level; the later ones at the application level [13]. Standards for applications become an issue in its complexity significantly surmounting the private sector. Further on, standardization has to be seen with a broad focus including several issues: establishing a common understanding of processes, building on widespread administrative concepts, ensuring interoperable platforms, having a workable administrative domain ontology, defining formats for data interchange. Standardization is a huge task. Yet in the long run, all partners involved (public agencies, software industry, private companies) will gain. There are already some advanced fields such as e-procurement, however, the core administrative processes are still far away from that.

A common governmental mark-up language has to be developed acting as a means for defining governance-specific content. Among others, this is a prerequisite for the transport of data from back offices and from the distributed information repositories serving them, to both (virtual and physical) front offices which deliver the services produced elsewhere. Especially for cross-border e-government having such definitions is a must! These standards will be built on XML combined with domain ontologies. For domain ontologies a rich kit of methods for knowledge representation already exists (taxonomies, semantic nets, semantic data models, hyperlinks etc.). Present deficiencies in this field are a problem of praxis often caused by lack of commitment.

4.3 A *conditio sine qua non*: Safeguarding Trust, Security and Privacy

Quite similar to last issues, differences occur at the higher level. Requests are more strict since the e-identity is needed in all administrative transactions and since wrong passports may have more serious consequences than bouncing checks. In addition, taking the point of the users, informational guarantees and the trust in the system becomes crucial. Delivering electronic services will largely depend upon the trust and confidence of citizens. For this aim, means have to be developed covering a sole purpose: achieving the same quality and trustworthiness of public services as provided by the traditional way. Regarding the level of systems design, fundamental requests have to be met:

- Identification of the sender of a digital message
- Authenticity of a message and its verification
- Non-repudiation of a message or a data-processing act
- Avoiding risks related to the availability and reliability
- Confidentiality of the existence and content of a message.

4.4 Don't Forget the User – Enhance Usability

Speaking on portals a long catalogue of shortcomings has been listed. Usability is a main concern and it can be improved in several ways. One is building on past experience (and common sense as well). Some examples that even plain rules will benefit are below:

- The prime obligation is: "Stress usability - not alone visibility".

- "Less is more" and "Keep it straight and simple" are sayings that can be applied to design. They will match because overloaded or too complicated presentations are a nuisance.
- Further on it may be wise neglecting a drive to perfection. Designers have to avoid the widespread mistake of shifting too much burden to the client.
- In addition design will be successful when using more analogies.

Nothing against folks wisdom and common sense, but there are complicated interaction processes needing a deeper analysis. Citizen contacting agencies for advice in complex cases is such an issue needing closer inspection. Often the concrete situation is so that design has to resolve rather conflicting demands:

- the citizen's requests are commonly posed in a rather urgent situation,
- there may occur need for an in-depth explanation in an unambiguous way,
- the explanatory capabilities of the system are limited,
- interactions are connected with a high translation effort (i.e. transforming demands of the everyday world in the legal-administrative jargon and vice versa).

In case of the example of giving advise to citizens, design has to use several means. One would construct program clarifying dialogues, and describe illustrative scenarios. Also detailed knowledge (on both, on the field in question and on the interaction) can be embodied in software agents. All this works in actively helping users in accomplishing their tasks. Finally, very advanced future design will result in intelligent multi-lingual and multi-cultural personal assistants being integrated in electronic public services portals.

4.5 Change Management – The Key to Success

Change management is the key to success urging for cooperative efforts of a wide range of actors from administration and software industry. Particularly for the public sector, a quantum leap in the innovative capacity is asked. It starts on the political level with a strategic thinking and creating advanced infrastructures. Other critical success factors include best practice-evaluations and guidelines derived from ground-breaking projects. Competent change management means empowerment of staff and starting a remarkable qualification initiative. Cultural change and dissemination of know-how become crucial as well: the old egotistic behavior of shielding information, knowledge and process know-how has to be cast off. Eventually a new way of thinking will emerge with information sharing and cooperation as guiding stars.

5. THE ROADS AHEAD ARE FAR AWAY FROM BEING SMOOTH

Those who travel the road have to overcome many obstacles: bureaucratic attitudes and historical legacies, inertial institutions and impeding regulations, time and budget constraints. Yet those traveling the roads will be rewarded when they closely perceive the impending e-transformation of society. For them, the journey might become an overwhelming experience. There is chance and opportunity – we have to take advantage of the kairos of the moment.

REFERENCES

1. Galindo, F., Quirchmayr, G. (eds.), Advances in Electronic Government. Proceedings of the Working Conference of the IFIP WG 8.5, Univ. of Zaragoza, 2000
2. Grönlund, A. (ed.), Electronic Government - Design, Applications and Management, Idea Group Publishing, 2002
3. Leitner, C. (ed.), eGovernment in Europe: The State of Affairs, Report of the European Institute of Public Administration, Maastricht, 2003 (to appear)
4. Reinermann, H., von Lucke, J. Speyerer Definition von electronic Governance. Forschungsinstitut für Öffentliche Verwaltung Speyer, Online Publikation, 2002, http://foev.dhv-speyer.de/ruvii/SP-EGvce.pdf
5. Schedler, K., Proeller, I., New Public Management, Verlag Haupt, Bern, 2000
6. Traunmüller, R. (ed.), Electronic Government. Second International Conference EGOV 2003. Springer, LNCS # 2739, Heidelberg et al, 2003 (to appear)
7. Traunmüller, R., Lenk, K. (eds.), Electronic Government. First International Conference EGOV 2002. Aix-en-Provence. France. September 2002. Springer, Heidelberg 2002
8. von Lucke, J., Reinermann, H., Speyerer Definition von Electronic Government, Forschungsinstitut für öffentliche Verwaltung, Speyer, 2000, http://foev.dhv-speyer.de/ruvii/Sp-EGov.pdf
9. Wimmer, M. (ed.), Knowledge Management in e-Government. Proceedings of the 3rd International Workshop of the IFIP WG 8.5, Schriftenreihe Informatik 7, Trauner Verlag, Linz, 2002
10. Wimmer, M. (ed.), Knowledge Management in Electronic Government. 4th IFIP WG 8.5 International Working Conference Proceedings, Springer LNAI # 2645, Heidelberg et al, 2003
11. Wimmer, M., Tambouris, E. Online One-Stop Government: A working framework and requirements. In Traunmüller (ed.), Information Systems: The e-Business Challenge. Proceedings of the 17th World Computer Congress of IFIP in Montreal, Kluwer Academic Publishers, Boston et al, 2002, pp 117 - 130
12. Wimmer, M., Traunmüller, R., Lenk, K. Electronic Business Invading the Public Sector: Considerations on Change and Design. In Proceedings of the 34th Hawaii International Conference on System Sciences (HICSS-34), 2001
13. Wimmer, M., von Bredow, B. A Holistic Approach for Providing Security Solutions in e-Government. In Proceedings of the 35th Hawaii International Conference on System Sciences (HICSS-35), Hawaii, 2002

Chapter 2

REDUCING NORMATIVE AND INFORMATIVE ASYMMETRIES IN FISCAL MANAGEMENT FOR LOCAL ADMINISTRATIONS

M. Carducci, M. A. Bochicchio, A. Longo

Abstract: Fiscal incomes are vital for Governments, both for central and local agencies, therefore on-line fiscal services will play a key role in the e-Government perspective. The creation of citizen-centered fiscal e-services, however, requires a new citizen –centered institutional and juridical context to be effective. The current Institution-centered scenario, based on the authoritative approach, is in fact unaware about the active citizens' role in the e-Government perspective, as well as about the deep impact of the information and communication technologies in the citizen s' everyday life. The paper describes how the extensive application of the regulative approach can be synergic to the extensive adoption of information technologies and user centered e-services to reduce these "normative-informative asymmetries".

Key words: Institutional change, IT-driven modernization of public governance, user centered design

1. INTRODUCTION

In recent years the diffusion of communication networks and distributed applications allowed the development of new interaction paradigms in Public Administration, under the collective name of e-Government, meant also as a way to organize public governance for better serving citizens and enterprises on a comprehensive scale. As a consequence, the basic outline of an e-government vision has recently emerged and governments have taken promising steps to deploy e-government services both in USA and in Europe [1], even if much remains to be done if this vision is to be broadly realized, to create innovative services within a coherent system of juridical and economical rules based on these new technologies and concepts.

Innovating the Public Administration, in fact, is not just the union of various tasks (innovating institutions, reengineering administrative processes and using innovative technologies) performed in isolation, because of the additional complexity coming from their coupling [1]. The crucial point is to foster the consciousness of the holistic approach to integrate and extend models, design methodologies and techniques to face the new e-government challenges. This paper is focused on the innovation in Local Public Administrations (LPAs in the following) of federal (or regionalist) countries, like Italy, Spain, Germany and Brazil, to give an integrated and trans-disciplinary answer to the following questions:

1. how to promote an effective e-government approach in LPAs, in the current context of “institutional uncertainty” about the role of the LPAs toward the citizen’s community?
2. in the current devolution panorama, which new options are imposed and which new tools are given to the LPAs to improve their relationship with citizens?

So the present paper aims at three main goals:

1. to analyse how and how much the normative autonomy of LPAs can improve the efficiency and the effectiveness of their administrative processes and services to citizens adopting suitable e-government approach;
2. to develop an institutional know-how oriented to the holistic innovation, able to actively involve the institutional stakeholders;
3. to demonstrate how to organize an effective e-Government solution based on the “regulatory approach” and on the “information” of citizens and Administrations.

We decided to orient our approach to the Local Administration because of their very poor performance in several key sector: in Italy the official evasion figure of estate tax is around 30% while the fiscal contentious between citizens and LPA has reached a critical level [5]. Moreover, small and medium LPAs are often overwhelmed by outdated and bureaucratic employees culturally unable to manage the complexity of the new institutional and technological scenario. We chose to develop an egovernment framework to achieve better performances in their core services (e.g. fiscal services, Municipal-knowledge based services, governance services etc.) rather than only to create new auto-referential services solely based on novel technologies.

The research experience proposed here comes from the collaboration among the University of Lecce, the Municipality of Taviano and a local private partner in the South-East of Italy, with the aim at supporting the LPA to reduce the normative and informative asymmetries in local taxation. The research activity, synthetically described by "*e-government as new-government*", has been positively evaluated by the Italian Ministry of Innovation and Technologies (Agreement Protocol signed in Lecce, on Dec. 23, 2002).

The paper is structured as follows. In Section 2, we present the juridical and technological context and the motivations at the root of our research. Section 3 discusses the reference framework we adopt and the main related tools. In Section 4 we present some results we achieved applying this framework in a Municipality. Section 5 concludes the paper and depicts some further research developments.

2. CONTEXT AND APPLICATIONS

2.1 The change in normative techniques

In the perspective of the institutional dimension of technology innovation, the unclear distinction, wide spread in the European legal tradition, between law and regulations creates uncertainty [6]. In fact, in the Law the decisional process takes place as authoritative form in the circle of command, with informal and uncontrollable behaviours. The result is a "mono-directional construction" aimed at reaching specific goals. On the other hand, the regulation is characterized by its practice, so it has conditional and "multidirectional" nature. It is based on the "what if" principle, with hypothetical not executive clarifications and it depends on the position of the single actor.

In the Italian institutional history about the "administrative simplification", from Law 241/1990 (regarding administrative proceedings) to Law "Bassanini" (law 59/1997, 127/1997, 50/1999), Local Public Administrations have tried to be modernized in regulations. However especially LPAs keep on acting in informal ways, through uncontrollable behaviours, inherited from consolidated traditions or from formal authoritative procedures, inspired to the mono-directional discipline, little sensitive to the information access and control. The "multi-purpose" nature of LPAs is the cause of the decisional uncertainty of LPA about e-government. E-government implies two risks for LPAs: the "auto-reference" of technology innovations, if operating in an unchanged institutional context, disciplined by authorities [7], and additional administrative decisional charges with the subsequent overlap of old behaviour with those determined by the use of new techniques[1]. In Italy, the "Analysis of Regulation Impact" (A.R.I.), introduced by Law 50/1999, answers to the first risk, and the "simplification risk", introduced by Law 59/1997, answers to the latter.

2.2 The technological scenario in LPAs

The use of ICT can produce radical improvements in administrative procedures, if it is supported by the optimization of the procedural iter. The normative techniques of regulation and simplification pursue the goal of "*regulation in simplification*" [8], which means:

- eliminating useless procedural steps unnecessary to decision making;
- reducing functional interferences between procedures regulated by different norms, unifying them in a unique procedural flow with one regulation;
- rationalizing the communication processes among the figures involved in decisions and creating a unique and consistent interface with the citizen;
- promoting the widespread access to decisions and their effects.

The user centered approach in designing e-services and on-line applications is the corresponding facet in the software development community.

Actually, providing citizens with e-services is a hot research topic today, in particular researchers focus their attention on the implementation of a single point of access to public services and information, the development of integrated platforms, which will allow the public sector to provide citizens, businesses and other public authorities with information and public services structures. Examples of research and business applications and methods for integrating heterogeneous legacy information systems are extensively described in [3]. Nevertheless, the scenario is not homogeneous, and from an informal survey of the current context in the Southeast of Italy, we found:

- inconsistent and uncoordinated organizational growth with hundreds of processes overlapping in the years, with outdated organization solutions;
- incompatible ICT solutions and a correspondent inefficient usage of ICT
- lack of horizontal and vertical communication in the institutional structure.

Other considerations related to the local taxes' management process contribute to better understand the poor quality of data owned by the Local Administration. In summary, our experience is that:

[1] An example is the "computer protocol" that coexists with the "paper protocol" and the plurality of offices that rule it

- Several sources may provide data to the system with different quality, according to the source (manual data input, data copied from system to system, etc.);
- Data is often duplicated either due to the poor definition of the business logic (or bad bureaucratic processes) or to technical reasons, or simply for convenience.
- The semantic relationships among information don't exist, making difficult to enforce the integrity constraints and to guarantee the value correctness.

3. V.I.O.L.A.: THE VIRTUAL INCOME OFFICE FOR LOCAL ADMINISTRATION

The holistic approach in e-government creates new processes and situations, as they are highly knowledge intensive and they rely on the strict interaction between people and IT. Moreover it is related to the Knowledge enhanced government. These points imply IT must reach the heart of the administrative work: taking decision. So the management of legal/ administrative knowledge becomes a decisive driver in governance. At the same time, designing governmental applications and services touches two issues: coupling concepts and systems and making use of standards. So the methodology must combine Legal Drafting and engineering techniques, developing normative and technological tools in an integrated fashion to reduce the gap of the normative asymmetry, cause of the informative asymmetry.

Our research starts from the upsetting of *authoritative discipline* experimenting the impact of innovation as *decision regulation* in real LPA cases. As experiment a single mission has been chosen, derived from the Constitutional reform of the Constitutional Law n. 3/2001 which amended the whole Title V of the Constitution: this is a "constitutional innovation", legitimating the related institutional, organizational and technological innovation. According to the new Title V of the Italian Constitution, Municipalities have normative autonomy in the organization of powers and functions and in the standardization of process adequacy in providing services. Moreover for the first time in Italian history Municipalities have taxation autonomy to finance their own functions and to "measure" the fiscal capacity of the territory. This incentive is particularly relevant for ICT.

Before the Constitutional reform the LPA situation (see [10]) in technological innovation consists of two main models: *insourcing* or *outsourcing* ICT services . In the first approach the use of technological tours has been subjected to unmodified informal behaviors and to authoritative bureaucratic formalism. In the latter case externalisation made providers de-facto owners of data and knowledge, acquiring dominant position in IT innovation. In both cases LPAs lacked of any independent position in the entrance of new technologies and they have used (or have been used by) unidirectional approaches without carrying about the delivery and the economic exploitation of information access.

Nowadays Municipalities, to be autonomous, must "know themselves", they must be conscious of their heritage of experiences, usual procedures and functionalities in delivering services, of their ability in understanding citizens needs, of checking their performance, just in order to change.

3.1 The Local Taxation

The single mission dimension chosen for the research is local taxation, because taxation is a decisional field where the use of innovative technology can be widely spread, but the resistance to change is strong. Moreover the new art. 119 of the Italian Constitution make the decisions of local taxation dependent on the Municipal area, which becomes a relevant informative factor both to make the decisions legal [11] , and to define and calculate the "citizen's fiscal capacity" (art. 119 line 3), as parameter of State subsidies for the equalized regulation.

So the information about the taxpayer and his relationship with the his territory becomes the *trait d'union* in the local fiscal discipline and in the management of Municipal data archive to allow State interventions. In conclusion, the Local Town Council should regulate its internal process, in order to create the data archive for itself and for the State.

3.2 The tools based on the framework

The considerations developed so far have led to the creation of an integrated group of legal and technological tools to enable an effective local taxation management. To integrate them and in order to teach institutions how to use them in LPAs' everyday life, a related organizational structure is needed, a *Contact Center*, where the fiscal problem is faced through different disciplines, in homogeneous legal and informative "environment", with the support of two complementary instruments called:

- TUnifET (*Testo Unificato delle Entrate Tributarie*, that is Unified Sheet of Fiscal Incomes)
- SIFET (Sistema Informativo della Fiscalità e dei Tributi, that is Informative System of Taxation and Tributes)

3.2.1 TunifET

TUnifET is a legal code to regulate fiscal proceedings, that are identified through activities, structures and information for managing relationships with taxpayers. So its "subject/object" is not the tribute as citizen's unilateral duty, but it is the fiscal relationship between Administration and citizen, built through the acknowledgment of the right to be informed and of the data which administrative decisions are based on. The TUnifET normative structure is open, thanks to the use of adaptable clauses. Using the TunifET regulation the *Contact Center* works as the institutional place of the informative interaction between user and supplier in the fiscal relationship. The TunifET foundation lies in the above mentioned article 119 of Italian Constitution. The interaction between the taxpayer and the *Contact Center* enables to

- correct data held by Administration, to avoid mistakes;
- know the expectations and good faith of Taxpayer about fiscal questions concerning him, in pursuance of Law 212/2000;
- remove informative asymmetries between Administration and users, which can create useless burdens or wrong expectations.

In other words, also in the perspective of *Legal Drafting,* the *Contact Center* puts the self-correction approach of *FMEA* (*Failure Mode and Effect Analysis)* in action. As the criteria of good faith and legitimate expectation can't be traced back to taxpayer within predefined control schemas, because such templates are founded on the interaction between human beings and events and events are often the consequence of activities or behaviors of Administration, the regulation must distinguish its objects from activities: the first are represented by fulfillments and regulated conditions according to the law, the latter by relationships that spontaneously take place within the "events" happened in the specific taxpayer's case.

These relationships are not one-way disciplined by Administration, but they represent the occasion to personalize the appliance of the regulation on the basis of the specific case, enriching the experience and the institutional learning in building relations between administration and citizen.

The subsequent effects can be summarized as follows:

1. self regulation of the fiscal relation on the field of information reciprocity between the Administration and the taxpayer inside a unique regulative process schema
2. elimination of useless, ineffective activities in taxpayers requirements from the normative discipline *(regulation in simplification)*
3. tuning of the process with respect of actual users' behaviors

4. knowledge of Administration's clients, represented by local taxpayers

So, it is possible to estimate the attitude of people to pay taxes, not exclusively through juridical factors, which is particularly important in the panorama of the new Constitutional autonomy. In fact the criterion of "fiscal capacity per inhabitant", recalled by Italian Constitution, requires the knowledge of the taxpayer on the basis of his incomes and personal situations, activities and events. In other words, the tuning of TunifET has "diagnostic" aims instead of regulatory, because it promotes the attitude to pay of taxpayer according to reasonable criterions by mutual consent of information and data sharing.

3.2.2 SIFET

SIFET assures the homogeneity of the informative settings. In order to know itself and its territory and to build the tools to measure its performance, the Municipality must be conscious of information and knowledge heritage owned inside, eventually completing and complementing it with external sources. The goal is to expose information owned inside, asking citizens and enterprises to complete it. New technologies (Internet, mobile, etc.) are very effective for this task, because they support the development of real-time/near-time services. To enable this instant access/instant response to take place, a blending of Ubiquitous Web applications with data from legacy archives is needed, requiring design methodologies borrowed from both database and hypermedia communities. With reference to local taxation, the ICT goal has been the design and the implementation of the *Virtual Office of Incomes,* a virtual center of real time aggregation of Municipal incomes, to give it the chance to know its actual entrances, to plan the budget on "certified" data and to start creating the Municipal database (an integrated virtual archive storing all information belonging to the Municipality, its citizens, its estates). The system for managing and facilitating the heterogeneous data transformation and integration is based on techniques of database integration design, while the design of the Web application for the management of ICI – citizen side - is based on the UWA framework (borrowed from the hypermedia communities) [2]. Different design methodologies are used according to the application views: the new services, based on UWA conceptual framework, are designed in a user centered perspective, whilst we adopted suitable models and technologies to face database integration and process design issues. The architecture to clean and integrate information sources (fiscal data, demographic data, cadastral data, phonebooks, etc.) is designed around the idea of pushing data through a pipeline of pre-defined processing blocks.

The interconnection path among blocks represents the steps to solve the specific problem of cleansing Italian fiscal data. The integration strategy is based on a logical integration of sources, keeping physically distinct the databases. The user centred design methodology, generally recognized as very effective in improving the usability of application, aims at reducing the informative gap necessary to approach the tools. Data integration is constantly monitored by the Contact Center according to users'/citizens' requests in the TunifET regulatory design. Beside, through the contact center the digital gap is reduced, because users/citizens are initiated to use multichannel communication techniques with the Administration, in order to make the informative certainty easier , which is necessary condition to reach the income certainty.

4. VALIDATING THE APPROACH: THE TAVIANO PROJECT

The town of Taviano (about 12.600 citizens) has 11.500 buildings and 17.500 lands. The results of our approach impact both the Municipality's management and citizen services. Since summer 2001 the main aims of the present Administration have been the citizen right's protection, fiscal equity, autonomies development, financial resources' review, reform of the Municipal autonomy's resources.

To meet these goals the Municipality has singled out technology and juridical innovation as an opportunity of autonomy exploitation and citizen relationship development. The first step was to design and support an effective tax management process (the *Virtual Office of Incomes*), in order to give the Municipality the chance to know its actual incomes and to plan the budget on "certified" data. We used the tools described in the previous section and at the end of 2002, the results achieved have been the following:

- the reduction of financial advance in 500 K€
- savings in financial charges in 30 K€ (from 70 K€ in 2001 to 40 K€ in 2002)
- higher entrances of 100 K€ (10% better than in 2001), without any increase of taxes' rates
- savings of 35 K€ by direct management of taxes.

From a technical perspective, to support efficient and effective services to users (both outside and inside the Municipality), the first issues we run into have been the evolutionary maintenance of legacy systems towards the new services. System assessment showed the unfeasibility of integrating the new services within the old systems and required deep data quality assessment evaluation and the development of a systematic approach to clean and integrate information, since data necessary to set up the Virtual Office of Incomes had to be extracted from paper and unreliable flows coming from internal legacy applications or other institutions .Several sources have been cleaned and integrated to understand and assess the fiscal position of each citizen and his own estates.

Due to the nature of local taxes, the integration step has been two folded:

- given a year information flows coming from different sources have been integrated to get the picture of the citizen's ownerships and payments (vertical integration)
- given a tax, several years of the same flow have been integrated to assess the citizen tax position

Proper cleaning methods [see 4], customized on the Italian case, have simplified the clerical review, reducing the need of manual inspection and the related costs by 70%. From the Administration point of view, this tool is very powerful to discover fiscal evaders and misalignments among data owned by the Public Administration about the citizen and his estates, highlighting history and discrepancies in data.

The left side hand of Fig.1 shows a sample of the user interface of the demonstrator supporting clerks in matching different information flows for the discovery of tax evaders: given a year and a street, all the estates are displayed. Selecting one, integrated data about the estate and its owners, from the Land registry's office, from the ICI statement, from the National Electrical Agency, are displayed. The same set of information is the basis to build innovative online services to citizens. The user centered design for Ubiquitous Web applications had highlighted the taxpayer's requirement for a synthetic and clear view of his position. The right hand side of Fig 1 reproduces a screen shot of ICI online Web application enabling citizens to manage ICI tax. Once logged in, the taxpayer can display its ownership in the Municipal territory, and securely communicate a variation of his position.

We used these prototypes to test the whole framework and to evaluate the feasibility of the technical choices (the network, security, development environment, programming languages).

From the normative side, the regulative environment offered by TunifET has allowed the construction of a relationship between users and Administration not founded on duties and prohibitions but on actions of "incentive regulation" and legitimization of taxpayers' position, according to the information they provide in order to correct Municipality's internal data and to tune its internal processes.

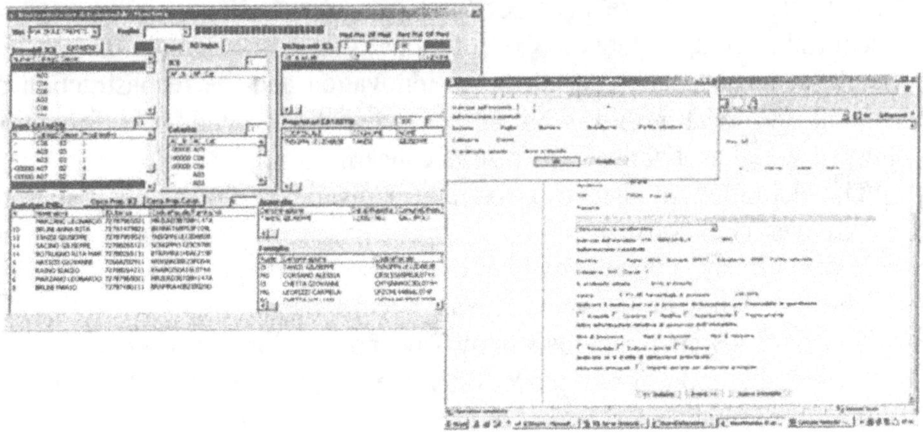

Figure 1: Some interfaces of ICI online services from the clerk's perspective and from citizen's perspective

In the Municipality of Taviano two important figures to measure the application of V.I.O.L.A. are the *fiscal trial* for the introduction and the usage of the tools (TunifET e SIFET) in the "Contact Center", which is zero, and the increasing diffusion of A.D.R. (Alternative Dispute Resolutions) methods in the Administration. In particular the use of "*istanze di autotutela*" (autoprotection instances), submitted by taxpayers, has been largely improved, most of all to request the data correction to Municipality.

Briefly, in Taviano the usage of self correction presents these figures: in the fiscal period of 2001, 1120 "istanze di autotutela" were submitted to correct data out of 1370 and only 120 out of 1120 had soundness. In 2002, this rate is even more significant, because 330 instances have been presented out of 866 tax assessments done. Among these nobody protested against the legitimacy of regulatory processes in the taxation relationship with payers. In both 2001 and 2002 no litigation against the Municipality was undertaken related to the regularity of its decisional processes on local taxation. Up to April 2nd, 2003 the results show 563 fiscal controls, without any fiscal litigations or "istanze di autotutela".

5. CONCLUSIONS AND FURTHER WORKS

In the interdisciplinary experience of Taviano project, V.I.O.L.A. has promoted the institutional, normative and organizational learning process. In this e-government experience Taviano has not been "buyer of innovation", but "author and actor of institutional innovation" in the normative and informative scenario of technology usage. So, instead of bearing the increase of knowledge, learning, discipline and adaptation costs, the Municipality of

Taviano finds out the relativity of the e-government impact in an ongoing transforming context, like the Constitutional Italian one, and, in the specific sample of local taxation, it combines innovation and the reconstruction of trust in the relationship with citizens, on the ground of information circulation and availability. The results concern

- The definition of clear lexicon and semantic in the regulations with taxpayer's relationship
- Positive financial benefits without additional costs
- The elimination of unwanted effects, like fiscal trial

The main advantages perceived by Taviano's Major have been:

- learning by doing for employees involved in the project, and knowledge transfer from the project team to the rest of employees;
- better aptitude to innovation;
- savings and reorganization of local taxation.

We are planning to extend the empirical validation of the model and the tools in order to tune and customize them to different contexts and sizes. Moreover this experience of holistic model can be extended from the local taxation to other functions in the LPA, having already verified the citizens' goodwill to pay according to indicators of regulation and reorganization of informative processes.

6. REFERENCES

1. Lenk k., Traunmüller R., Electronic Government: Where Are We Heading?, in Traunmüller R., Lenk K. (Eds), EGOV 2002, LNCS 2456, pp. 1-9, 2002, Springer-Verlag, Heidelberg 2002.
2. UWA (Ubiquitous Web Applications) Project: http://www.uwaproject.org.
3. Bochicchio M. A., Longo A.: "An Effective Approach to Reduce the "Avalanche Effect" in the Management of Fiscal Data in Local Public Administration. ICSM 2002: 560- 567
4. Bochicchio M., Longo A.: "Data Cleansing for Fiscal Services: the Taviano Project", in ICEIS 2003, Angers, April 2003
5. Ministero dell'Interno, *Analisi Gestionale e Finanziaria degli Enti Locali* (Ufficio Studi della Direzione Centrale della Finanza Locale), Settembre 2002.
6. Graham G.A., *Regulatory Administration*, in Graham G.A., Reining jr. H.F. (eds.), *Regulatory administration*, Wiley and Sons, New York, 1943, 16
7. Dipartimento della Funzione Pubblica, *Proposte per il cambiamento nelle pubbliche amministrazioni*, Rubettino, Soveria Mannelli, 2002
8. OECD, Regulatory quality and public sector reform, Puma/reg (97), 1, Paris, 1997
9. Salmon P., *Decentralization as an Incentive Scheme*, in Oxford Review of Economic Policy, vol. 3, n. 2, 1987, pp. 24-43
10. Morgan G. *Images*, Trad. italiana, Franco Angeli, Milano. 1995
11. Giarda P., Le regole del federalismo fiscale nell'articolo 119: un economista di fronte alal nuova Costituzione, JEL Classification. H7, Working Papers No. 115/2001, Jan. 2002, 11

SECTION 2
Business Models of e-Applications

Chapter 3

WHO ARE THE INTERNET CONTENT PROVIDERS?

Identifying a realistic taxonomy of content providers in the online news sector

Cornelia C. Krueger, Paula M.C. Swatman

Faculty of Informatics, University of Koblenz-Landau, Germany

Abstract: The Internet continues its growth as a medium for the sale of goods and services – and yet, although it would seem that digital content was one of the most obvious products for sale, content providers continue to struggle to find business models which will bring in adequate revenue. In this paper we initially review the literature on business models for digital content providers and note the wide variance in perspective amongst those writing in this field. Building on our work for the European Commission SimWeb project, we then consider the ways in which content providers in the online news industry actually operate and the ways in which they contribute to the content value chain, suggesting a framework of our own which suggests a continuum of business model types for content provision – ranging from "pure" content provision at one end to an extension beyond content itself to the provision of the technology needed to read content at the other end. We then elaborate on this framework, discussing the ways in which companies are making money from content provision in the online news industry, on the basis of a number of cases within the European market. Finally, we discuss the possible directions other content providers might wish to take in this environment.

Key words: Internet content providers, business models, value chain, online news

1. INTRODUCTION

Despite the continuing growth of the Internet as a commercial environment, online content providers are still looking for successful ways to sell their content over the Internet. The initial hope, that advertising would finance most of their business, simply did not eventuate – Gaffney (2001)

notes, for example, that click-through rates for banner ads are averaging only 0.3%. Advertisers are also cutting their marketing budgets in response to tight economic conditions, both in terms of products and of job advertisements and, since advertising is a major source of revenue for both print and online newspapers, a decline in advertising volume and value has a serious impact on newspapers' profitability. As an illustration of this decline, the regional and local job advertising market in Germany declined by 45%, the real estate advertising market by 18.4% and automotive advertisements by 10.8% between 2001 and 2002 (Die Zeit, 2002).

Although this loss of advertising revenue affects newspapers and magazines of all types and sizes, from the news agencies to national dailies to the smaller regional papers, those which wish to be successful must adapt their business model(s) to make use of the new technologies (Schneider, 2001). Some content providers concentrate on their core business, while others try to widen their business activities and endeavour to gain expertise in neighbouring fields such as technology – and thus have evolved from being content providers in the narrow sense of news and associated information, to being the providers of content in a much broader sense.

In this paper we suggest an appropriate taxonomy for content providers on the Internet, with a particular focus on the online news sector. This taxonomy covers the full gamut of content provision – from the narrowest sense in which "only" digital content is provided, to the widest sense in which the provider does everything, from the creation of content to its digital distribution. We begin with a survey of the literature on existing Internet business models for content providers, and then investigate the ways in which the providers of news make use of these models. Are the Internet business models described in the literature used by online news providers? Or can we find other, perhaps more successful, types of business models for this industry sector by studying the models our case studies have adopted? Finally, we discuss the advantages of content provision in the broader sense.

2. BUSINESS MODELS FOR ONLINE CONTENT

Afuah and Tucci (2001) define an Internet business model as "the method by which a firm plans to make money long term using the Internet". Research into appropriate eBusiness models has grown significantly over the past few years, with authors taking both theoretically and empirically-based approaches to the development of definitions and taxonomies of business models suitable for the New Economy.

Timmers (1998) is one of the earliest and most widely cited of the authors writing on B2B eBusiness models and, like Schuster and Weiss

(2001) he points out that business models in general can be based on fragmentation and combination of the value chain, or on value chain de- and re-construction. The value chain is decomposed into components and, by means of new contact points and new markets, these parts are recombined in different (and often more profitable) ways. A variety of potential inter-organisational architectures can be constructed by combining interaction patterns with value chain integration – Rappa (2002), for instance, points out that a business model shows how a company makes money by identifying its place in the value chain.

Picard (2000) believes that a business model includes the concept of the value chain, that is, the value that is added to a product or service at each step of its acquisition, transformation, management, marketing and sales, and distribution. Porter (1999) had earlier divided value chain activities into "primary" and "supportive" – and pointed out that limited resources, increasing competitive pressure and a shorter innovation cycle led to a concentration on particular value chain processes and to a combination of these processes with other companies' complementary processes. Such an emphasis on inter-organisational complementarity in turn can lead to the evolution of a value "network" (Parolini, 1999), from the traditional value chain, as more and more cooperating business partners, at every possible level, become involved. One goal of such a value net could be to offer new or more broadly-based products, another to create economies of scale. A characteristic of value networks is the fact that each member company is responsible for its own part of the value creation process, but planning and organisation of the value chain activities occur overall (Loos/Scheer, 2002).

In applying value chain concepts to the digital media industry, it is clear that the nature of the product (essentially information-based) lends itself particularly well to the idea of a value net and, indeed, many media producers made use of a network of resources which relate very neatly to the value net concept. Nonetheless, for the purposes of this paper we retain the value chain terminology because of the ease of distinguishing the stages of product development which it offers. Woessner (2001) differentiates the media products value chain from the more general B2B eBusiness value chain, noting that it relates to the generation of ideas, the editing of those ideas in order to obtain utilisable content, the production of media products, the copying of these products and their distribution. Content providers, he adds, are: "those firms that provide users access to content of interest including news, information and entertainment, leisure activities, and other material". Overall we can break the value chain as it currently exists in the online content market into four component parts: the creation of content, the adding of value to content, the distribution of content, and its consumption (figure 1).

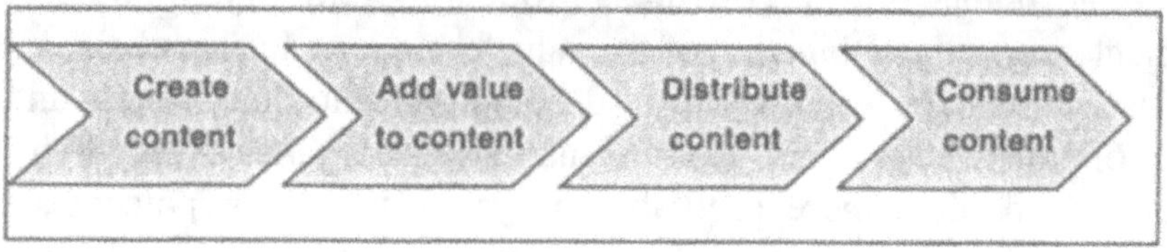

Figure 1. The content market value chain (Krueger et al. 2003)

This chain appears comparatively simple at first glance – but the definition of a content provider within this industry sector is not quite as simple as it might appear. Content providers are currently defined in a variety of different ways by the major authors in this field. Table 1 summarises the best-known of these approaches, and provides a foundation for our later discussion of business models in the context of online news provision.

Table 1. Comparison of business models for content providers

Authors	**Catchword**	**Value chain**	**Critical success factors**	**Sources of revenue**
Bartussek	Newsfilter	Producer and Intermediary	- Identify and serve customers' needs; - Traditional strengths; - New income resources	Selling product Advertising
Farhoomand/ Lovelock	Content Provider	Intermediary		Subscription fees
Niewiarra	Content Network	Network	Networks	Revenue sharing
Rayport	Content Business	Network	Revenue sharing between content provider and online service	Revenue sharing
Weill/Vitale	Content Provider	Producer	Branding; Recognised as best in class; Network	Fees from third parties or allies
Wirtz	E-Information	Intermediary	Differentiation	Pay per transaction; Subscription

Bartussek (2001) calls content providers "newsfilters" and argues that newspapers need to recognise and extend their existing core competency as filterers of real-world news into the digital world; and use this to create a competitive advantage. He sees providers as being simultaneously producers of content and intermediaries which add value to content (whether home-grown or purchased). For Weill and Vitale (2001), by contrast, content providers are purely producers of content. According to these authors, content providers are firms which create and provide content in digital form to customers using third party intermediaries. The physical world analogy of a content provider is a journalist, recording artist, or stock analyst. Typical offerings include software, electronic travel guides, digital music and video. A content provider offers expertise and leadership in a niche market and it is clearly important that such a provider understands customers' needs and wants, so that s/he can create and price content appropriately.

Both Farhoomand/Lovelock (2001) and Wirtz (2001) position the content provider in the value chain as an intermediary. Farhoomand/Lovelock (2001) describe a business-to-consumer (B2C) eCommerce business model which they call "content provider" – a "web-based data host and electronic publisher of newspapers and magazines". Web-based data hosts gather a variety of information and organise them into electronic databases, with revenue coming from subscription fees. Online newspapers and magazines seldom charge for general content, but rather change small fees for archived news and special services. Wirtz (2001) has a slightly more production-line approach, describing a content business model which consists of collecting, selecting, systematising, packaging and providing content on a company-owned platform. The user must have ready access to the content required, which may (or may not) be personalised. The E-Information business model depends on the content offered, for example E-Politics, E-Economics, which is focused on political, economic or social content.

For Niewiarra (2001) and Rayport (1999) only network can be successful in the content market. According to Niewiarra (2001), networks are the key to success in content distribution. Content is used cross- and multi-media. The increasing fragmentation of the market demands new preparation and new ways of offering content to the customer. Rayport calls his model "content business". He believes that the period in which everything on the Internet was free is now over. Users are increasingly accustomed to paying for Internet access and this revenue should be shared between the content provider and the online service provider previously, the way in which such revenue was shared had to be negotiated).

What is clear about the majority of these views of the online content-creation sector is that the authors take a value-chain oriented approach to identifying the roles and responsibilities of the content providers. Rayport alone has a customer-oriented view of the content creation process, seeing the days of free Internet access as numbered (a very farsighted view for 1999).

3. RESEARCH APPROACH

The Internet continues to change the way content providers are positioned within the value chain and, with each new technology innovation, the opportunities available to content providers becomes more complex (and potentially more interesting). The literature overview in the previous section shows that although these is a fairly general agreement on a value-chain orientation as the appropriate standpoint for viewing the digital content market-place, there is no consensus about either the definition of a content provider, or about the role of the content provider within the value chain. This lack of unanimity led us to consider the following research questions:

- Is there a single definition of the term "content provider", or can a content provider enlarge its business model according to its needs – especially as the Internet itself evolves?
- How do content providers develop their business models?
- How do content providers position their company in the content value chain?

Our interest in these questions stemmed from our involvement in the European Commission project SimWeb (Exploring Innovative eBusiness Models using Agent Simulation), which is building multi-agent simulation models of the online news and online music markets to assist companies in making informed strategic decisions. As the group developing "sector surveys" of these market-places, we have spent considerable time trying to understand the business models of online content providers in both these sectors, although in this paper we focus on the online news industry.

Our data gathering has thus far been concentrated on the German experience, but we also wanted to understand the situation in the US and in other European countries (such as Portugal, Spain and Italy). We have made use of 'secondary data research' (Jarvenpaa 1991; Neuman 1997), which involves the synthesis of existing data collected for other purposes than the specific research project concerned; and with drawing inferences from those data. This has meant our scanning the Web for information related to online content provision in the news industry, and using data from interviews undertaken as a part of our SimWeb data gathering activities.

As part of our work for the SimWeb project we interviewed several stakeholders in the online news market, including: ddp, the second largest news agency in Munich/Germany; RZ-Online, a very innovative regional newspaper in Koblenz/Germany; FAZ.net, the online company of a national German high quality newspaper; and Publico.pt, a very well regarded Portuguese online newspaper.

We also analysed the Web sites of a number of newspapers and magazines which had interesting business models, most of which restrict their "charged for" content to specific foci. These newspapers include: The Wall Street Journal (public.wsj.com), the New York Times (www.nyt.com), Il Corriere della Sera (www.rcs.it) and El Pais (www.elpais.es).

Finally, we analysed the Web sites of some of a new type of content provider. These companies offer newspapers and magazines a technical platform through which they can sell their online editions to their Internet customers, and include: Newsstand (www.newsstand.com), Tecnavia (www.tecnavia.ch) and In a Daily (www.inadaily.com).

4. FINDINGS FROM THE CASES

Wössner (2001) suggests that the value chain for media products such as online news can be divided into different stages: The party which generates an idea, the party which adds value to this idea, the party which turns the idea into a product and, finally, the party which multiplies and distributes the content.

The positioning of online content providers along this value chain, according to the literature analysed in the second section of this paper, however, differs greatly from one author to another. Bartussek (2001) sees the content provider as being both a producer of content and an intermediary. For Weill and Vitale (2001) the content provider is solely a producer, while for Farhoomand/Lovelock (2001) and Wirtz (2001) s/he is only an intermediary. Niewiarra (2001) and Rayport (1999) see content providers primarily as part of a network. Such a divergence of views suggests that there may be different way of viewing this marketplace.

What does the online news market itself say? Can one say that a content provider merely produces or adds value to digital content? Our research shows that it is important to differentiate the content value chain more accurately. Figure 2 illustrates the different types of content providers existing in the Internet market-space today:

<u>Content Creator</u>

These are the firms which concentrate on producing content, such as the big news agencies or even journalists. Production is their main focus. For example journalists create and deliver content for Web sites. They can either work as freelancers, or as employees of a news agency or a media company or a portal, etc.

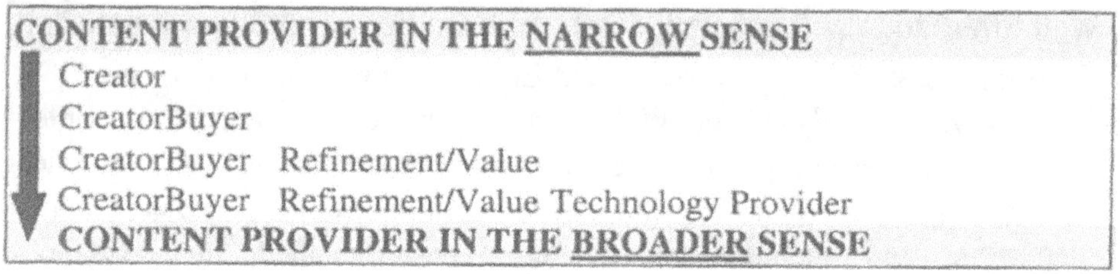

Figure 2. Taxonomy of Internet content providers

Content Creator and Buyer

These are the companies which create and buy content to sell through their Web presence. This is the case for smaller news agencies which create, for example, national, business and cultural news but have to buy sports news in order to make their offering complete. News agencies like ddp purchase articles from journalists or other news agencies, but also produce articles and content in-house.

Content Creator, Buyer and Value Adder

This could be a portal that is creating, buying and adopting content for their customers. Most newspapers and magazines fall into this category (for example, Publico.pt or the Wallstreet Journal). They create some of their content in-house with the help of employees and/or freelancers, and most of the time this is also where their core competence can be found. Then they buy specialist material to round out their offerings – for example the Wallstreet Journal purchases sport news (Dutta, 2002) because this is something a full-service newspaper must have, but it would be too expensive to produce it in-house. Finally this group also adds value to both the content they have purchased and the content they created in-house, for example by offering a daily newsletter to customers containing personalised information, or by offering premium services such as financial news analysis, as ft.com does (Dutta, 2002).

This group of online content providers does not offer Internet access or application services, but cooperates with companies which supply these services to the newspaper's or magazine's customers. A good example of this is RZ-Online, which creates regional and local news in-house as its core competence. To increase its income the company not only sells advertisement space and produces a facsimile version of the print edition but, in cooperation with an Internet service provider, it has created a subsidiary company which sells ISP and ASP services (Krueger and Swatman, 2002).

These examples make it clear that the boundaries between Content Creators, Buyers, Value Adders and Internet service provision are not well defined. The digital distribution of content and the changing market-place forces telecommunication companies, Internet providers and media companies to work together. Only those with a good understanding of their customers' needs and those who choose the right partners will be successful (Gregg, 2001).

Content Creator, Buyer, Value Adder and Technology Provider

Media companies have the choice of purchasing technological infrastructure, finding a technology partner or becoming a technology provider themselves and simultaneously offering appropriate technological infrastructure and content to their own customers. They can even go a step further, as the New York Times did with its NewsStand (www.newsstand.com) subsidiary and offer their technology platform to other newspapers and magazines worldwide such as The Australian, The Harvard Business Review, or Le Monde (Figure 3). NewsStand makes use of a proprietary, Acrobat-based reader to present the Quark Express print-ready output of 36 (as at the date of writing) newspapers and magazines. The NYT has developed a highly successful new business model out of technology which is intrinsic to the newspaper, together with a modification of one of the world's most widely-used presentation platforms.

NewsStand (www.newsstand.com) is typical of a development which is already very well known in other industry sectors such as music, and which has now been adopted by entrepreneurial companies in the news industry. In a bid to get to know the customer's habits ever better, the music industry makes use of software like Windows Media Player, Real Player or Liquid Audio which the user needs to stream or download music from a Web site. This concept is itself not original to the music industry, which has copied the idea from Internet Access Providers who place their software on new PCs prior to purchase. The music subscription service Pressplay is now bundling its software with new Gateway computers (www.spiegel.de, 2002) and, although the online news industry has not yet started to bundle modified browsers with PCs, such a move cannot be far away. NewsStand's use of a modified version of the Acrobat viewer can readily be compared to Liquid Audio, for example. The online newspaper reader must first download the software before s/he can actually download, for example, the daily edition of The Boston Globe. The software provides online companies with an efficient tool to acquire customer data and to analyse customer behaviour – as well as to promote their latest offerings. Customers can easily be segmented and can be targeted with customised offerings (Preissner, 2001).

NewsStand is not the only company using a mixture of technology and content provision, of course. Other good examples of this combination include Tecnavia and inadaily.com. Both make use of software which must be downloaded before the reader can get to the content of the newspapers Il

Corriere della Sera and El Pais. Inadaily.com states frankly on its Web site that it is aiming to sell its Premium News Service after a 1 month free trial .

The really innovative feature of inadaily.com's offering is that the customer can "cherry pick" among the content of all the partner newspapers offered through this delivery portal, which include major European dailies such as El Pais, FAZ, Le Figaro etc. That means the reader can combine the sports news from El Pais with the business news from FAZ and the political news from Le Figaro and avoid all the other news s/he isn't interested in. This not only attracts readers who see the benefits of this selection facility, but also shifts the work of customising the product from the content provider to the customer him/herself.

The technological extension of the business model of a content provider has a lot of advantages for both the customer and the media company, as Table 2 shows:

Table 2. Effects of the technological extension

Advantages for the content provider	Advantages for the customer
Independence of Internet providers	Customer can be mobile
Customer Data of high quality	Customer doesn't have to be online
Customised/personalised offers	Customer only needs one software for different content providers
Network => greater potential customer base	Customer chooses his/her own offer
Technological platform can become a portal	

5. CONCLUSIONS

Earlier in this paper we identified three research questions which we believed were relevant to online content provider business models:

- Is there a single definition of the term "content provider", or can a content provider enlarge its business model according to its needs – especially as the Internet itself evolves?
- How do content providers develop their business models?
- How do content providers position their company in the content value chain?

We have shown that the existing literature is taking such an eclectic (and, indeed, contradictory) view of content provision than no consistent theme emerges from the many authors working in the definition of business models for this group. Our work with the SimWeb consortium has, however, given us the opportunity to study content provision in the online news and online music industries in perhaps more depth than is normally possible. In this paper, we have shown that the online news sector content providers, at least, can be classified in a logical fashion from those who offer content in the narrowest and most literal sense of this term, through to those who are combining technology and core news-provision competence to create a new form of content provision.

It is obvious that not every content provider can afford to set up its own technological platform, of the sort offered by NewsStand or inadaily. But such a step is not necessary. By joining one of the existing platforms, any company can profit from the advantages mentioned above.

Even though content provision in the broader sense seems to be the most promising online business model for content providers at the moment, each of the other categories also offer possibilities for survival. But the closer companies come to the narrow definition of content provision, the more they must concentrate on their core competency in order to achieve a competitive advantage. It would therefore appear that there are two types of promising Internet business models:

1. Those Internet business models which integrate the creation, the acquisition, the value-adding and the digital distribution of content with the help of a software platform and therefore profit from the network effect of linking all these offerings.
2. Those Internet business models which concentrate on what the news providers know best, their core competence, and where the companies concerned can find the right partners to support this strategy.

Perhaps it is even possible to combine those two ways successfully. But this would need further research. We plan to extend this investigation to the online music industry and, through our on-going investigation of both sectors across Europe, to further enhancing our understanding of online content provision.

REFERENCES

1. Allan Afuah, Christopher L. Tucci, Internet Business Models and Strategies, 2001
2. Joerg Bartussek, Vom Newspaper zum Newsfilter, in: Andrej Vizjak, Max Ringlstetter, Medienmanagement: Content gewinnbringend nutzen, Trends, Business-Modelle, Erfolgsfaktoren, Wiesbaden, 2001, pp. 63-73
3. Ciaran Byrne, FT.com says 17,000 now paying, in: http://media.guardian.co.uk/newmedia/story/0,7496,765179,00.html, 2002
4. Dave Chaffey, E-Business and E-Commerce Management, Prentice Hall, Essex, 2002
5. Soumitra Dutta, Zeitungen gehen online: Eine Herausforderung für alte Medien, in: Thomas Bieger, Nils Bickhoff, Rolf Caspers, Dodo zu Knyphausen-Aufsess, Kurt Reding (Hrsg.), Zukünftige Geschäftsmodelle, Konzept und Anwendung in der Netzökonomie, Berlin, Heidelberg, New York, 2002, pp. 117-150
6. Ali Farhoomand with Peter Lovelock, Global e-Commerce, TEXT and CASES, Singapore, 2001

7. John Gaffney, The Battle over Internet Ads, Business 2.0, July 25 2001, available online: http://www.business2.com/articles/web/0,1653,16546,00.html
8. Stefanie Gregg, Die Konvergenz: Telekommunikationsanbieter und Medienunternehmen – Wettbewerber oder Partner?, in: Andrej Vizjak, Max Ringlstetter (Hrsg.), Medienmanagement: Content gewinnbringend nutzen, Trends, Business-Modelle, Erfolgsfaktoren, Wiesbaden, 2001, pp. 37-44
9. Sirkka Jarvenpaa, "Panning for Gold in Information Systems Research: 'second-hand' data", pp 63-80 in: Nissen H-E, Klien HK and Hirschheim R (Eds) 'Information Systems Research: Contemporary Approaches and Emergent Traditions', North-Holland, 1991.
10. Cornelia Krueger, Paula M.C. Swatman, Regional Online Newspapers: paths to glory, or the road to ruin?, in: The Proceedings of the 6th BIT Conference, Manchester, 2002
11. Cornelia C. Krueger, Kornelia van der Beek, Paula M.C. Swatman, SimWeb Sector Reports of the European Online Music and Online News Market, 2003
12. Peter Loos, Christian Scheer , Internetbasierte Geschäftsmodelle. Neue Möglichkeiten der Wertschöpfungsorganisation in der Internet-Ökonomie, available online at : http://chris.scheer.bei.tonline.de/docs/scheer02_wertschoepfung_paper.pdf , 2002 [accessed 17.03.2003]
13. Neuman WL, Social Research Methods: Qualitative and Quantitative Approaches, 4th ed, Allyn and Bacon, Boston, 2000.
14. Manfred Niewiarra, Der Rahmen: Content Management in der dritten Generation der Telekommunikation, in: Andrej Vizjak, Max Ringlstetter, Medienmanagement: Content gewinnbringend nutzen, Trends, Business-Modelle, Erfolgsfaktoren, Wiesbaden, 2001, pp. 45-59
15. Cinzia Parolini, The Value Net, John Wiley & Sons, 1999.
16. Robert G. Picard, Changing Business Models of Online Content Services, Their Implications for Multimedia and Other Content Producers, in: JMM – The International Journal on Media Management, Vol.2, No.II, 2000, pp. 60-68
17. Michael E. Porter, Wettbewerbsstrategien, Methoden zu Analyse von Branchen und Konkurrenten, Frankfurt/New York, 1999
18. Andreas Preissner, Marketing im E-Business, Online und Offline – der richtige Marketing-Mix, München, Wien, 2001
19. Jeffrey F. Rayport, The Truth about Internet Business Models, in: http://www.strategy-business.com/press/article/?art=14631&pg=0 , 1999
20. Michael Rappa, Business models on the Web, http://digitalenterprise.org/models/models.html , 2002
21. Karsten Schneider, Geschäftsmodelle in der Internet-Ökonomie, in: Dieter Ahlert, Jörg Becker, Peter Kenning, Reinhard Schütte, Internet & Co. im Handel, Strategien, Geschäftsmodelle, Erfahrungen, 2. Auflage, 2001
22. Michael Schuster, Michael Weiss, Ecosystems – Ein Paradigma in der Medienindustrie, in: Andrej Vizjak, Max Ringlstetter, Medienmanagement: Content gewinnbringend nutzen, Trends, Business-Modelle, Erfolgsfaktoren, Wiesbaden, 2001, pp. 109-121
23. Paul Timmers, Business Models for Electronic Markets, in: http://www.commerce.net/research/reports/1998/98_21_n.html , 1998
24. Peter Weill, Michael R.Vitale, Place to Space, Migrating to eBusiness Models, 2001
25. Bernd W. Wirtz, Medien- und Internetmanagement, Wiesbaden, 2001
26. Mark Woessner, Die Medien: Eine traditionsreiche Branche am Scheideweg, in: Andrej Vizjak, Max Ringlstetter (Hrsg.), Medienmanagement: Content gewinnbringend nutzen, Trends, Business-Modelle, Erfolgsfaktoren, Wiesbaden, 2001, pp. 19-25
27. PC mit Pressplay, www.spiegel.de/netzwelt/netzkultur/0,1518,225902,00.html , 2002
28. Noch in dieser Woche…, in: Die Zeit, No. 27, 2002, pp. 11-14

Chapter 4

NET MARKET MAKERS IN THE AUSTRALIAN B2B E-SPACE

Mohini Singh

Abstract: This paper discusses the role of net market makers as intermediaries in the emergent Australasian B2B e-space. The discussion and findings of this paper are from a research project that investigated the business and operational issues of these intermediaries as highly volatile business entities. The findings include business opportunities arising from technology, revenue sources from intermediary services, factors contributing to success and the challenges of operating in an evolving and dynamic industry.

Key words: B2B E-Business, net market makers, intermediaries, e-procurement, e-auctions, e-marketplace

1. INTRODUCTION

B2B e-business in Australia is largely dependent on intermediation from net market makers either private, consortia based, national or international, to facilitate e-procurement, B2B exchanges, e-supply chain management and information sharing between business partners. B2B exchanges include various categories of market spaces, including vertical market portals, horizontal exchanges, hubs and various types of online auctions. Intermediaries generally referred to as net market makers or e-markets bring together exchange partners, organise and manage auctions, generate e-catalogues and other auxiliary services pertinent and relevant to efficient exchanges in the B2B e-space. Since 1998 Australian businesses have increasingly adopted e-procurement largely depending on net market makers to provide intermediary services and facilitate the exchange. As a result net market makers proliferated at an astounding rate because of the business opportunities in facilitating B2B exchanges. For buyers these intermediaries lower purchasing costs while reaching new suppliers. For suppliers they lower sales costs and help the supplier reach new customers.

However, as reported by Frew (2002) and NOIE (2001) many of the net market makers in Australia either collapsed, did not get off the ground or consolidated. It is now believed that as many as only 100 e-markets (net market makers) exist in Australia (Frew, 2002).

Although the Internet offers great opportunities to net market makers, operational and business issues for success and survival as business entities and current and future challenges have not been addressed. In this paper I present the findings of a research project that highlights the business practices, success factors and challenges faced by Australian net market makers. Net market makers and e-markets in this paper mean the same thing and are used interchangeably.

2. LITERATURE REVIEW

Electronic intermediaries provide an information infrastructure by which traders can realise commerce over electronic networks (Lee and Clark, 1999). Deganais and Gautschi (2002) describe net market makers as organizations that own and operate the e-marketplace. Characteristics of net market makers according to Deganais and Gautschi are that these owners and operators may be companies that are fully independent (that is, are not also buyers or sellers) or could be major and dominant participants in the exchange as buyers and sellers. It is important for net market makers to adopt a business model that will allow them to operate the exchange and make a profit. These net market makers also depend on ancillary services to complete the exchange with support from electronic payment agencies, security system providers, and in some cases the engine (technology) to host auctions.

A business-to-business electronic marketplace has several buyers and several sellers. It is an arena on the Internet where a trusted intermediary (e-market) offers trading functionality to registering companies (Swedish Trade Council, 2001). Archer and Gabauer (2002) describe an electronic marketplace to be a virtual marketplace where buyers and suppliers meet to exchange information about prices and product and service offerings, to collaborate, and to negotiate and carry out business transactions. Gallaugher and Ramanathan (2002) suggest that electronic marketplaces provide the basic infrastructure to allow suppliers and buyers to interact in an online environment.

Based on the above definitions, in this paper the terms net market makers and e-markets are terminologies used interchangeably meaning the same thing.

E-markets (net market makers) offer services that facilitate transactional service needs to buyers and suppliers. Weill and Vitale (2001) and Zwass (2000) advocate that e-Markets enable easy search of products and services, information on product specifications which reduce communication costs for both buyers and sellers, dynamic pricing based on demand relationships, sales transactions that include payment and settlement, product delivery, market surveillance for stock market, auction results and enforcement of proper conduct by buyers and sellers.

The large amount of bid, order and transaction management for the B2B procurement of parts and supplies usually require assistance from auxiliary services such as e-markets (Thomson and Singh, 2001, Laudon and Traver, 2002). An 'e-market' functions as a trusted intermediary whose well-integrated business procedures and technology save costs and streamline the purchasing and sales processes (Swedish Trade Council, 2001). They provide customers with buy and sell services to enhance busines, efficiencies and competitiveness.

There are generally different value propositions for buyers and suppliers to participate in a trading exchange or an industry consortium. Archer and Gebauer (2000) advocate that e-markets can reduce costs of acquiring and communicating information about prices and products, as more organisations join the marketplace benefits to participants increase, e-markets can impose significant switching. Palmer (2002) suggests that buyer benefits gained from e-markets are easier and faster buying process, reduced transaction cost and time, reach to global community of sellers, easy access to a large amount of information, reduced search costs, streamlined business processes, a wider choice, quick order fulfilment and less chance of errors. For supplier benefits Palmer includes cost saving and tighter inventory control, reduced transaction time and cost, broader customer base, information privacy and security, savings in time and capital, streamlined business processes, easy to update prices, availability and other product information, enable smaller regionalised suppliers a level playing field with larger suppliers and consolidation of orders onto a single invoice.

2.1 The Role of Net Market Makers in E-Procurement

Business and government organisations in Australia and around the world are increasingly moving from paper-based traditional procurement to e-procurement, to capitalise on the efficiencies of price transparency, online catalogues, easy search for products and suppliers, and to capitalise on the benefits of technology and automated processes. While EDI has been the most common method for automating procurement, its extent was limited by its substantial cost that made it only accessible to large firms with recurring volume purchases (Pavlou and Sawey 2002). The more ubiquitous Internet, which is also economically accessible to small-scale B2B exchanges, has further advantaged e-procurement. Developments in B2B e-business in Australian has seen businesses and the government, both at the State and Federal levels adopting Web-based e-procurement to achieve volume purchase, dealing with a wider choice of buyers and suppliers, lower costs, better quality, improved delivery, and reduced paperwork and administrative costs (Singh and Thomson, 2002).

Businesses buy a diverse set of products and services, ranging from paper clips to computer systems, from steel to machinery. At the broadest level these purchases have been classified by Kaplan et. al. (1999) into manufacturing inputs and operating inputs. Business purchases are dominated by systematic sourcing and spot sourcing of goods and services. Systematic sourcing, buying through pre-negotiated contracts with qualified suppliers, is relationship oriented and contracts are long term. Spot sourcing is fulfilment of an immediate need, typically of a commoditised item for which it is less important to know the credibility of the supplier (Thomson and Singh, (2001) and Chaffey, (2002) and Christiaanse, et al (2001)).

E-marketplaces or e-hubs that enable B2B purchases have been categorised by Kaplan et al (2000) as MRO (maintenance, repair, operating) hubs, horizontal markets that enable spot sourcing of operating inputs for manufacturing, labour, and advertising. Vertical exchange markets that enable spot sourcing of manufacturing inputs and commodities and vertical catalogue hubs that enable systematic sourcing of non-commodity manufacturing inputs.

E-markets support e-procurement by providing technology and governance solutions, auctions, electronic catalogue content, aggregating supplier input, industry news and standardised data access to buyers (Archer and Gabauer, 2000). Archer and Gabauer further explain that e-markets face the challenges of a balancing task as they set up solutions that satisfy suppliers and buyers ensuring that their costs do not outweigh the overall benefits they will receive from the arrangement, the need of a particular business model to determine which buyers and suppliers to recruit as participants and compete with other techniques that suppliers and buyers use to meet their needs.

2.2 Net Market Makers in Australia

In the year 2000, many market makers (e-markets) proliferated in Australia due to e-procurement being the most important B2B e-business application. However, in the year 2001 consolidation of some e-markets took place in Australia due to online exchanges being new and unproven roles of these intermediaries (NOIE, 2001).

Standing and Stockdale (2002), who did a preliminary study of the net market makers in Australia report that initiatives such as 'try before you buy', free transactions, site tours and lengthy FAQ sections, strongly addressing security information an important feature of their web sites. They also indicated that the most important means of revenue generation was a charge on transaction costs. Standing and Stockdale also suggest that Australian ownership of e-marketplaces is generally based on first mover advantage operating as intermediaries. These are either horizontal or vertical hubs enabling catalogues, auctions, exchanges, storefronts and negotiations.

Bryant (2002) is of the opinion that in Australia and New Zealand there are too many market makers for the size of the industry in this region. According to Bryant (2002), most of these net market makers have similar offerings and flawed revenue models. Therefore to survive and expand business e-market makers need to build specialist businesses to support B2B e-business and look beyond their current business model.

Net market makers in Australia do have an important intermediary role in e-procurement matching buyers and sellers, ensuring trust among participants, supporting market operations and transactions, ensuring quality control and aggregating buyer and supplier information. However, with evolving e-business models and technological developments, consolidation and liquidation of a large number of these market makers have taken place. It is not clear from literature, the challenges, revenue models, success factors, management and operational issues that affect these intermediaries and their roles in B2B exchanges. Therefore to investigate business practices, governance issues, challenges and success factors research was carried out with five Australian e-markets (net market makers). The method of investigation, findings and discussions are presented in the next section.

3. RESEARCH APPROACH AND METHODOLOGY

This research project was exploratory in nature. Profiles were gathered from 5 companies operating as net-market makers in the Australian B2B e-space via interviews. The interviews explored company demographics, business model, practices and governance, success factors and challenges of these intermediaries operating as business entities. The companies were selected on the basis that they had been identified as successful e-markets from their presence on the Internet and other industry based literature. The 5 organisations investigated are those that agreed to participate in the project. Initial contact was made by email with the person who headed the organization, emphasising the importance of this research and agreeing to share the findings with these organizations. Five organizations were considered adequate for an exploratory study.

The interviews were semi-structured because of the exploratory nature of the research. Open-ended questions were used to guide discussion about each firm's business practices, successes and challenges. With the permission of the interviewees all discussions were recorded on tape and later transcribed. The questionnaire included seven sections to extract information on company demographics, business practices, strategic issues, infrastructure issues, services offered, success factors and challenges. Data collected are summarised in Table One. The method of analysis was qualitative the findings of which are discussed in the following section. For reasons of confidentiality, names of companies discussed are not identified. In this paper they are referred to as Companies A to E. One of the respondents was a large organization that provided the infrastructure for online auctions and exchanges only. Therefore some of the analysis is drawn from four companies only.

4. FINDINGS AND DISCUSSION

Company demographics, business practices, challenges and success factors identified from the findings are discussed in the following section.

4.1 Company Demographics

Organisations investigated were less than five years old, set up to provide intermediary services in B2B exchanges. Except one all were small organisations with less than 20 employees. Company D was a large international organisation which did not provide intermediary services only the infrastructure for auctions, bidding and payments. Companies B, C and E were privately owned while companies A and D were subsidiaries of international organisations. Companies B and E facilitated vertical hubs providing services to the construction and the steel industries, A and C facilitated horizontal and vertical hubs providing services to specific industries as well as exchange of commodities. Company D provided support services to other locally based B2B intermediaries.

Table 1.

	Co A	Co B	Co C	Co D	Co E
Demographics	New, small, international parent co., Melbourne based	New, small, local, Melbourne based	New, small, Melbourne based	New, large, Australasian based with international parent co.	New, small and privately owned
Business Motivation	Technology opportunities	Business & technology	Business opportunity	Technology opportunity	Business opportunity
Bus Model & Governance	Vertical & horizontal	vertical	Vertical & horizontal	None	Vertical
Services	Auctions, catalogues, RFQ, participant selection	Forward auctions	Reverse auctions, search, catalogues	NA	e-catalogues, tendering, search assistance
Revenue Model	Flat fee & charges for exchange services	Transaction fees	Auction & service fees	Software lease fees	Software, training users, negotiations
Strategic Issues	Holistic solutions	Niche applications	A focused business model	New alliance	New services
Infrastructure Issues	Bidware	Developed in-house	Developed in-house	Commerce One	JAVA/LINUX

The respondents' positions ranged from senior consultant with the company to partner and director. Core services provided by these organizations included document tracking, online catalogues, strategic sourcing of suppliers, technology for online auctions, prequalification of suppliers and facilitation of product sales.

4.2 Motivation to Act as an Intermediary

Business opportunities from the Internet and e-procurement being the largest component of B2B e-business were the main motivating factors for these organisations to act as intermediaries. Company A saw a window of opportunity in the Australasian B2B E-Business, thus decided to expand its operations in this region. Core services provided by these intermediaries included provision of technology for online bidding and auctions, sourcing of solutions and suppliers, document and transaction management and online catalogues. Company E provided services specifically to the Construction industry and Company B capitalised on technology to reach out to global buyers of Australian Steel. Company D being a supplier of ERP systems in

Australia and New Zealand was pressured by large users of these systems to provide an infrastructure to support e-procurement that could be integrated to its back end ERP systems.

It is apparent that opportunities arising from technology are a strong driver of net market makers to provide intermediary services and to operate as business entities. Another compelling reason to operate as a net market maker was first mover advantage in the emergent Australian B2B e-space.

4.3 Business Models and Governance Issues

Companies A, C and D operated as neutral third parties supporting both vertical and horizontal exchanges, Companies B and E operated as vertical intermediaries supporting two specific industries the construction and steel. All five organizations had over 100 buyers and sellers on their database and the value of trade facilitated ranged from half a million to 100 million dollars. Vertical exchanges took advantage of the knowledge of the industry to target suppliers and buyers. Horizontal operators used advertising both formal and informal and allowed 40 % of the potential suppliers, based on prequalification and buyer selection to participate in any event. In most exchanges, the suppliers were known to the buyers due to the limited number of suppliers of some goods and services in this part of the world.

Value proposition to the buyers and sellers included transparency of information and prices, business efficiencies resulting from reduced costs of purchases, administration accuracy of data transmitted, reduced time for acquisition of products, scalable and reliable applications and improved and standardised business processes.

Findings of this research reveal that business models of intermediary operations in the Australian B2B sector of e-business are similar to those identified by Kaplan (2000), discussed earlier.

4.4 Services Provided

Intermediary services included provision of directories listing suppliers, their profiles and ratings; inventory listing; bulletin boards and online information browsing capability; information about the site and customer service, buyer guides and news. Auctions, reverse auctions, online negotiations, RFQ's and document management and after event reports and payments were some of the other important services provided to buyers and sellers. Training users to use technology to bid online, access information and to negotiate online are also provided by these e-markets.

4.5 Revenue Model

The most common revenue model for services included a flat fee and a percentage of transaction fees charged to the buyer. Other sources of revenue included charges for technology for bidding and auctions, preparing e-catalogues and other small technology solutions. Company D collected a licensing fee for the use of its software and Company E servicing the construction industry included a subscription fee for participating organizations. All charges were levied on the buyer, except in the case of Company B, which supported the suppliers of steel. Prevalent payment methods used were traditional, although all had the infrastructure to support e-payments. Most received payments for their services 15 to 20 days after the event. Two respondents indicated that the fee charged covered costs but was inadequate. Low fees were maintained to attract business.

4.6 Strategic and Tactical Issues

Main strengths of their business included open source technology, skilled staff, a focused business model, technology for niche applications, holistic solutions encompassing services and technology, leveraging global knowledge base, applications of technology and first mover advantage in the industry. Competitive threats common to these organizations were other net market makers operating in the region and in the specific industry, as well as consulting organizations. Consulting organizations such as Peat Marwick and A T Carney provided B2B solutions as well as the engine for auctions. Business strategies for these e-markets focussed on business expansion, capitalising on technology, new and profitable business models, packaging of disparate issues for simplified solutions, new partnerships for business growth. To form new alliances with organizations that can support software development, content, catalogues, payment services and technology.

The findings indicate that competition for these net market makers is stiff, although their strategy is focused on growth and expansion. Technology can be capitalised on for differentiated services and integrated solutions to lock in customers.

4.7 Infrastructure Issues

Infrastructure for managing an electronic marketplace varied for all the respondents. Company D specialised in e-marketplace technology and supported intermediaries, buyers and suppliers setup and create an e-marketplace for bids, product information, payments, security and transactions. This organization used Commerce One operating system, a friendly user interface, 'on ramp' solution to allow companies to participate in the online exchange and a 'plug and play' solution to enable them integrate the e-marketplace to its back office. Company A, which is a subsidiary of an international net market maker owned its e-marketplace platform called 'Bidware' which supports global markets and hosted on a server in Pittsburg. Companies B and E supported vertical exchanges with privately owned technology either developed or modified in house to support the online exchange. Company B's platform required participants to make their bids from their personal computers, and Company E was based on open source JAVA and LINUX.

From the above it is clear that e-market services are not technology specific and can be operated from different platforms. The need for a seamless integration of procurement data and the back end system is essential for online real-time processing of information. Information technology is the backbone to B2B e-business exchanges.

4.8 Success Factors

Factors that contributed to the success of net market makers included skilled and motivated employees, a focused business model, first mover advantage, open source technology, e-market technology compatibility with internal business processes, knowledge of the industry, a good value proposition to participants, and Australia's limited supplier base.

Technology and business knowledge of the market makers, need for support in B2B exchanges with the increased application of the Internet, and a well planned business model clearly leads to first mover advantages in innovations such as e-procurement and e-business.

4.9 Challenges

Challenges faced by net market makers identified from this research project are clearly more than the benefits. The most common ones being the small size of the industry in which they are operating, fragmented market, a lack of belief in the industry that e-market will work, undoing the legacy of existing markets, old inflexible back end systems, credibility issues due to the small size of the company, limited access risk capital in Australia, fear of change to a new way of doing business, negative perception created by the collapsed and unsuccessful market makers, a lack of trust, education and awareness of the benefits of e-marketplaces.

The above findings indicate that overcoming conservatism, finding a champion in the organization to promote the role of e-market services in B2B exchanges, the need for a leader company in the industry to support net market maker services, breaking existing supplier and buyer relationships, convincing buyers and sellers of the benefits achieved from intermediary services, the need for standards to support trust and security, winning investor confidence and overcoming internal resistance are hindering the growth of net market operations in Australia and New Zealand.

5. CONCLUSION

From this research it is apparent that technology opportunities gave birth to a new type of business, the intermediary services. Business models of e-markets are predominantly neutral third parties operating in the horizontal, vertical or both horizontal and vertical markets. Leading intermediary services include e-catalogues and auction management. It is a very competitive business which is indicative of the need for new, innovative and integrated services and business strategies for revenue expansion and future survival. The evidence that intermediary business is possible from disparate technology platforms reveals that for technology savvy people this is a business opportunity without a huge outlay. It is also evident that technology skills, industry knowledge and the implementation of e-procurement contributed positively to the business of net market makers. Challenges faced by these business entities are the same as those of most innovations including change management and leadership issues, overcoming resistance to change, creating trust to win investor confidence and the need for standards to support the marketplace.

Further research issues arising from this research are identification of future business trends in the B2B e-space, impact of new technologies in e-procurement, a sound business model and establishing the value of ancillary services of logistics management, e-supply chain management. Determining the issues for developing standards for secure e-marketplace negotiations and transactions is also required.

REFERENCES

1. Archer, N. and J. Gebauer (2002), B2B Applications to Support Business Transactions: Overview and Management Considerations. Business to Business Electronic Commerce: Challenges and Solutions, M. Warkentin. Hershey, Idea Group Publishing, pp 22-50.
2. B2B in Australia, New Zealand http://www.ebizchronicle.com/backgrounders01/january/bcg_autralia.htm
3. Bryant, G., 2002, Online Exchanges: Buoyancy Test, http://www.brw.com.au/stories/20010206/8767.asp
4. Chaffey, D., 2002, E-Business and E-Commerce Management, Prentice Hall, Great Britain.

5. Christiaanse, E., Sinnecker, R. and Mossinkoff, M., 2001, 'The Impact of B2B Exchanges on Brick and Mortar Intermediaries: The Elemica Case', 9th European Conference on Information Systems, Bled, Slovenia, June 27 –29, pp 422 – 432.
6. Dagenais, T. and Gautschi, D, 2002, Net Markets: Driving Success In The B2b Networked Economy, McGraw-Hill Ryerson, Canada.
7. Frew, W., 2002, The Trouble with E-Markets, http://www.brw.com.au/stories/20020502/14540.asp 9 August 2002
8. Gallaugher, J. M. and S. C. Ramanathan (2002). Online Exchanges and Beyond: Issues and Challenges in Crafting Successful B2B Marketplaces, Business to Business Electronic Commerce: Challenges and Solutions, M. Warkentin. Hershey, Idea Group Publishing, pp 51-70.
9. Kaplan, Steven, Sawhney and Mohanbir, 2000, 'E-Hubs: The New B2B arketplaces', Harvard Business Review, May/June 2000, Volume 78, Issue 3, pp 97.
10. Lee, H. G. and Clark, T. H., 1999, 'Electronic Intermediaries: Trust Building and Market Differentiation', Proceedings of the 32nd Hawaii International Conference on Systems Sciences.
11. NOIE Report (2001, 'B2B E-Commerce: Capturing Value Online', Commonwealth of Australia http://www.noie.gov.au/publications/NOIE/B2B/start.htm March 2003
12. Palmer, J. W., 2002, Electronic Market Mechanism to Support Business to Business Opportunities in Emerging Economies, Online Report, 27/11/2002.
13. Pavlou, P. A. and O. A. El Sawy (2002), A Classification Scheme for B2B Exchanges and Implications for Interorganizational eCommerce, Business to Business Electronic Commerce: Challenges and Solutions. M. Warkentin. Hershey, Idea Group Publishing, pp 1-21.
14. Raisinghani, M., and Hanebeck, H. L., Journal of Electronic Commerce Research, Volume 3, No 2, pp 86 – 97.
15. Singh, M. and Thomson, D., 2002, 'An E-Procurement Model For B2B Exchanges: An Australian Example', 15th Bled Conference on eReality: Constructing the eEconomy, June 17 – 19, Slovenia, CD ROM.
16. Standing, C. and Stockdale, R., 2002, 15th Bled Conference on eReality: Constructing the eEconomy, June 17 – 19, Slovenia, CD ROM.
17. Swedish Trade Council
18. http://www.emarketservices.com/about_emarket_places/main.htm
19. Thomson, D. and Singh, M., 2001, 'A Macro Level Business Model For E-Enabled Procurement', CollEcTeR, December 3-4 pp 227 – 237.
20. Weill, P. and Vitale, M., 2001, Place to Space, Harvard Business School Press, USA.
21. Zwass, Vladimir, Foundations of Information Systems, McGraw Hill www.netstoreusa.com/cbbooks/069/0697133125

Chapter 5

THE SUCCESS STRATEGIES FOR HYBRID BUSINESS MODEL

Savanid Vatanasakdakul, Eugene Lee Boon Kiat and Joan Cooper

Abstract: The advent of Internet technologies has impacted the way we do business. It was once believed that the Internet would bring about a business revolution and all business would be e-business. However, the revolution of the Internet has not yet turned out to be anything we once thought it would be. There is evidenced indicating that not all businesses have transformed to online business and not all consumers prefer to engage in online activities. In fact, some consumers still prefer to engage in traditional business processes. The reality is a hybrid model, which is convergence of traditional and online methods. This paper aims to analyze hybrid consumer behavior and develop a theoretical model, which serves as a guideline to better understanding hybrid consumer behavior and their needs.

Key words: E-commerce, business model, business strategies, and consumer behavior

1. INTRODUCTION

Since the reported failure of dotcom companies, the argument that e-commerce will replace traditional business model is not supported (Steinfield *et al.*, 2002; Wind *et al.*, 2002). While dotcom companies are struggling to gain competitive advantages in the non-linear and rapid change of business environment, much attention has shifted from pure play Internet based business model to hybrid business model. The hybrid business model is also known as "multi-channel commerce", "bricks-and-clicks" or "clicks-and-mortar". It focuses on maximizing the competitive advantage from the combination of both the web channel and the physical channel (Steinfiled *et al.*, 2001).

Despite the attention focused on the hybrid business model strategies, there is little research that addresses the relationship of hybrid consumer behavior and how it affects the building of a hybrid business model. Research shows that consumers do not change as quickly as the rapid change of technology (Wind *et al.* 2002, p.291). Evidence exists that not all consumers prefer to engage in online activities and some consumers still prefer to engage in traditional business processes. In fact, a consumer can use both online and offline methods. Today, the reality is a "*hybrid consumer*", which is the convergence of the traditional consumer and cyber consumer (Wind *et al.* 2002, p.xiii).

While the "hybrid consumer" has been defined, there has been little research that identified the behavioral characteristic of such a consumer (Wind *et al.* 2002, p.xiv). However, one piece of research conducted by *Jupiter* showed that 68 percent of American did research online but purchased in store and 54 percent of them did research in store but purchased online (Wind *et al.* 2002, p.xv). The questions are how can we understand the behavior of such a complex behavior, what are the criteria that consumers rely on when choosing their preferences between online and offline methods and how does this affect the development of an organization business strategies?

This paper aims to examine hybrid consumer behavior. By understanding hybrid consumer behavior, organizations can understand both traditional and technology requirements. This will help organizations better understand customer requirements and can be used to model its business according to its customers needs. Thus, this paper proposed a theoretical model to analyze hybrid consumer behavior, which serves as a guideline to better understanding hybrid consumer behavior and their needs.

Three research methods were used to conduct this study namely literature review, observation and interview. The authors spent two months working closely with one of the top local IT companies in Malaysia in order to understand the real world situation and to identify the existing problems and their needs. Firstly, the authors reviewed the existing theories and frameworks related to failure of online business and the emerging of hybrid business model. While the literature identified many of the same problems that were being experienced by the IT company interviewed, there was a gap in the literature with regard to how to solve these problems. We concluded with a theoretical model to examine the hybrid consumer.

2. FAILURE OF DOTCOM AND EMERGING OF HYBRID BUSINESS MODEL

During the Internet bubble, there was a proliferation of e-commerce model and the exponential growth of the Internet as a commercial medium has resulted in the development of a number of frameworks that seek to enable a better understanding of what businesses are doing on the Web (Timmers, 1998). Much attention was paid to the Internet as the only channels that will save cost and reach global markets. Many new online business models were created such as dell.com with its online customization business model, priceline.com with its discount business model and ebay.com with its auction business model. Researchers even predicted that in the long run Internet stores would replace traditional stores (Armbruster, 2002).

In contrast, there is now evidence showing that online businesses are struggling to find a recipe for success (Krishnamurthy 2003, p.4). Some businesses survive such as E-Bay, Homestore.com, Yahoo.com, Amazon.com and Priceline.com (Anajana, 2003). But many failed such as Toysmart.com (Morehead, 2000), eToys (Wired News, 2002), Pets.com (Sullivan, 2000), Violet.com (Glasner, 2000) and Boo.com (Grose, 2003). According to *Internetweek*, more than 500 online businesses, that have obtained more than 1 million of investment, had to stop their operations since January 2000 (Kauffman *et. al.*, 2001). The research conducted by *Merrill Lynch* showed that 75 percent of all e-commerce start-ups fail (Grose, 2003).

The failure of dotcom companies is a widely debated topic. Much research has been undertaken to analyze the causes of failure. Some experts said that online business model failed as they faced problems in cost of setting up and maintaining an online store, requirements of physical distribution, poor financial management and lack of trust from the consumers (Armbruster, 2002 and Grose, 2003). Moreover, one of the most interesting points that caused the failure is misunderstanding the value of consumer (Krishnamurthy 2003, p.4). The literature identified two main causes of failures.

Firstly, in information and communication technologies (ICTs), in particularly, Internet related products and services, the value chain for consumer has increased its importance. The success or failure of the technology providers depends on how well they understand the user perspective (Barr, 2000; Neal 2002, p.4). Many technology providers failed because they tend to start from what they know about technology and what they can offer rather than trying to understand what consumers want (Barr, 2000; Barua, 2001). Thus, participants who wish to succeed in offering these technologies based products and services need to understand social and culture dimension of the consumers (Barr, 2000).

Secondly, many of dotcom companies proposed new online business model to provide new ways of doing things such as online reservation, online banking, online payment, and online shopping. However, persuading customers and suppliers to change to new ways of doing things is not a simple task because customers do not change as quickly as the rapid change of technology (Wind *et al.* 2002, p.291; Krishnamurthy 2003, pp.4-5). Thus, companies should not "underestimate the ease with which human behavior can change" (Krishnamurthy 2003, p.4).

Even though, many dotcom companies fail. Still, the rapid population growth of the Internet has attracted many people to use Internet as a medium of communication. As the result, many businesses are still seeing the opportunities that underlie the Internet. However, as there is no solid business model when a new technology is launched, many of the remaining online businesses are struggling to sustain themselves in the dynamic and non-linear business environment (Kauffman *et al.*, 2001; Krishnamurthy 2003, pp.14-15).

A paradox situation occurs when the online business world is struggling to find a recipe for success and traditional business are trying to identify their needs to implement online strategies. This has resulted in the emergence of a new business model known as a "Hybrid business model". The hybrid business model is the combination of business strategies between online and offline strategies. This type of business makes use of traditional commerce and e-commerce to give them competitive advantage from different market places. Today, some businesses choose a business model that integrates physical and online channels because they realized the benefits from both channels.

The success story of a hybrid model, for example, can be seen in the case of Tesco. Tesco uses a hybrid strategy to compete with the past pioneers of online grocer such as Peapod and Webvan. Tesco implemented online service bases on the existing business platform of the retail stores in the UK. By mid 2001, Tesco was handling 70,000 orders per week and increased $400 million in sales the year before (Wind *et al.* 2002, pp. xiv - xv).

Integrating e-commerce with a physical channel is a challenge as well as integrating traditional commerce with online channel (Stienfield, 2002). Many theorists have long recognized the channels conflict problem, which may occur when there are alternatives paths for delivering products to end consumers (Stern and Ansary, 1992). Inappropriate channel mix strategies may create even more conflict potential. For example, an e-commerce strategy can cannibalize sales from an organization own physical operation (Stenfiled, 2002).

It is obvious that many businesses are now facing problems in finding a right business model. The questions rise in what should be an optimal solution in integration online and offline strategies for a business. In the other words, should a business use online or offline strategies, should a traditional business transform to online business, should a business use hybrid strategies and to what extend that they should implement online or offline strategies. In responding to these questions, one of the factors to be considered is to understand consumer behavior.

3. CONVERGENCE OF CONSUMER BEHAVIOUR

The advent of Internet technologies has impacted not only business processes but also on the consumer. Using Internet technologies in business processes, consumers have more channels to interact with companies (Peter *et al.* 2001, p.5). As a result, the behavioral of consumer has also changed. With the Internet, a new kind of consumer has emerged known as "cyber consumer". Unlike the traditional way of doing business where customers have to go to physical store, cyber consumers prefer to conduct buying activities through companies' websites.

Before Internet technology was integrated into the business model, consumers conduct buying activities by using conventional ways such as phone, fax and store front. But after the Internet was integrated into business models, some consumers prefer to use online methods to interact with companies. It can be seen that the Internet has lead to the diversity of human behaviors. This phenomenon has created the divergence of consumer belief "which states of almost constant warfare with each other" (Neef 1999, p. 114).

Belief and attitudes towards a specific subject can lead to a specific type of behavior (Fishbein and Ajzen, 1975). The divergence of consumer belief can lead to the creation of two schools of thought (Neef 1999, p.114; Wind *et al.* 2002, p. xiii) that are traditional oriented group (non Internet-adopter) and technology oriented group (Internet adopter) (see Figure 1).

Two schools of thought,

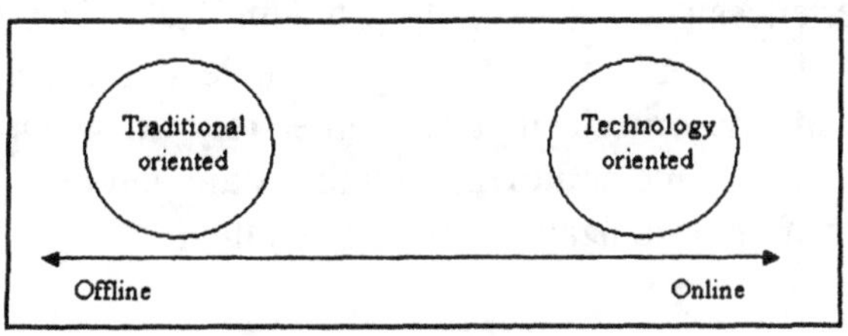

Figure-1. Divergence of consumer belief and behavior

The traditional oriented group is the group of people who do not believe in technology. This might be because of fears of changing, unawareness of technology and prefer physical contact. They prefer to use traditional methods. This group is also known as "traditional consumer" (Wind *et al.* 2002, p. xiv) or "high touch group" (Neef 1999, p. 114). On the other hand, technology oriented group is a group of people who has a strong belief in technology. They believe that technology can solve anything. They tend to be an early adopter of new technology. This group is also known as "cyber consumer" (Wind *et al.* 2002, p.xiv) or "high tech group" (Neef 1999, p.114).

If consumers really could be classified as one or the other, it would not be too difficult for companies to predict consumer's needs. However, a number of problems occur when the real world cannot clearly separate people into two groups. Even though, an online strategy is viewed as a very efficient communication channel, an individual may use both online and offline methods in a buying process. Thus, today the reality is a "hybrid consumer", which is the convergence of the traditional consumer and cyber consumer (wind *et. al.* 2002, p.xiii) (see Figure 2).

Two schools of thought,

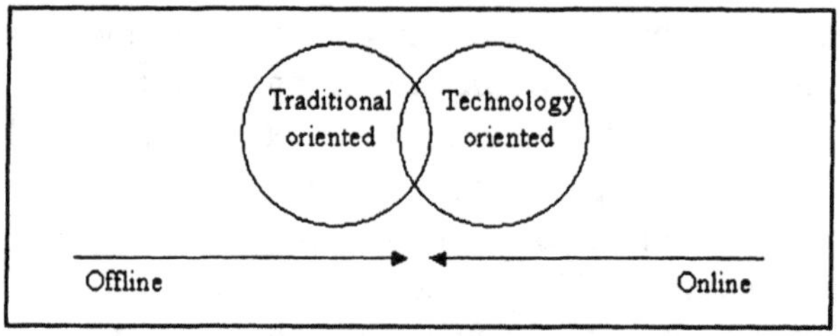

Figure -2. Convergence of consumer belief and behavior

There are several reasons to explain why consumer behavior is moving towards convergence. For example, some people adopt online methods because of the increase in accessible to computer and increase in technological knowledge. On the other hand, some people can move experience from online back to offline. For instance, consumers may have bad online experiences such as security problem, lack of trust and poor user interface design. The problem is how can we understand such a complicated behavior, what are the criteria that consumer rely on when choosing their preferences between online and offline methods are and how does this affect the development of an organization business strategies.

4. THEORETICAL FOUNDATION

To be able to generate an explicit explanation of hybrid consumer behaviors, the theory of reasoned action (TRA) and the theory of planned behavior (TPB), which are among the widely supported theories, were chosen. These two theories form a basis on the concept of attitude towards behavior, which is often applied in marketing research and many consumer theories were derived from (Arnould *et al.* 2002, p.427; Davis *et al.,* 1989 and Schillewaert *et al.,* 2000).

4.1 Theory of reasoned action

In 1967, Professor Martin Fishbein developed a theory of reasoned action. It is an expanded version of the initial Fishbein model. Feshbein suggested that the behavior of intention could be predicted from knowledge of consumers' attitudes, social normative belief, personal belief recognized. Behavior could be predicted from knowledge of behavior of intentions and the judged influence of extraneous events such as time constraint, technology constraint, competitive promotion and economic constraint. Consumers' actual behaviors may not be the same as their intentions. This model awakened a strong influence that other people can have on a person's behavior. In some circumstance, social influences can be as strong as or stronger than the person' own attitude (Arnould *et al.* 2002, pp.474 – 475). This model can be represented as follows (Arnould *et al.* 2002, pp.474 – 475).

$$BI = f(A, NBs * MC, NBp) \quad (1)$$

$$B = f(BI, EE) \quad (2)$$

where,
B = Behavior, A = Attitude, NBs = Social normative beliefs
MC = Motivation to comply, NBp = Personal normative beliefs
BI = Behavioral intention, EE = Judge influence of extraneous events

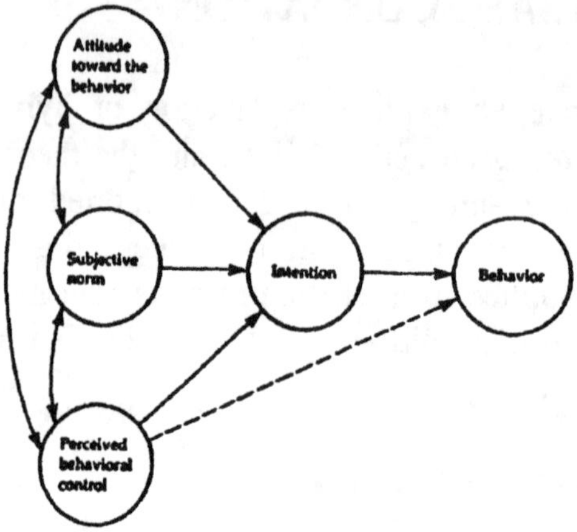

Figure -3. Theory of Planned Behavior

(Source: Ajzen, 1991)

4.2 Theory of planned behavior

The theory of planned behavior (TPB) is a well-established general theory of social psychology developed by Professor Icek Ajzen. In 1988, the Theory of Planned Behavior (TPB) was added to the existing model of reasoned action theory to address the inadequacies that Ajzen and Fishbein had identified through their research using the TRA. TBP asserts that specific salient beliefs influence given behavioral perceptions and subsequent actual behavior. Following TPB, there are three types of beliefs in the TPB that impact three perceptual constructs: behavioral beliefs that influence attitudes, normative beliefs that affect subjective norm, and control beliefs that shape perceived behavioral controls (Ajzen, 1991).

5. A MODEL FOR ANALYSIS HYBRID CONSUMER BEHAVIOUR

To understand the behavior of hybrid consumer, this paper proposed a conceptual model, which serves as a guideline to analyze hybrid consumers. This model applies a basis concept of TPB and TRA. These theories suggest that "attitudes could explain human actions" (Fishbein and Ajzen, 1975). The authors have recognized that the behavior of hybrid consumer can be explained by belief and behavior. The main purposes of this framework are as follows:

1. To understand hybrid consumer behavior both actual and intention in adoption of online and offline channels in buying process;
2. To identify the hidden potential target market for online and offline strategies;
3. To identify the strategies that companies can use to avoid channel conflicts in online and offline channels.

According to the proposed theoretical framework, hybrid consumer behavior both intention and actual behaviors are guided by three main variables of consideration: influence factor; attitude & beliefs; and extraneous.

Influence factors are the consumers' background and experience that formulate consumers' attitude and beliefs. Influence factors could be from external source such as social influence and subjective norms and it is also from internal source, which is self-interested.

Attitude and Belief is the feeling and expectation of consumers toward a specific object that could lead to the formation of behavioral intention. It explains the likelihood of consumers' intention and it predicts the potential of behavioral.

Behavioral intention represents the plan of what consumers have in their minds before the actual process. Behavioral intention explains what consumers' want or/and need is.

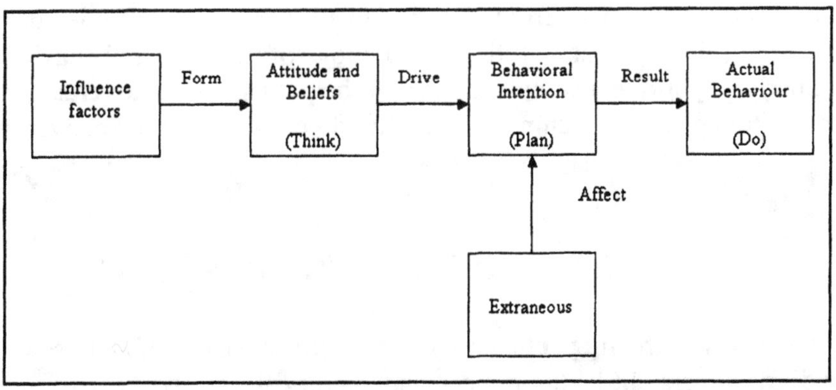

Figure -4. Proposed theoretical model to analyze hybrid consumer behavior

Actual behavior is the fact-based information. It indicates the reality of consumer behavior. Information regards actual behavior can be obtained by research survey such as questionnaire, interview, and observation with consumers. Consumers' intention behavior may be different from the actual behavior.

Extraneous is the factors that affect consumers' behavioral intention and it may lead to different actions. It explains why consumers' actions are not always as they originally plan. For example, extraneous could include time constraint, technology constraint, economic constraint or competitive promotion (Arnould *et al.* 2002, pp. 474-475). Table 1 summarizes the explanation of key terms in this model.

Table -1. Construct and Measurer

Construct	Definition	Question
Influence Factors	The factors that form attitude and belief.	What are the factors that form consumers' attitude and belief?
Attitude and Belief	Feeling and expectation toward a specific object.	What do consumers think about it?
Behavioral intention	The behavior that plan to perform.	What do consumers plan to do?
Actual Behavior	The actual performance of behavior.	What is the actual behavior?
Extraneous	The factors that make the difference between intention behavior and actual behavior.	What make consumers behave differently from their original plans?

This model can be used to evaluate hybrid consumer behavior towards the adoption of online (through Internet) and offline (without Internet) activities. One can benefit from applying this model to better understanding the current situation and future prediction of consumer behavior, in particularly, hybrid consumer. This will help businesses to create right business strategies for reaching the new hybrid consumer in the convergence era.

6. CONCLUSION AND FUTURE RESEARCH

As there is a growing reliance on the hybrid business model as well as hybrid consumers, little theoretical and empirical research exists on this topic. The theoretical model to analyze hybrid consumer behavior is proposed to fill this gap. This model is based on the underline concept of belief and behavior derived from TRA and TPB. The outcomes from this model is seeking to understand hybrid consumer behavior both actual and intention in adoption of online and offline channels in buying process to identify the hidden potential target market in building online and offline strategies. Future research is currently being conduct to test the efficiency of this model.

ACKNOWLEDGEMENTS

We would like to thank Mr. Eric Sito, General Manager of the Dataprep Group, Malaysia, for his assistance in the development of this paper.

REFERENCES

Anajana, C. S.. (2003), *Internet Business Models*, Available: http://stylusinc.com/website/business_models.htm [Accessed 2003, March 19]

Ajzen, Icek. (1991), The Theory of Planned Behavior. *Organizational Behavior and Human Decision Processes [Online], Vol. 50*, pp. 179-211. Available: http://www-unix.oit.umass.edu/~aizen/publications.html [Accessed 2003, January 10]

Armbruster, Karin. (2002), *Hybrid Strategies in consumer retail markets*, Available http://e-business.fhbb.ch/eb/publications.nsf/id/126 [Accessed 2002, December 28]

Arnould, Eric J. & Price, Linda and Zinkhan, George (2002), *Consumer*, 2nd end, McGraw-Hill, USA

Barr, T. (2000), *User perspectives of the future of the Internet*, paper deliver at ALIA (The Australian Library and Information Association), 24-26 October 2000, National Convention Centre, Canberra.

Barua, Anitesh & Konana, Prabhudev & Whinston, Andrew B. and Yin, Fang. (2001), Managing E-Business Transformation: Opportunities and Value Assessment, *Sloan Management Review, Vol. 34(1)*, Fall 2001, pp. 36-44

Boar, Bernard H. (2001), The art of strategic planning for information technologies, 2nd end, Wiley and Sons, USA

Davis, F. D., Bagozzi, R. P., and Warshaw, P. R. (1989). User Acceptance of Computer Technology: A Comparison of Two Theoretical Models. *Management Science, Vol. 35*, pp. 982-1003.

Fishbein, M., and Ajzen, I. (1975). Belief, Attitude, Intention and Behavior: An Introduction to Theory and Research, Reading, MA: Addison-Wesley.

Glasner, Joanna. (2000), Dead Startups: Shrinking Violets. Available: http://www.wired.com/news/business/0,1367,36069,00.html [Accessed 2003, 20 January]

Greenstein, Marilyn (2000), Electronic Commerce: Security, Risk Management and Control, McGraw-Hill, USA

Grose, Thomas k. (2003), *Wanna Be an Internet Star?*, Available www.time.com/time/europe/specials/ eeurope/process/dotcomstar.html [Accessed 2003, March 16]

Janger, Wander (2000), Modelling consumer behaviour (PhD thesis, University of Groningen), [Online]. Available: www.ub.rug.nl/eldoc/dis/ppsw/w.jager/titlecon.pdf [Accessed 2002, December 27)

Kauffman, Robert J. and Wang, Bin (2001), *The Success and Failure of Dotcom: A Multi-Method Survival Analysis*, Proceedings of 6th Conference on Information Systems and Technology, Florida

Krishnamurthy, Sandeep (2003), *E-commerce Management: Text and Cases*, Thomson, Canada

Morehead, Nicholas (2000), Toysmart: Bankruptcy Litmus Test, Available http://www.wired.com/news/business/0,1367,37517,00.html [Accessed 2003, January 15]

Neal, Cathy M. (2002), *Consumer behavior: implications for marketing strategies*, 3rd edn, McGraw-Hill, Australia

Neef, D. (1999), *A Little Knowledge is a Dangerous Thing: Understanding Our Global Knowledge Economy*, Butterworth Heinemann, Melbourne

Peter, Paul and Olson, Jerry C. (2001), *Consumer Behavior*, 6th edn, McGraw-Hill, USA

Schillewaert N., Ahearne M., Frambach R. & Moenaert R. (2000), The acceptance of information technology in the sales force, ISBM Working Paper Series.

Steinfield, Charles & Bouwman, Harry and Adelaar, Thomas (2001), *Combining physical and virtual channels: Opportunities, Imperatives and Challenges*, Proceedings of 14th Bled Electronic Commerce Conference on e-Everything: e-Commerce, e-Government, e-Household, e-Democracy, Slovenia

Steinfield, Charles. (2002), Understanding Click and Mortar E-Commerce Approaches: A Conceptual Framework and Research Agenda, *Journal of Interactive Advertising, Vol. 2*, Number 2, Spring 2002

Steinfield, Charles & Adelaar, Thomas & Lai and Ying, Ju. (2002), *Integrating Brick and Mortar Locations with E-Commerce:Understanding Synergy Opportunities*, Proceedings of 35th Hawaii International Conference on System Sciences – 2002, IEEE, Hawaii

Stern, L. W. and A. I. Ansary (1992), *Marketing channels*, Englewood Cliffs, NJ: Prentice Hall.

Sullivan, Tom (2000), *Dot-com shakeout clears landscape, executives say*, Available http://archive.infoworld.com/articles/hn/xml/00/11/09/001109hnbea.xml [Accessed 2003, January 15]

Timmers, P (1998) 'Business Models for Electronic Markets', Journal of Electronic Markets 8(2), July: pp. 3-8.

Wind, Yoram & Mahanjan,Vijay & Gunther, Robert E. (2002), *Convergence Marketing: Strategies for reaching the New Hybrid Consumer*, Prentice Hall, USA

Wired News Report. (2002), Toys Were Us, Available http://www.wired.com/news/business/0,1367,56144,00.html [Accessed 2003, January 20]

SECTION 3
Innovative Structures in the Internet

Chapter 6

INFLUENCE OF ELECTRONIC BUSINESS TECHNOLOGIES ON SUPPLY CHAIN TRANSFORMATIONS

Wojciech Cellary, Sergiusz Strykowski[1]

Abstract: Electronic business technologies enable to separate the information flow from the commodity flow in the supply chains. As a consequence, the information role of intermediary links in supply chains may be highly reduced. Only if customers are familiar with electronic business technologies, a manufacturer may conduct business processes electronically without the need of intermediaries as information providers. In the paper, a model with both real and virtual warehouses and direct shipments is analyzed. It is shown how such model may help to evolutionary shift business from a traditional to electronic one. A system of e-procurement *e-MAX-ML* deployed in Philips Lighting is discussed as an example of successful application of the described model.

Keywords: logistics, supply chain optimization, supply chain transformation, electronic business, virtual warehouse

1. INTRODUCTION

During the 1990s most of the companies deployed the ERP systems resulting in the optimization of their internal business processes. A motivation for such optimization was improvement of the competitiveness of a company versus other companies of the same profile. As the possibility of achieving further improvement in this way attained its limits, the next step was thinking about competition in terms of whole supply chains. A critical issue within the supply chain optimization is sharing sensitive data between all chain components. Due to development of information and

[1] author responsible for correspondence

telecommunication technologies, the technical problems of that issue have been overcome. Unfortunately, mutual trust among companies composing a common supply chain that is necessary to share sensitive information between them appears to be very hard to be established. The main reason for that is participation of a particular company, for example, a retailer, in several supply chains being often the rival ones. Under such circumstances it is hard to be sure that sensitive information about one supply chain will not be passed to the rival one. All this means that on the contrary to theoretical good wishes, in practice, chain optimization is hard to achieve.

In this paper we propose a different approach to the problem of supply chain optimization. This approach consists of supply chain transformation due to the application of electronic business technologies. It is based on the observation that electronic business technologies enable to separate the information flow from the commodity flow. While the commodity flow remains unchanged, because of unavoidable transportation delays, the flow of digital information is significantly different from the flow of information on a paper carrier. Digital information is available immediately after its creation for all the supply chain links independently of their geographical dispersion, which are authorized to access it. In this paper we analyze the consequences of the separation of the digital information flow from the commodity flow. In particular we show the changed role of intermediaries in supply chains, which may be gradually eliminated from the information flow.

The reminder of the paper is organized as follows. In Section 2, two main cases of traditional supply chains are analyzed: (1) a supply chain with a direct distribution realized by a manufacturer, and (2) a supply chain with an indirect distribution realized by intermediaries. In Section 3, the role of intermediaries is analyzed in a new environment provided by electronic business technologies. Section 4 is devoted to possible supply chain transformations. In Section 5, a supply chain with virtual and real warehouses and direct shipments is proposed for a manufacturer of mass commodities of small prices. In Section 6, the case of Philips Lighting is studied. Section 7 concludes the paper.

2. TRADITIONAL SUPPLY CHAINS

In the traditional economy there are three fundamental forms of the supply chain on the side from a manufacturer to ultimate customers:

- manufacturer → wholesaler → retailer → customer;
- manufacturer → retailer → customer;
- manufacturer → customer;

The above list of the supply chain forms can be reduced to two generalized forms:

- a supply chain with direct shipments performed by the manufacturer;
- a supply chain with distribution performed by intermediaries.

In the traditional approach to supply chain design, the direct shipments performed by the manufacturer is advised only in the particular cases like when the single purchase order is large, a product has short consumption or usage deadline, or a manufacturer has to control product's parameters till the very moment of transaction, because it will be responsible for them after the transaction is committed.

Much more advised is the distribution performed by intermediaries:

1. an intermediary represents certain community of customers; it stays in touch with them and can offer them conveniently located place of purchase – an intermediary is able to effectively promote products in a certain area or for certain community of customers;
2. an intermediary accepts orders of the size too small and thus not acceptable by the manufacturer – an intermediary aggregates a number of small orders in a large one that may be accepted by the manufacturer;
3. if orders are collected by an intermediary in another place than product storage, then an intermediary becomes responsible for steering the flow of orders from a customer to a manufacturer and supervising product shipments from the manufacturer to the customer;
4. an intermediary takes part in exchanging and sharing information among other links of the supply chain;
5. intermediaries instead of the manufacturer cover costs of maintaining distribution infrastructure – then the manufacturer perceives only variable costs depending on distribution scale;
6. intermediaries purchase products in batch orders which reduce transportation costs and let the manufacturer to shift inventory from its warehouses to intermediaries' ones;
7. an intermediary sells products of a manufacturer in a composition of complementary products;

The preference in the traditional economy for the supply chain with intermediaries results from the fact of tightly coupled flows of information and commodities. In the traditional supply chain, a manufacturer is located at the beginning of the commodity flow and at the end of the information flow. The manufacturer needs an intermediary to pass its products down towards the ultimate customer, and to pass information up – from ultimate customers towards itself. For this reason, the main values added by intermediaries are: promotion of products passed to customers and marketing information collected from the customers.

In the traditional economy, to some extend, a manufacturer is forced to use intermediaries. If a manufacturer likes to be in a direct contact with ultimate customers, i.e., to be located at the beginning of the information flow and gain additional, extraordinary profit, it should either establish its own points of sales at the local markets, or it should apply direct shipment strategy. However, establishing and running points of sales at the local markets requires know-how about buy-and-sell business and knowledge about these local markets. A manufacturer usually does not have the required knowledge, and does not want to invest much in points of sales either. Moreover, customers are used to visit points of sales that sell products from more than just one line, and within each line – products of more than just one brand. Thus, own manufacturer points of sales are generally not justified economically.

3. ANALYSIS OF THE ROLE OF INTERMEDIARIES IN DIGITAL ECONOMY

Let us recall that the essence of digital economy consists in [2]:

- replacing everywhere possible physical products and services with digital products and services provided through the network;
- introducing new digital products and services which do not have corresponding physical equivalents, because they did not meet required conditions on cost, time, etc.

In case of a supply chain of physical commodities, the above statement concerns digital information flowing down and up a supply chain. Characteristics of digital information compared to information on a physical carrier: low costs, unlimited access time, 24 hours per 7 days availability, unlimited geographical range, arbitrary capacity, and possibility of automatic processing, cause the existing business rules to change. In this section, for analysis reasons, we assume a perfect digital economy, where all supply chain links, from a manufacturer to ultimate customers, are well acquainted with electronic business technology, they have unconstrained and unlimited access to the network, and they accept using these technologies in everyday private and business life. Under such conditions we now reexamine one by one argument for the role of intermediaries in supply chains.

1. *An intermediary represents certain community of customers whom it can effectively sell products to.*

This argument is meaningless in the perfect digital economy, where customers may directly access any information about products, prices, delivery conditions, etc. via network.

2. *An intermediary aggregates a number of small orders into a larger one that may be accepted by the manufacturer.*

It is true that every order entails certain fixed cost, independent from the value of this order – fulfillment of orders below certain value is thus unprofitable for the manufacturer. Although, comparing to the traditional economy, in the digital economy the profitability threshold is much lower. For example, processing a banking transaction in a brick-and-mortar bank costs $1.25, while processing the same transaction in an Internet bank costs only $0.01. We conclude that in the digital economy the role of an intermediary as the aggregator of small orders becomes significantly reduced or even removed as unneeded.

3. *An intermediary steers the flow of orders from customers to a manufacturer and supervises product shipments from the manufacturer to customers.*

In the digital economy, where business processes are conducted through the network using appropriate IT systems, digital orders may be placed by customers directly at the manufacturer site. On the Internet, both the manufacturer and a customer are equally far away from each other – the distance is "one mouse click". A manufacturer does not need help of an intermediary to electronically manage a fulfillment of these orders (i.e., to steer the flow of commodities), even if the customer picks up products at the intermediary's location.

4. *An intermediary takes part in exchanging and sharing information among other members of the supply chain.*

Digital information can be very quickly and cheaply shared between any numbers of partners despite their geographical locations. Intermediaries are unnecessary in information flow and even harmful – in the digital economy an entity that only passes information on does not add any value but only generates delays.

5. *Intermediaries instead of a manufacturer cover costs of maintaining distribution infrastructure*

In the digital economy, distribution infrastructure at local markets becomes simpler in comparison to the traditional economy which reduces holding costs. In the traditional economy, this infrastructure is designed for storing physical goods and for conducting business processes to establish and perform transactions. In the digital economy, the second function disappears, because business processes are conducted through the network, so they do not require any infrastructure at local markets. Moreover, the digital nature of business processes makes it possible to process them in a centralized IT system at the manufacturer site. Such system enables global optimization and cost reduction of distribution infrastructure at local markets. As a result, the only unavoidable costs are those of rented space in a cheap, local warehouse, where customers may pick up ordered goods, or where a delivery company may pick them up and deliver to a destination indicated electronically. Such centralized IT system makes it also possible to detect a situation when aggregation of many independent orders goes beyond a threshold that justifies direct shipment from a manufacturer's facilities to locations chosen by customers.

In the digital economy, the holding cost of distribution infrastructure at local markets is no longer as important as it was in the traditional economy.

6. *Intermediaries purchase products in batch orders, which reduce transportation costs and let the manufacturer shift inventory from its warehouses to intermediaries' ones.*

The electronic business technologies enable all customers to place their orders despite their geographical locations at one centralized IT system. Knowledge about all the placed orders permits a manufacturer to optimize the whole distribution. The manufacturer can use distribution strategies like aggregating shipments or cross-docking at the global level and thus fulfill a number of small orders at the same costs as one big order in the traditional economy. Moreover, the manufacturer does not have to fulfill each order itself – certain ones can be passed to intermediaries that hold warehouses nearby the customer who placed the order. In this case, the manufacturer steers the flow of information and commodities, while only shipment is performed by the intermediary.

7. *An intermediary sells products of a manufacturer in a composition of complementary products.*

In the digital economy, a manufacturer can attaint this goal easier than in the traditional economy. For example, a manufacturer of lamps can agree with a manufacturer of lighting accessories to apply affiliate business model [7]. In such case, both of them put mutual references on their web portals. In more advanced business models, they mutually accept orders for partner's products.

4. SUPPLY CHAIN TRANSFORMATIONS

As follows from Section 3, in the digital economy importance and range of the roles of intermediaries change a lot. This change is a result of the separation of the information flow from the commodity one. The integrated information-commodity function of the intermediary, natural in the traditional economy, undergoes a split into two independent functions in the digital economy. In the perfect digital economy mentioned in Section 2, the information function is performed exclusively by a manufacturer. In the current transitory situation, the information function may be electronically performed by a manufacturer only for technologically advanced customers. For customers who do not have sufficient skills and/or technical possibilities this function has to be performed by the intermediaries in the traditional way.

The commodity function changes in the digital economy not as radically as the information function, though still significantly. As mentioned in Section 3, a manufacturer may aggregate small orders and organize a single direct shipment to several customers who placed their orders independently. A manufacturer may also buy or rent warehouses located nearby local markets and manage them remotely. Finally, for the customers who are not yet ready to conduct business processes electronically, traditional warehouses run by intermediaries have to remain.

The new forms of the supply chain created as the result of electronic business technologies consist in purging certain components:

- from the flow of information (orders) only;
- from both the flow of information and the flow of commodities.

As there are two supply chain components between the manufacturer and the ultimate customer, the following ones may be purged:

- wholesaler,
- retailer,
- both wholesaler and retailer.

Purging a real wholesaler or retailer component from the flow of information (orders) is equivalent to creation of a virtual warehouse or a virtual retail outlet, respectively. Purging the wholesaler or retailer component from the flow of commodities is followed by either the strategy of direct shipments from manufacturer's facilities to the next level (after purging the wholesaler – to a retailer; after purging a retailer – to the ultimate customer) or shipments from distributed local warehouses belonging to or rented by the manufacturer and managed remotely.

These observations have led us to propose five transformation models of the traditional supply chain as the result of electronic business technology application:

- Model 1: The virtual manufacturer's warehouse with storage and distribution provided by a real wholesaler.
- Model 2: The virtual manufacturer's warehouse with direct shipments from manufacturer's facilities.
- Model 3: The combined Model 1 and 2 equipped with a switch depending on order conditions.
- Model 4: The virtual manufacturer's retail outlet and the real warehouse providing commodity storage and transportation.
- Model 5: The virtual manufacturer's retail outlet with the strategy of direct shipments to the customers.

The virtual warehouse and virtual retail outlet are the IT system running on the manufacturer's servers accessible via the Internet. Authorized wholesalers, retailers, and ultimate customers using web browsers can manage business processes related with their orders through the Internet directly from their own facilities.

Everything what is needed to place orders at a virtual manufacturer's facility is a computer connected to the Internet and a web browser. Clients start working with logging into their accounts. To prevent from unauthorized interception of data, the whole transmission between the clients' computers and the manufacturer's server is encrypted. Because clients must log in to their individual accounts, the manufacturer can track their individual behavior patterns. The available OLAP software permits to investigate not only the behavior and pattern of a single customer but also analyze data cross-sectional in a freely chosen manner. For example, the manufacturer can analyze purchase patterns of retailers from the same region and then seeks for correlations among them. In the traditional economy, where the manufacturer is located at the end of information flow, such analysis made by it would be impossible.

5. SUPPLY CHAIN WITH VIRTUAL AND REAL WAREHOUSES AND DIRECT SHIPMENTS

The model described in this section is devoted to a large manufacturer of cheap commodities. The model is presented in *Figure 1.* The red dashed arrows indicate the flow of orders; the black solid arrows indicate the flow of commodities.

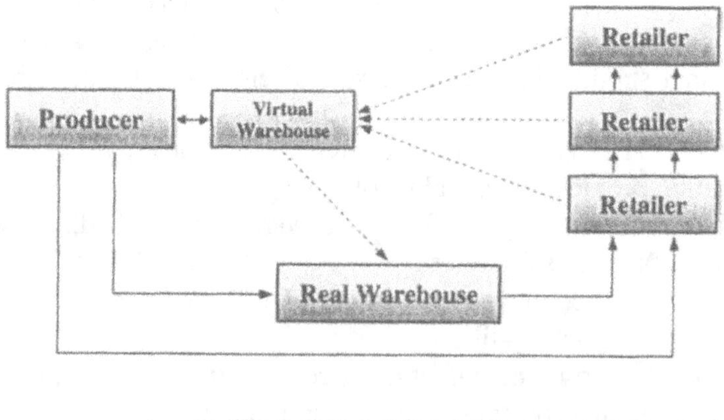

Figure 1. Supply chain with virtual and real warehouse and direct shipments

The manufacturer collects orders from retailers in its virtual warehouse. The manufacturer services the information part of order fulfillment. The physical distribution is either executed autonomously by the manufacturer applying direct shipment strategy or it is passed to a chosen real warehouse.

The term "real warehouse" refers here to:

- a warehouse being an independent enterprise that purchases commodities from the manufacturer; or
- a warehouse being an independent enterprise where the manufacturer rents space, stores and manages its own commodities; or
- the remote manufacturer's distribution center.

The distribution decision for a particular order depends on two main conditions: cost analysis (first of all transportation costs) and potential profit. Typically, if the manufacturer manages to aggregate shipments to several retailers that placed orders independently, then the manufacturer applies direct delivery. Because the real wholesaler has been purged from the supply chain and transport has been optimized, the manufacturer takes the wholesaler's profit; it also turns into its own profit the wholesaler's overheads and variable costs. However, if appears that direct shipment is unprofitable, then the order is passed to the one of warehouses located nearby to the point of delivery.

The main advantages of the above model are as follows:

1. The relationship between a manufacturer and wholesalers reverses. Now a wholesaler does not select the manufacturer as one of available suppliers, but the manufacturer selects one of available wholesalers to fulfill an order. This of course improves the manufacturer's position in the negotiations with wholesalers.
2. The direct and straightforward contact with retailers – the manufacturer moved one position down along the information flow. The manufacturer does not need any longer an intermediary to obtain information from further links. Moreover, now the manufacturer provides the wholesaler with information.

The additional advantages of the proposed model are as follows:

1. In the model, real warehouses are not entirely purged from the supply chain starting with a given manufacturer, so the warehouses do not feel abandoned and forced to move to competitive manufacturers.
2. A manufacturer presents itself to real warehouses as a supplier of orders.
3. A manufacturer passes orders to those real wholesalers which signed the best agreements with it.
4. A manufacturer negotiates with real wholesalers more profitable agreements in exchange for the promise of passing orders to them.
5. A manufacturer and real wholesalers fulfill the orders which are too small to be fulfilled independently by the manufacturer.
6. A manufacturer take over the fulfillment of the most profitable orders which leads to taking over partial profits of real wholesalers and to

turning into the profit the wholesalers' costs due to purging one link from the supply chain.

7. A manufacturer flexibly steers the development of self-served market due to market conditions – readiness of certain manufacturer departments, transport availability, etc.
8. A manufacturer is protected from a price shock provoked by a competitive manufacturer that using electronic business technologies could offer lower prices to retailers.
9. A manufacturer is protected against independent creators of competitive virtual warehouses which instead of the manufacturer could take over profits following from purging real wholesalers from the flow of commodities.

6. CASE STUDY – PHILIPS LIGHTING

The model presented in Section 5 has been successfully implemented at Philips Lighting Poland.

Philips Lighting is the world leader in the lighting industry. It runs several factories worldwide, including eight in Poland. The headquarters of Philips Lighting Central and Eastern Europe Division is located in Pila city in western Poland. Their production is a very wide range of lamps and lighting accessories sold only to large wholesalers which sell them farther to smaller wholesalers or to retailers. In the remainder of this section, the term "manufacturer" refers to Philips Lighting Central and Eastern Europe Division, the term "customers" refers to its customers, and the term "products" refers to lamps and other lighting accessories manufactured by Philips Lighting.

The manufacturer sells products manufactured in its own plants and in other plants belonging to the corporation including ones from abroad. The catalog of products contains about 3,000 items. Each customer has several places of product delivery nationwide. Most of these places operate permanently – they are usually warehouses or distribution centers. Some of them are open temporarily for a few deliveries only – this happens for example in case of direct shipments to constructions. In average, each customer places one order containing over 100 lines every three days. Quantity of a typical line is hundreds of items. Prior to placing orders, each customer signs a contract with the manufacturer. The contract deals with product delivery conditions, credit conditions, etc.

Before deploying the model presented in Section 5, Philips Lighting serviced customers in traditional manner. The detailed analysis of that supply chain of Philips Lighting led to the following observations:

1. One category of wholesalers' customers is mass-retailers. Their typical order is suitable to be serviced directly from the manufacturer facilities.
2. Characteristics of orders placed by smaller wholesalers and retailers at large wholesalers' are such that they may be serviced directly by the manufacturer if advanced distribution strategies were applied.
3. The market as the whole is not yet ready to conduct all business processed exclusively electronically, but large and medium enterprises are developed enough with regard to this issue to partially shift to electronic business processes.

The model implemented at Philips Lighting is a mixed one – it contains both the digital economy elements (the virtual warehouse) and the traditional economy ones (the real warehouses). The model lets Philips Lighting introduce electronic business solutions in the evolutionary way depending on market development. The failure of Internet rush at the turn of the century has proven that the complete transition from the traditional to Internet business style will last at least several years. The chosen model does not make real wholesalers turn away rapidly from the Philips Lighting which could be a case of other models. The experience proves that there will be many customers who will prefer the traditional not electronic manner of conducting business processes. For this reason, it does not seem that it will be possible to completely purge intermediaries; although their role will be decreasing as the time passes.

The deployment process of electronic solutions finally ended in 2001 with development of the Internet e-Order Management System *e-MAX-ML*. Currently, this system reaches 28 countries from Central and Eastern Europe, Middle East, and Middle Asia. Almost 70% of company turnover in the professional channel is serviced by the system.

7. CONCLUSIONS

Digital economy enters a phase of necessary deployment of new business models basing on electronic business technologies. The case of Philips Lighting described in this paper proves advantages of evolutionary approach to supply chain transformations rather than revolutionary one. Particularly efficient are models which, on the one hand, follow evolution of customer readiness to use new technologies, and on the other hand stimulate customers to develop. The model presented in this paper, which consists of both virtual and real warehouses, as well as direct shipments, appeared successful while not provoking conflicts between the new and the traditional distribution channels.

It is worth noticing that the new business models and electronic distribution channels were so well accepted on emerging markets. It may be explained by the fact that traditional means of conducting business on emerging markets are not perfect, so managers from those markets are very ready to improve their business deploying new solutions.

REFERENCES

1. Amor, D., The E-Business (R)evolution: Living and Working in an Interconnected World, Prentice Hall PTR, 2000.
2. Cellary, W. (editor), Poland and the Global Information Society: Logging on, UNDP, Warsaw (Poland), 2002.
3. Cellary, W., S. Strykowski, Deploying the E-Commerce Application System in Wholesale Commerce, Proc. of 4th International Conference on Business Information System, Springer-Verlag, pp. 221 – 217, Poznan (Poland), 2000.
4. Cellary, W., S. Strykowski, L. Remesz, M. Toboła, Supply Chain Evolution in Electronic Economy, (in Polish), in: K. Rutkowski (editor), On-Line Logistics, PWE, Warsaw (Poland), 2002.
5. Kalakota, R., M. Robinson, e-Business 2.0: Roadmap for Success, Addison-Wesley, 2001.
6. Kaminsky, P., D. Simchi-Levi, E. Simchi-Levi, Designing and Managing the Supply Chain – Concepts, Strategies, and Case Studies, Irwin/McGraw-Hill, 1999.
7. Rappa, M., Managing The Digital Enterprise, Department of Technology Management, North Carolina State University, USA, 2002.
8. Shapiro, J. F., Modeling the Supply Chain, Duxbury Press, 2000.
9. Strykowski S., Ph.D. Thesis: Supply Chain Transformations under the Influence of Electronic Business Technologies (in Polish), submitted for evaluation, PUE, Poznan (Poland), 2003.

Chapter 7

PRODUCT PLATFORMS FOR THE MEDIA INDUSTRY

Lutz Koehler, Markus Anding, Thomas Hess

Abstract: The concept of product platforms has already been successfully applied in various industries such as automobile and software. Advantages reach from cost savings due to the re-use of product components to the simplified individualization of products. In the media industry, the application of product platforms promises ample benefits due to the industry-specific first copy cost effect and the move towards production and distribution of digital goods. However, product platforms for the media industry have so far not been widely discussed. The paper on hand proposes a framework for the platform-based production and distribution of digital goods, which is customizable for different media business models. The heart of the platform is a repository, storing content modules and content meta data. Further, we distinguish different components for the input and output of content and the platform management, which can be assembled according to the requirements of an individual business model.

1. INTRODUCTION

The media industry has been evolving rapidly in the recent past. Advances in information and communication technologies create new opportunities for the production and distribution of media products. At the same time, recipients become more demanding and expect ever new and individualized offerings. However, the creation of media products is a sophisticated process, which so far has not yet been highly elaborated in media related research. In other industries, product platforms have proven to be a successful concept for the cost effective, fast and individual production of different products. So are automobiles assembled of a well defined set of components and software is made of different modules on the basis of specified communication interfaces. Building on the idea of product modularization, product platforms enable and reinforce the re-use of product components like car chassis in the automobile industry and software modules in the software industry. Due to the first copy cost effect, the re-use of content modules on the basis of platforms can bring about substantial cost savings for the media industry and opens up a new field of research. In the paper on hand, we transfer the concept of product platforms to the media industry and propose a conceptual framework for the platform based production of media products. Thereby, we build upon first works of Meyer and Lehnerd [MeLe97]. We first give an introduction to the concept of product platforms in general and focus on the specificities in the production and modularization of media products in chapter two. After deriving requirements for product platforms in the media industry in chapter three, we develop a framework for the platform-based production of media products, which supports the re-use of content modules and is easily customizable to the needs of different media business models.

2. BASIC CONSIDERATIONS

In the following we discuss the principles of product platforms. Then we introduce a concept for the modularization of media content which can form the basis for product platforms in media companies.

2.1 Product platforms

Platform concepts focus on the similarities of products and processes, and do no longer consider products and technologies of a company as independent from each other [Sawh98, pp. 54]. Theory distinguishes product platforms, brand platforms, processing platforms, global platforms and customer platforms as five different types of platforms [Sawh98, pp. 56]. Most widely discussed are product platforms which support the planning and realization of complex products and production concepts and can be compared to a construction kit that enables a company to create differentiated offerings based on modularized products. Products are created by combining the modules on the basis of product architectures. Accordingly, Meyer and Lehnerd define a product platform as a "set of subsystems and interfaces that form a common structure from which a stream of derivative products can be efficiently developed and produced." [MeLe97, p. 39]

Using multi-purpose modules in multiple products is possible through a sweeping unification of different product architectures in a company, resulting in a significant advantage of efficiency compared to monolithic approaches in both product innovation and production. In monolithic approaches, products are developed and produced as a single unit and not as modules. The exact extent of the efficiency advantage depends on the modules' multiple usability and is greatest when it is possible to reproduce modules very cheaply or almost for free; for example software modules. The individual products of a product platform sum up to a product family as they depend on the same technologies and belong to the same group of products.

A first approach towards using product platforms for information products can be found in Meyer and Zack and Meyer and Lehnerd [MeZa96, pp. 46; MeLe97, pp. 209]. The product platforms for information products introduced in these works are part of an architecture for information products and consist of a repository in which contents and structures are saved. The repository in this model assists in the manufacturing process of information products. In their work, the authors outline the concept of product platforms on a basic level. The framework to be introduced in this paper picks up these approaches, substantiates and develops them. The concept furthermore provides a starting point for the construction of a product platform prototype for media companies.

In most cases, product platforms use the principle of modularization, an idea as old as the idea of division of labor. Tasks or problems are no longer treated as monolithic devices but are fractionized in sub-tasks or sub-problems, which possess clear-cut limited functions and interfaces, and can be looked at and worked with independently and can be combined with each other. The individual parts are called units or modules, the underlying principle is called modularization (process-related view), or modularity (condition-related view). Product platforms transfer these principles to products and create product architectures based on modules that are designed independently but still function as an integrated whole. However, the procedure is only beneficial if the partition is precise, unambiguous and complete. For this purpose the procedure is divided into visible design rules and hidden design parameters [BaCl97, p.86]. The visible design rules include all aspects that affect subsequent design decisions and therefore it is useful to establish all of them in an early stage of the design process and to communicate them to those involved. The visible design rules are split into the three categories: architecture, interfaces and standards. The architecture specifies the modules of a system and their functions while the interfaces describe in detail how the modules interact, including how they fit together, connect and communicate. The standards describe how the conformity of a module can be tested and how their performance can be measured. The hidden design parameters contain the decisions that do not affect the design beyond the local module. Hence modules can even be changed, chosen or substituted at a late stage in the design process without the necessity to communicate to anyone beyond the module design team.

2.2 Modularization of media products: modules and construction plans

A starting point for the modularization of media content is the media value chain, which, in three generic stages, describes the production, bundling and distribution of media contents [ScHe02, p. 9]. A form of modularized products already seems to exist when we relate these stages to the stages of modularization: media contents are first produced and then bundled to products to be distributed. Therefore, regarding media products, we can already talk about modularized products, which admittedly have a low degree of modularization because the products on the step of production are usually already marketable. This could, for example, be single TV shows which are distributed in a bundled form for economic as well as technical reason together with other products, for example a whole TV program. Therefore, a modularized production, taking a view from the bundling perspective, already exists. Modularization in the production stage, which will be discussed from the view of modularization in the following, is traditionally very low, mainly because of technical restrictions. In this context we will talk about monolithic content production. The main effort on the production step is clearly the creation of new original content, the so called "first copies", which are bundled and distributed on the succeeding steps and whose production costs are called "first copy costs" [ShVa99, pp. 20]. New technical means, such as the digitalization as a basis of media independent data management, low-priced storage and duplication of content, allow a modularized production of the former monolithically produced content. Consequently, the first copy and therewith the first copy costs split into two components, according to the two steps of modularization. Original content modules (for example text modules of a book, which are not marketable on their own) are generated during stage one and bundled during stage two to create original content products.

We can split the creation of first copies into two steps. While the activity of producing the "first" first copy is the same as in the production of original content (for example writing a text), the activity to produce the "second" first copy is a bundling activity. Compared to monolithic content production, there is a higher degree of modularization. While in monolithic content production the major effort lies in the production of original content, in the modularized production of content the effort is shifted towards the bundling of content modules and the effort for production of original content is reduced. A higher re-use of modules accompanies a rising degree of modularization. The effects on the first copy costs are obvious. Minimizing costs to a level significantly below the cost of monolithic content production can be attained by an ideal proportioning of efforts on both steps of the content production.

In the following discussion of the content platform, the term 'module' will be used with respect to media content being stored and processed by the platform. Moreover, it is illustrated how these modules can be bundled to products and how "construction plans" have to be designed for this purpose.

In other industries, modules are built by decomposing products into their individual parts. This decomposition can take place in different steps by deconstructing every part as often as possible. As a result, hierarchies develop between the modules on these different steps. Due to this procedure, modularization of products ends with screws and nuts for manufacturing goods or with single bits for digital goods. In this case, modularization of digital media contents could be extended to single bits. However, from an economic point of view such an extensive modularization is not reasonable. Therefore, the modularization of media products should not be based on a technical perspective; it should take place with respect to the human perception of content.

The particular size of single modules, such as a single audio stream, can be determined according to the economic value of the module. This value results on the one hand from the difference between additional costs of producing an original module (which can be saved by re-use) and the costs of finding a re-usable module in a database, and on the other hand, by the value customers assign to the module. For the technical representation of content, the markup language XML would be appropriate as it distinguishes between content, semantics and layout of media contents (see figure 1).

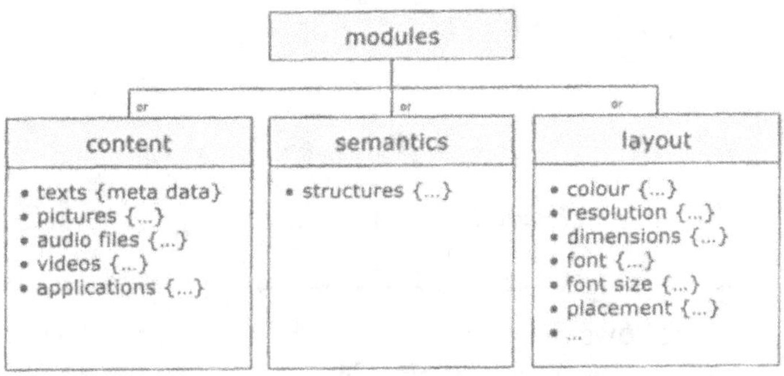

Figure 1: modules of media content

Following this approach, media content like texts, pictures, audio/video files and applications are content modules and classified using meta data. For consumption, the content modules have to be bound to a semantic module and a layout module. The semantic module describes the structure of the content and can be compared to a grammar. A semantic module, for example, describes that a specific content consists of a headline, an abstract, a body text and a number of graphics. These descriptions are stored in semantic modules, which in turn consist of meta data. To display the content, a layout is needed, which, with the help of layout modules, describes features like the font, font size, resolution or placement of a headline. Different layout modules results from different layout guidelines for different structures.

Each of these modules is exactly described by meta data. The meta data could comprise names or descriptions of people and objects on a photograph, the date of the shooting, the name of the photographer and the copyright of the picture. Copyright information is particularly important meta data for media companies because it enables the company to track the holder of a specific copyright and whether a module is free to be published or not.

According to the procedure described earlier, modularized media products consist of content-, layout- and semantic modules. We further want to discuss how these single modules can be assembled to a marketable content product. The assembly is based on construction plans, which determine the modules contained by the product, the way these modules are linked and the methods to be used for editing the product and the modules (see figure 2).

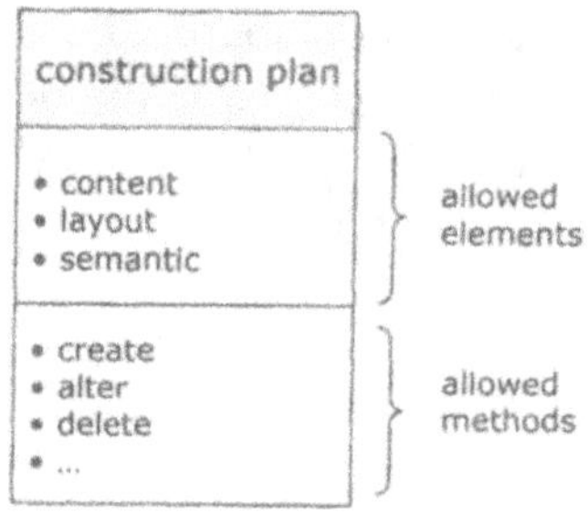

Figure 2: construction plans

The concept of construction plans resembles object-oriented software development and follows the class-principle, which contains the data used as well as the methods used to edit the data. Further, hierarchies can exist among different construction plans, similar to class-hierarchies in object-oriented software development. Construction plans on a higher hierarchy level define the spectrum of modules and methods, which can be used in construction plans on a lower level. For example, the construction plan of a product line defines the modules and methods that can be used in the construction plans of products in this product line. The construction plan of the product in turn defines the methods and modules that can be used to produce an issue of this media product (e.g. an issue of a newspaper). The construction plan of an issue finally contains all elements and methods that are used in this issue. Therefore, it is reasonable to name the construction plan of an issue "logic first copy". The construction plan of an issue again can be compared to an object (or instance) of the product class. Later on, the "physical first copy" is assembled on the basis of the construction plan for the issue. The physical first copy in turn serves as a master copy for copies of the media content.

3. PRODUCT PLATFORMS IN MEDIA COMPANIES

In the following, requirements for a product platform in media companies are analyzed, and, based on the results, a framework for the specific development of product platforms in media companies is introduced.

3.1 Requirements for product platforms in media companies

The media industry is made up of several sectors with different companies using different media and running different business models. Because of these differences, it is not possible to develop just one product platform that fits all companies. Instead, it is useful to compose a framework, which allows the design of customized product platforms, aligned to the specifics of each media company. To allow the design of individual platforms, the framework needs a modular design with independent system components that can easily be combined using standardized interfaces.

As discussed in the former section, not only the product platform should have a modular design, also the media content stored and processed by the platform needs to be organized in a modular way to benefit from all advantages of modularization. Hence, the product platform needs to support the modularization of media content based on the concepts introduced before. Besides the storage of media content, product platforms should also offer functions to support other activities. We will discuss some of these activities starting from the media value chain with its three steps: production, bundling and distribution. The framework for product platforms in media companies has to offer individually applicable system components that support these activities, even if some companies do not cover all value activities in their business model and do not need all of the system components. By analyzing the different value chain activities it is possible to identify more detailed activities, which are contained in the value chain activities but which should be offered independently by the framework.

Production, the first activity of the value chain, can be divided into purchase, creation and integration of media content depending on the business models of different media companies. Some companies just create their own media content, others only purchase media content from producers or syndicators. Some integrate media content from outside the company, e.g. media content from customers for other customers, and again others use several ways to produce media content. Because of these differences in the business models, the framework should offer independent components which support each function and which can be chosen independently.

The second value activity represents the process of bundling separate content modules to products. For modularized media content as described before, we can differentiate between bundling single modules to larger units (e.g. combining pictures and texts with the help of construction plans to articles) and bundling these larger units to complete products (e.g. different articles to a whole newspaper). The media content produced and bundled within the first and second activity is stored in the repository of the product platform. In the case of modularized media products, content-, semantic- and layout modules as well as the construction plans have to be stored separately from each other.

The third value chain activity covers all activities linked to the distribution of media content. These are activities of copying, marketing, distributing and syndicating media content as well as after sales services. Each of these activities has to be represented by a distinct system component within the framework. In this framework, the system component for copying media content supports the production of a first-copy and the duplication of this first-copy. Further, marketing-, distribution- and syndication components support the delivery of the media content to customers. Finally, the component for after sales services assists activities after the delivery of content, such as updates. In addition to these activities, there are activities that belong to or influence all steps of the value chain. Product innovation, for instance, influences all activities, because the process of product innovation determines the construction plans of product groups, product lines, or products and also whether modules acquired from other firms or created within the company or whether the products will be syndicated or not. Hence, product innovation not just affects one activity, it influences all activities. As a further activity relevant for the whole value chain, a controlling function collects data about all processes and units and generates information to manage all processes and units. According to the specific needs of media companies, product platforms should also support these functions by providing independent system components.

Finally, further components are needed to manage the product platform itself. With the help of these management components, system components can be integrated into the platform, updated or removed. In addition, management components support the administration of users and workflows.

3.2 A framework for product platforms in media companies

To enable the design of individual product platforms aligned with the specific needs of a media company, a framework to develop product platforms is discussed in this chapter. This framework is composed of independent system components, which can be combined and which interact using standardized interfaces. Within this framework, it is distinguished between the product platform in the narrow sense, which solely represents the repository of the platform, and the product platform in the broader sense, which – besides the repository – comprises additional system components for further functions.

3.2.1 Product platform in the narrow sense

The product platforms in the narrow sense consist of the content repository, which includes the content-, layout-, and semantic modules of a company and the construction plans of product groups, product lines, products etc. The repository itself is divided into a multimedia database and a XML database. While the first stores different content modules, the second holds layout- and semantic modules as well as the meta data describing content modules in the multimedia database. The product platform in the narrow sense only provides the storage of media content modules. To extend the functionalities of the product platform, different system components have to be integrated using standardized interfaces. By integrating system components, the product platform can be individually adjusted and designed, according to the needs of a specific media company.

3.2.2 Product platforms in the broader sense

The product platform in the broader sense is designed according to the principles of component based software development and comprises the product platform in narrow sense and different integrated system components to expand the functionalities. The various system components can be categorized into three groups, which resemble the various value chain activities of a company (see figure 3). The first group of components supports the input of media content into the platform. The second group supports the output of media content while the third group provides components for the platform management. In the following the different components are further discussed. Each group consists of several components, which can be combined according to the needs and the business model of a company. One of the input components is the purchasing component, which supports the process of buying media content from producers or content syndicators and which includes several functionalities, e.g. a billing system, a retrieval system and a system to generate relevant meta data for the purchased media content. The next component supports the creation of new content by the company and provides editing functionality for different media, systems to split content into modules as well as content mining systems. A further component enables the bundling of various modules with the help of construction plans. The bundling does not need to take place manually by editors, but can be an automated process, based on personalized product configurations. These, in turn, could be generated manually by the customers or automatically by collaborative filtering systems. Another component to input media content could support the integration of media content from outside the company. This media content may be advertising content from advertising customers or media content such as comments produced by consumers.

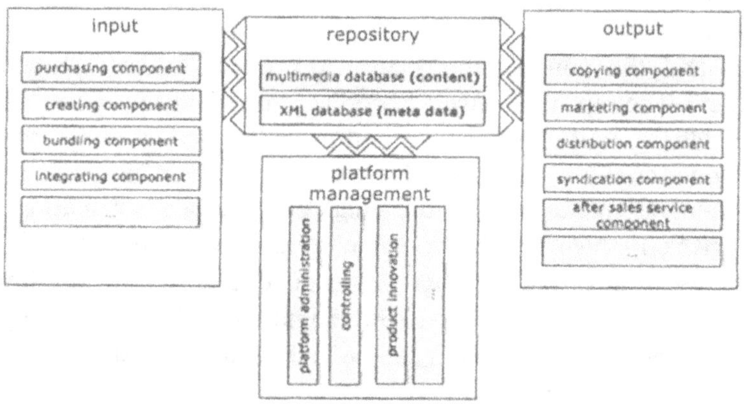

Figure 3: framework for product platforms in media companies

The group of output-components also provides various functionalities. The copying component allows the physical production of the first-copy, linking the construction plan of an issue to a medium. This component also supports the duplication of this first-copy. The marketing component supports the advertisement for and the sale of the product. Functionalities for analyzing customer target groups or producing the advertisements can be integrated in this component. The distribution of products to the consumers is supported by a distribution component, enabling the customization of the products (e.g. to fit the customers' technical equipment). The B2B-distribution of products to other companies can be supported by a specific syndication component, which is customized to the particular features of these deals. Another component may support the after sales services, including content updates after the sales process.

The third group of components supports the management of the product platform. To administer the platform, these components offer functionalities to plan and design the platform as well as to integrate, update and remove components. The components also deliver functionalities for user- and workflow administration. The second part of this group supports the overall activities of the company, such as product innovation or controlling. The component for product innovation offers functionalities to design construction plans for product groups, product lines and single products. It also supports market testing of new products. The different groups can be extended by new components, which may be customized for a specific company if the functionality of the standard components is insufficient.

4. SUMMARY AND OUTLOOK

Starting point of the paper was the question, how product platforms as a very successful concept in various industries can be applied in the media industry. We have discussed the modularization of products as the basic concept of product platforms with respect to media content and have extended the first copy approach by a distinction of content modules and content products. This allows the definition of a platform for media content and shows that product platforms unfold a huge potential in the media industry due to the reinforced re-use of content modules. The proposed platform framework covers a content repository for storing content modules and meta data as well as supportive components for content input, output and the platform management. Due to the many different types of media companies to be supported, the proposal puts emphasis on the flexible configuration of the platform depending on the functions needed by specific media business models. While each customized version of the platform framework for a specific media business model represents a platform for the production and distribution of media content, the platform framework itself can be considered a meta platform for the definition of specific platforms.

Further research is needed on the definition of platform components and the XML-based representation of content semantics and layouts in the repository (e.g. Document Type Definitions for different layouts). Furthermore, the demand for an easy configuration of the platform requires interfaces between the repository and the supportive components to be well considered in order to allow smooth access. After these refinements, a prototypical implementation should deliver a proof of concept of the proposed framework and pave the way for its application in a real world environment.

REFERENCES

[BaCl97] Baldwin, C.Y.; Clark, K.B.: Managing in an age of modularity, Harvard Business Review, September – October 1997, pp. 84-93.

[MeLe97]Meyer, M.H.; Lehnerd, A.P.: The Power of Product Platforms: building value and cost leadership, New York, 1997.

[MeUt93]Meyer, M.H.; Utterback, J.M.: The Product Family and the Dynamics of Core Capability, Sloan Management Review, Spring 1993, pp. 29–47.

[MeZa96]Meyer, M.H.; Zack, M.H.: The Design and Development of Information Products, Sloan Management Review, Spring 1996, pp. 43-59.

[Sawh98]Sawhney, M.S.: Leveraged High-Variety Strategies: From Portfolio Thinking to Plattform Thinking, in: Journal of the Academy of Marketing Science, 1998, pp. 54-61.

[ScHe02]Schumann, M.; Hess, T.: Grundfragen der Medienwirtschaft, Berlin, 2002.

[ShVa99]Shapiro, C.; Varian, H.R.: Information Rules: astrategic guide to the network economy, Boston, 1999.

Chapter 8

DYNAMIC MANAGEMENT OF BUSINESS SERVICE QUALITY IN COLLABORATIVE COMMERCE SYSTEMS

Bob Roberts, Adomas Svirskas

Abstract: The importance of e-business services quality continues to increase, as the use of e-commerce to support business activities becomes a routine practice for many enterprises. Companies need robust, predictable and efficient services that they can rely upon. This paper explores ways of how e-business quality can be established, monitored, reported and managed. A review of the literature considers the work recently undertaken in both business-level quality of service (QoS) and the QoS issues at the infrastructure level, as well as the relationship between these two areas. From a practical research perspective, the work within the framework of the EU-funded LAURA project is presented. The key goal of this project is to facilitate interregional zones of adaptive electronic commerce using the potential of the ebXML architecture. A Dynamic QoS Management framework is proposed to inform the implementation of QoS and SLA concepts within the LAURA project.

Key words: e-business, e-commerce, virtual organisation, service level agreement, SLA, quality of service, QoS, ebXML, Web Services

1. INTRODUCTION

The original motivation of the research discussed in this paper was to ensure an adequate level of business services quality in Request-based Virtual Organisations (RBVO). This task is very important for dynamic ad-hoc formations of enterprises, taking different roles in many different business processes at the same time. Quality of service (QoS) needs to be specified, agreed upon, measured and monitored. Parties should be compensated for deviations from a service level agreement (SLA). These provisions promote trust between business parties and make e-business services predictable and manageable. The further organisation of this paper is as follows: Section 2 serves as a literature overview concerning Virtual Organisations, Quality of Service and Service Level agreements, Section 3 introduces the LAURA project, Section 4 explains QoS management within the chosen ebXML framework, raises some open points and discusses possible solutions, Section 5 summarises the overall effort and outlines future research.

2. REVIEW OF THE LITERATURE

2.1 Virtual Organisations and Enterprises

The concept of the virtual organisation (VO) is briefly discussed as our research on e-business is focused within this particular context. Tapscott (1996) discussed how companies are increasingly focusing on their core competencies and partnering with other organisations having complementary competencies. In this scenario companies may enter into multiple and ever changing partnerships of collaboration to achieve competitive success. There are, however, various definitions of VO that reflect different perceptions of the concept. Some broad definitions of VOs are:

> *A VO or company is one whose members are geographically apart, usually working by computer e-mail and groupware while appearing to others to be a single, unified organisation with a real physical location (VO 2002)*
>
> *A temporary network of independent companies that come together quickly to exploit fast-changing opportunities (Byrne 1993)*
>
> *An opportunistic alliance of core competencies distributed among a number of distinct operating entities within a single large company or among a group of independent companies (Goldman et al 1995)*
>
> *Less a discrete enterprise and more an ever varying cluster of common activities in the midst of a vast fabric of relationships (Sieber et al 999)*

A VO is described in most cases as a network among organisations and/or individuals. Another opinion is that VO's should not be viewed solely as networks among organisations or individuals but as a radical approach to management, or a strategic approach that leads to dynamically re-configurable enterprises (Saabeel et al 2002). In such cases, the inherent limitations of being able to plan in an uncertain environment are taken into account by creating high structural flexibility (Davidow & Malone 1993).

2.2 Service Level Agreements

In its most basic form, an SLA is a contract or agreement that formalizes a business relationship, or part of the relationship, between two parties. Most often, it takes the form of a negotiated contract made between a service provider and a customer and defines a price paid in exchange for an entitlement to a product or service to be delivered under certain terms, conditions, and with certain financial guarantees.

In many e-commerce contracts, the service provider agrees to guarantee a certain level of QoS for each class of service, and in return, each business agrees to pay the service provider for satisfying the QoS guarantees in serving its set of customers. Those contracts are based on a Service Level Agreement (SLA) between each business and the service provider that defines the QoS guarantees for a class of service, the cost model under which these guarantees will be satisfied, and the anticipated level of per-class requests from the customers of the e-business.

Per-class SLAs usually have clauses where the service provider gains revenue for each request satisfying the per-class SLA, and incurs a penalty for each request failing to do so. Hence, in order to maximize profits, one needs to pay attention to resource management issues, so that customers can be served according to the restrictions defined in the SLAs.

The concept of SLA is revisited, from a more technical perspective, later in this paper (Section 4) where it is discussed in relation to an SLA specification language and the ebXML context.

2.3 Quality of Service in E-Business

Quality of Service can be viewed as a collective measure of the level of service a provider delivers to its customers or subscribers. In telecommunications, for example, QoS can be characterized by several basic performance criteria, including availability (low downtime), error performance, response time and throughput, lost calls or transmissions due to network congestion, connection set-up time, and speed of fault detection and correction. Service providers may guarantee subscribers a particular level of QoS as defined by a service agreement (IDC 2000, IBM2001, Schmidt 2000). The overall definition of the QoS of an e-business service can be determined by consolidating the compliance with the detailed SLA. To allow a machine to compare the SLAs with the real world behaviour the SLAs need to be mapped into measurements that can be taken from the running system (Sahai 2001).

The growing use of e-commerce is creating demand for SLAs with financial incentives in which service provider revenues are determined by the number of completed transactions and where there are penalties for SLA violations such as exceeding response time guarantees. Diao et al (2002) introduce a simple profit model in which the service provider receives revenues for each completed transaction and where a cost is incurred if response times are excessive

Business-level QoS instrumentation often builds on top of the lower-layer QoS mechanisms. The closest layer in the e-business architecture stack is the application integration level, represented by Web Services, where QoS issues are quite intensively researched. For example, HP Laboratories have carried out various research projects on QoS and SLAs for Web Services. (Sahai 2001, Jin 2002, Pruyne 2000).

Sahai et al (2002) elaborate on SLA definition using XML schema-based model in an attempt to specify SLAs in precise and unambiguous manner as well as keep the specification flexible. Lammana et al (2002) propose the use of an SLA language (SLAng) to more thoroughly address the problem of SLA modelling and is further discussed below in section 4.2.

2.4 Relationship between QoS and Business Metrics

Wolter and van Moorsel (2001) discuss the effects of QoS degradations on the profitability of e-services. They characterise possible relationships between quality-of-service metrics (throughput, delay, availability) and business metrics (revenue, costs). This relationship is denoted as Q2B (quality of service to business). For IT and business managers there is a growing need to track the Q2B relationship at run-time in order to understand the consequences of QoS alterations on the bottom line. From a system management perspective, this implies monitoring both QoS and revenue or cost, as well as identifying the statistical correlation between the two. Based on such information, one may be able to tune a system or business process appropriately, thus bridging the gap between IT management and business management.

It is important to emphasize the conceptual similarity between the e-services and the RBVO underpinnings. This similarity is primarily defined by a possibility for the actors of both formations to issue requests for services upon dynamic discovery of these services. The actors can play roles of both service providers and consumers, thus enabling ad-hoc peer-to-peer value networks, depending on the circumstances. All interactions in RBVO are based on actors' requests; some of the requests might have associated monetary value, therefore the model proposed by Wolter and van Moorsel (2001) is relevant for QoS management in an RBVO environment.

Wolter and van Moorsel (2001) try to find answers to the following questions:

- Can a system of federated e-services (something conceptually similar to RBVOs) be instrumented, monitored and the Q2B relationship visualized?
- Can a business manager be notified if QoS starts influencing the bottom line?
- Can the business process be adapted to improve the overall gain in case of QoS changes?

To obtain the data necessary to correlate QoS with business metrics, e-services, which collect run-time data and visualise the above relationships are implemented using HP e-speak middleware, a proprietary service-oriented framework closely resembling Web Services architecture (HP 2001) and the HP ChaiServer framework for visualisation via Java applets (HP 1998). Visualisation of the business and QoS metrics allows a business manager to get the picture of the 'financial health' of the system and see if QoS influences this health parameter.

The conceptual and technological significance of their proposed approach has particular significance for the LAURA project discussed in section 3 because of the following attributes:

- The Service Oriented Architecture is used for the Q2B instrumentation framework. This fits with the concepts of Web Services and ebXML as HP 'e-speak' middleware is based on the same notions of unified description, registry-based publishing, dynamic discovery and on-demand invocation of services as Web Services and ebXML (HP 2001).
- It is based on a conceptual similarity to the RBVOs discussed in section 2.1 i.e. a dynamic network of business entities engaged in ad-hoc e-business activities with each other and the notion of federated e-services used to construct Q2B framework. RBVOs heavily depend on customer satisfaction and supplier flexibility, therefore dynamic monitoring, analysis of results and co-management of QoS and business processes is of great importance.
- It is a non-intrusive interceptor-based approach, orthogonal to the business e-services. This approach is used for monitoring, filtering, collecting and exchanging of QoS parameters. A chain of pluggable interceptors on either side of the service can be used to monitor different features of e-services. This technique is suitable for usage with many types of modern middleware tools. The interceptors can be implemented using different techniques (e.g. J2EE of .NET components, Web Services, etc.) for different nodes of heterogeneous network of e-services as long as they comply with the Q2B protocol
- The Q2B information exchange protocol is based on XML. This makes the information model of the Q2B framework as flexible as

XML and allows exploitation of many XML processing tools (transformation, data binding, messaging, etc.), which are available from commercial vendor and open source organisations.

Concepts from the Q2B framework are revisited in section 4 from the perspective of using the ebXML framework in the LAURA project context.

3. THE LAURA PROJECT

The authors of this paper are currently involved in an European Commission project sponsored by The Information Society Technologies (IST) Programme that is part of the Fifth Framework Programme for Research, Technological Development and Demonstration Activities. The project, called LAURA – 'Adaptive Zones for Interregional Electronic Commerce based on the concepts of Request-Based VO and sector-specific Service Level Agreements' (LAURA 2002, LAURA 2003) is directly related to the issues of e-business QoS.

LAURA is a project that innovates in terms of focusing on RBVOs, introduced in section 2.1, as a specific type of the VO taxonomy. RBVOs are clusters of partnering organizations that have replaced their vertical integration into a virtual one through collaborative networks between discrete business partners. Only some of the operations are within a particular organization - operations are now spread between separate organizations, which are linked to the original organization, to produce a new VO.

Managing and improving the service quality in both intra and inter-enterprise operations among collaborative network of enterprises, is virtually impossible in a reactive environment of a RBVO that does not provide a method of monitoring performance and measuring against SLAs. A VO has several individual enterprises-suppliers communicating with one another, fulfilling customer requests and/or triggering e-services that carry out their parts of some complex workflow of transactions. Without the right tools a VO has no way of knowing if it meets its commitments to the customer/user and supplier/user community.

Currently the described service is being implemented by the LAURA project team using an Open Source e-commerce framework called Open For Business (OFBiz). OFBiz is an e-business applications suite built on a common architecture using common data, logic and process components. This is an Open Source project with a rapidly growing user base characterized by cooperation between its creators, contributors and users. It is licensed under the MIT Open Source License (MIT 2003), which defines the rights users are granted to customize, extend, modify, repackage, resell, and use it (OFBiz 2003). The open nature of the OFBiz framework will allow required enhancements in order to accommodate both ebXML interoperability options and QoS management extensions.

4. PRACTICAL ASPECTS OF QOS AND SLA IN EBXML ENVIRONMENT

4.1 The ebXML Framework and QoS

The ebXML framework, among other things, includes declarative, executable languages to express e-business collaboration protocol profiles and agreements (CPP/CPA) in a non-proprietary, XML-based format. These specifications can be shared and ported between compliant implementations. The ebXML messaging services complements these by offering a very capable standards-based e-business messaging system (ebXML & OASIS 2000, ebXML(b) 2003). These features fit the LAURA project business requirements very well, so ebXML has been chosen as a key component for implementation of the LAURA project.

Collaboration protocol profile (CPP) describes the company capabilities, such as the supported business processes, transport, security and messaging protocols. The profile defines the functional and technical support for business processes and roles for the trading partner. A trading partner, therefore, can publish information about their supported business processes and specific technical details about their data interchange capabilities (Chiu 2002).

To a certain extent QoS is supported in the ebXML framework through Collaboration Protocol Agreements, which are based on the Collaboration Protocol Profiles of the parties. A CPA defines the system-level agreement for data interchange between the trading partners and describes all the valid, visible, and hence enforceable, interactions between the parties and the way these interactions are carried out. It is independent of the internal processes executed at each Party. Each party executes its own internal processes and interfaces them with the business collaboration described by the CPA and Process Specification document. The CPA does not expose details of a party's internal processes to the other party. The intent of the CPA is to provide a high-level specification that can be easily comprehended by humans and yet is precise enough for enforcement by computers.

4.2 Specification of SLAs

Collaboration protocol agreements (CPA), explained in the previous section define mainly conversational and protocol aspects of the business interaction between the trading partners. A CPA does not specify business level agreements between the partners, therefore a CPA is not an SLA. An SLA is an agreement between a hosting service or similar service and the clients, which are hosted on or use that service. It includes the CPA functions related to communication between the hosting service and one of its clients but is primarily for expressing the measurable aspects of the services that the hosting service is providing to its hosted clients. The SLA might well be a separate agreement that can be referenced by the CPA with the hosting service. In this case, a non-trivial question of SLA specification arises, which we will discuss here, based on some related research.

Lamanna et al. (2002) also introduce a two-dimensional SLA taxonomy in their model. Horizontal SLAs are contracted between different parties providing the same kind of service while Vertical SLAs regulate the support parties get from their underlying infrastructure. For example, a container provider can define an agreement with an ISP for network services. Once again, the resulting types of SLA differ in terms of their expressiveness, and SLAng defines them separately. Therefore, an important goal of an SLA definition language is to provide the means to accurately express features of a service in both qualitative and quantitative terms. Lamanna et al. also claim that other relevant aspects are *"the possibility to easily make comparisons between offers, to advertise and retrieve information about them, to reason about service proposals, understanding what one can offer and expect to receive, and to easily monitor QoS guarantees, both for fulfilling and claiming them......The main requirements for achieving these goals we had in mind while developing SLAng were parameterisation, compositionality, validation, monitoring and enforcement."*

The resulting SLAng is an XML language for capturing Service Level Agreements while the legal issues are addressed by embedding the SLA in an SLA contract that is essentially a framework containing one or more SLAs as well as the names of the two juridical persons contracting the agreement, together with their digital signatures.

Lamanna et al claim that XML proves ideal for the parameterisation of service level specifications that is supported at different system tiers, including vertical and horizontal agreements as indicated below.

The vertical SLAs are:

- **Application:** between applications or web services and components
- **Hosting:** between container and component providers
- **Persistence:** between a container provider and an SSP

- **Communication:** between container and network service providers.

The Horizontal SLAs that parties enter into by composing vertical SLAs are:

- **Service:** between component and web service providers
- **Container:** between container providers
- **Networking:** between network providers

(adapt. from Martinka et al 1998, Pruyne2000, Woolter et al 2001, Lammana et al 2002,)

However, Lamanna et al (2002) concede that the efficacy of SLAng does need further validation through assessing the benefits of inserting SLAng instances into standard XML-based deployment descriptors and also in testing the effectiveness of SLAng for monitoring compliance to SLAs. However, SLAng does appear to provide a language to capture SLA related information and is compatible with the ebXML framework that will serve as the basis for the LAURA project implementation (see sections 3 and 4.1 above).

4.3 Dynamic QoS Management Framework

SLAng-based service agreements can constitute the data model for the QoS and Q2B based business quality enforcement discussed in sections 2.3 and 2.4 above. Taking an eclectic approach to draw on previous work in the areas of QoS and SLAs provides a valuable and useful basis for the development of an initial framework to guide the implementation of these aspects into the LAURA project. Figure 1 reflects the structure of the framework and possible flows of information between the components while business interactions between the parties are to be carried out using ebXML-based implementation.

This initial proposed framework is based on the following main ideas that are of direct relevance to the implementation of QoS and SLAs within the LAURA project:

- Specification of the SLAs using a flexible XML-based language
- Monitoring of business services performance at runtime using middleware-specific interceptors and XML-based data structures combining metrics of both performance and business features
- Comparison of actual performance metrics against the parameters specified in the SLAs
- Calculating statistical correlation between the infrastructure-level QoS and the overall business performance, e.g. profit, through usage of business metrics associated with e-transactions
- Notification of users and managers upon certain conditions when level of QoS threatens business-level performance of the services

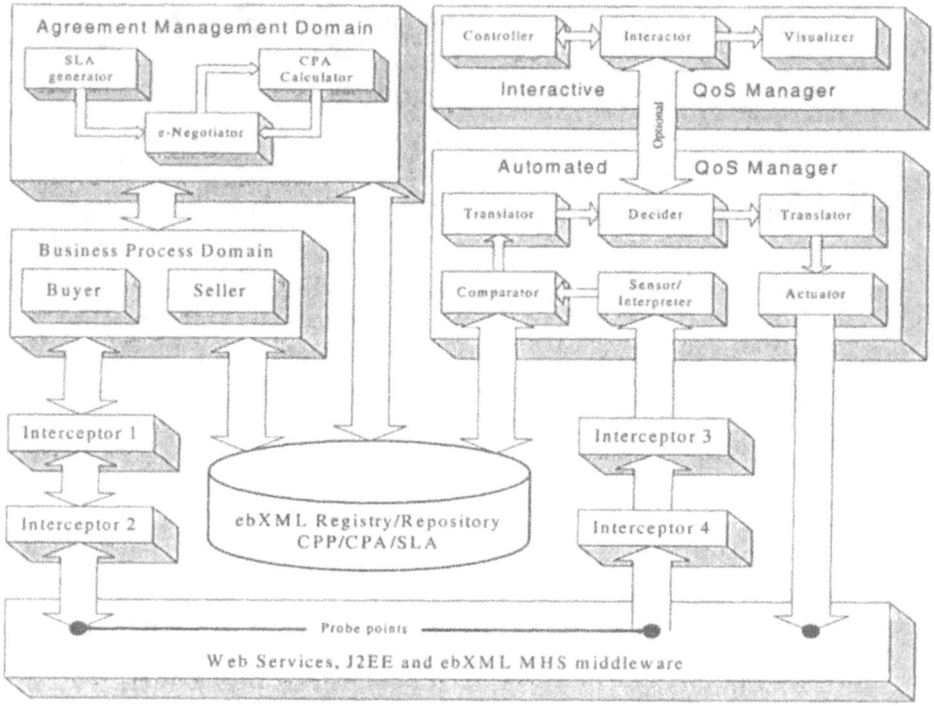

Figure 1 - Dynamic QoS Management Framework

- Dynamic adaptation and optimisation of business workflows according to the results of monitoring and comparison (with or without) human interaction
- Usage and enhancement of ebXML CPP/CPA mechanism to associate comprehensive SLAs to business collaboration
- Usage of ebXML registry to store the SLA data for run-time access

The LAURA project team currently is working on definition of flexible SLA templates adaptable to various industry sectors and geographical regions. The SLAs will provide the basis negotiate the terms of the business transactions and to measure business quality of service.

5. CONCLUSIONS

This paper discusses the need for QoS management in the context of RBVOs, the initial attempts to formalise business-level QoS and the early work of applying them in practice in the LAURA project. In general, there are ways to implement business level quality of service, however instrumentation and models need further elaboration as more parameters will be needed to support a realistic e-business SLA.

The ebXML framework will play a key role in supporting QoS management through Collaboration Protocol Agreements. The LAURA project partners are already progressing towards a 'proof of concept' sector-specific SLA for monitoring business service compliance to the SLA and other related aspects of business level QoS. The concept of a RBVO implies close interdependence and SLAs will be a fundamental component in ensuring that contractual obligations are monitored and adhered to.

We have referenced two important concepts of business QoS management: the specification of SLAs in a technology-independent manner and a dynamic service monitoring and management framework based on the interrelation between QoS metrics and tangible business-level parameters. These two concepts combined provide a good basis for an ebXML-compliant business QoS management framework. Another area of research in the near future related to this framework is the provisioning of pluggable negotiation implementation components to calculate a CPA from a set of CPPs, as the ebXML specification leaves this to implementers.

The ideas and work related to QoS and SLAs discussed in this paper will inform continuing research to building and testing the validity of a technology framework for the LAURA project to support the concept of RBVO based adaptive e-commerce zones.

REFERENCES

Byrne, J. A., (1993), The Virtual Corporation, International Business Week, 8 February 1993, pp. 36-41

Chiu, E., ebXML Simplified: A Guide to the New Standard for Global E Commerce, John Wiley & Sons; 1st edition (June 15, 2002)

Diao, Y. et al., Hellerstein, J.L., Parekh, S. Using fuzzy control to maximize profits in service level management, IBM Systems Journal, Vol. 41. No 3, 2002, p.p. 403-420

Davidow., D. H., Malone, M. S., (1993), The Virtual Corporation: Structuring And Revitalising The Corporation For The 21st Century, New York, Harper-Collins

ebXML & OASIS (2000), "ebXML Business Process Methodology Guidelines", http://www.ebxml.org/project_teams/business_process/wip/ccbp-analysis/eMethodologyGuidelinesV0_3a.pdf, accessed March 2003

ebXML (a) Collaboration-Protocol Profile and Agreement Specification, http://www.ebxml.org/specs/ebCCP.pdf, accessed March 2003)

ebXML (b) Technical Architecture Specification, http://www.ebxml.org/specs/ebTA.pdf, accessed March 2003

Frolund, S., Koistinen J. QML: A Language for Quality of Service Specification. Software Technology Laboratory, HP Laboratories, HPL-98-10, February, 1998

Foster, I., Kesselman, C. and Tuecke, S. The Anatomy of the Grid: Enabling Scalable Virtual Organizations. International Journal of High Performance Computing Applications, 15 (3). 200-222. 2001. www.globus.org/research/papers/anatomy.pdf

Goldman, S. L., Nagel, R. N., Preiss, K., (1995), Agile Competitors And Virtual Organizations: Strategies For Enriching The Customer, New York: Van Nostrand Reinhold, International Thomson Publishing

HP ChaiServer Framework http://www.hp.com/emso/products/chaisvr/index.html, 1998

HP Web Services Platform: a Comparison with HP e-speak, 2001, http://www.hpmiddleware.com/downloads/pdf/espeak_webservices.pdf

IBM, Quality of service: Evolving to the next generation, October 2001, http://www-1.ibm.com/partnerworld/pwhome.nsf/vAssetsLookup/QOS.pdf/$File/ QOS.pdf

IDC, Quality of Service for eBusiness: - The Impact of Infrastructure, An IDC Brief, December 2000, http://www-1.ibm.com/partnerworld/pwhome.nsf/vAssetsLookup/IDC_QOS.pdf/$File/IDC_QOS.pdf

Jin, L., Machiraju, V., Akhil Sahai, A. Analysis on Service Level Agreement of Web Services, Software Technology Laboratory, HP Laboratories, Palo Alto, HPL-2002-180, June 21st, 2002

LAURA project on the IST Web site http://www.cordis.lu/ist/projects/projects.htm, accessed September 2002

LAURA The LAURA project official Web site. http://www.lauraproject.org , accessed March 2003

Lamanna, D., Skene, J., Emmerich, W. SLAng: A Language for Defining Service Level Agreements. 2002, http://www.cs.ucl.ac.uk/staff/d.lamanna/tapas/SLAng.pdf

Manthou, V., Vlachopoulou, M., Folinas, D., The Supply Chain Perspective Of E-Business Evolution, (2002), in Towards the Knowledge Society: eCommerce, eBusiness, and eGovernment, pp229-242, Monteiro, J., Swatman, P. M. C., L. Valadares Tavares (eds.), The Second IFIP Conference on E-Commerce, E-Business, E-Government (I3E 2002), October 7-9, 2002, Lisbon, Portugal. IFIP Conference Proceedings 233 Kluwer 2002, 766pp, ISBN 1-4020-7239-2

Martinka, J., Pruyne, J., Jain, M. Quality-of-Service Measurements with Model-Based Management for Networked Applications, Software Technology Laboratory HPL-97-167 (R.1) September, 1998

Meade, L. M., Rogers. K. J., A Method for Analyzing Agility Alternatives for Business Processes, Proceedings of the Sixth Industrial Engineering Research Conference. Miami Beach, FL. 1997. pp. 960-965.

(MIT 2003) The MIT License, http://opensource.org/licenses/mit-license (accessed May 2003)

Muller Nathan J. (1999), "Managing service Level Agreements", International Journal of Network Management, 9

Pruyne, J., Enabling QoS via Interception in Middleware, Software Technology Laboratory HP Laboratories Palo Alto HPL-2000-29 February, 2000

Preiss, K., 1995. Models of the Agile Competitive Environment. Bethlehem, PA: Agility Forum.

(OFBiz 2003), The Open For Business Project, http://www.ofbiz.org (accessed May 2003)

Saabeel, W., Verduijn, T. M., Hagdorn, L., Kumar, K., (2002), A Model Of Virtual Orgasnization: A Structure And Process Perspective, Electronic Journal Of Organizational Virtualness, Vol. 4, No. 1

Sahai, A., Ouyang, J., Machiraju, V., Wurster, K. Specifying and Guaranteeing Quality of Service for Web Services through Real Time Measurement and Adaptive Control, E-Services Software Research Department, HP Laboratories, HLP-2001-134,

Sahai, A., Durante, A., Machiraju, V. Towards Automated SLA Management for Web Services, Software Technology Laboratory HP Laboratories, Palo Alto, HPL-2001-310 (R.1), 11 July 2002

Schmidt, H., Service Level Agreements based on Business Process Modelling, http://www.darmstadt.gmd.de/%7Ewombach/sem_proc_02/schmidt00service.pdf

Sieber, P., Griese, J., (1999), Virtual Organizations As Power Asymmetrical Networks, Organizational Virtualness And E-Commerce, 2nd International VoNet Workshop, Zurich

Tapscott, D., 1996. The Digital Economy: The promise and peril in the age of networked intelligence. McGraw-Hill, 342 pp. ISBN 0-07-063341-8.

http://www.opengroup.org/public/member/q202/documentation/forums/qos/tunney.pdf

(VO 2002) Virtual Organization – A whatis Definition http://whatis.techtarget.com/definition/0,,sid9_gci213301,00.html (4th Nov 2002)

Wolter, K., van Moorsel, A. The Relationship between Quality of Service and Business Metrics: Monitoring, Notification and Optimization

Chapter 9

SOFTWARE FOR THE CHANGING E-BUSINESS

Maria Alaranta, Tuomas Valtonen and Jouni Isoaho

Abstract: In this article, we first acknowledge the requirements for more rapid and cost-efficient development cycles and systems evolution for e-business software applications. Thereafter, we discuss the contemporary solutions used to meet the requirements. These include technological and organizational innovations, as well as commoditization. After that, we discuss attributes of modification of an e-business application, i.e. the depth of modification, the sophistication of the modification method, operational continuity, and freedom from errors. These attributes are combined into a framework that is then used to evaluate four common e-commerce applications, a spreadsheet application and a novel dynamic e-commerce platform, also presented in this article. The dynamic e-commerce platform is proposed to be the most favorable solution in cases where system specifications change frequently.

Key words: e-commerce, information system, modification, flexibility

1. INTRODUCTION

Due to the globalization of business and the evolution towards web-based systems, it is necessary to re-evaluate the way information systems are developed, modified, operated and maintained [1]. Changes in the global marketplace require frequent changes in software because firstly, globally used systems need to be locally adjusted [15]. Secondly, different industries – e.g. banking, insurance and stock exchanges in both Europe and also globally – are responding to increasing competition by mergers and acquisitions [12].

Hence, there is a demand for systems that evolve with and support the changing organization, facilitate business process redesign to better exploit the characteristics of IT, and fulfill the requirements for outward-facing information systems linked to networks of suppliers and customers [7].

These links include e.g. supply chain management (SCM), for which an increasing number of companies are using web sites and web-based applications [14].

These new applications, or changes in those currently in use, are called upon at a pace that requires significantly shorter development cycle times. Reducing both the cost and the time from idea to market – while ensuring high quality – is crucial for gaining a competitive advantage in the increasingly competitive IT market that is facing new entrants also from developing countries, such as China and India. [3], [15] Furthermore, the migration towards web-based systems makes time and creativity essential success factors as technology and demand change rapidly [5]. However, manifesting the apparent need for flexible software, as well as corrective and adaptive maintenance, accounts for a significant share of software activities in organizations, and erroneous concentration on the development project – rather than the whole product life cycle is a major cause of software problems [7].

In this article, we aim at answering the question: How can the ever-changing requirements for the software for e-business be met? In order to reach this objective, we (1) review contemporary solutions, (2) present a framework for analyzing the characteristics required for a solution aimed at fulfilling the requirements, (3) present a technically oriented concept in software development that aims at reducing evolution cycle time and increasing flexibility, and (4) analyze the novel concept, as well as some examples of contemporary solutions, against the framework.

2. CONTEMPORARY SOLUTIONS

Looking back at more than 50 years of history in software development, three main paths of trajectories of innovation can be observed. These are: (1) technical change, i.e. new programming languages, tools, techniques, and methods, etc.; (2) organizational change, i.e. new ways of managing the people and the process; and (3) substitution of standard products (generic packages) for custom building. [9]

Technological change manifests itself in the development of programming languages, starting from writing in machine code, all the way to 4G languages that have vocabularies and syntax very similar to natural language. After these, the technology advanced to e.g. "declarative systems", and structured techniques such as modularity and object-oriented (OO) design and programming. Object-oriented techniques provide significant possibilities for shortening the development life cycle, in addition to greater rigor and predictability. [9] Tools for supporting the development processes have also evolved, ranging from programmer aids – e.g. testing and debugging tools – to tools supporting the whole development life cycle, such as computer-aided software engineering (CASE) tools [9], as well as structured methods supported by CASE tools, e.g. information engineering (IE) [1]. Other technical innovations include the "cleanroom" approach in which the aim is to prevent the entry of defects during the development, and non-serial machine architectures, such as neural networks [9].

Organizational innovation aims at offering better tools, techniques and methods for the quality of development, supply and maintenance of software, as well as project management and the organization of work [9]. These include time-based software management [2], total quality management (TQM) [4], quality function deployment [5] and the Capability Maturity Model[SM] (CMM) [7]. Extreme programming (XP) is a team-based engineering practice that is suggested to be especially suitable for the high-speed, volatile world of web software development, which can also be combined with other innovations such as CMM [8]. Recent developments include the Model-Driven Architecture (MDA), which aims is to automate the transformations between the models and code [13], and the ISO 9001 [e.g. 15] quality standard that distinguishes between the technical and organizational aspects of software development.

The third development tendency is *commoditization*, which refers to the substitution of the process of custom building software for a software product or package. Packages should reduce uncertainty in the length of time and cost of development and ensure a predictable level of reliability and known quality, as bugs are identified by earlier users. However, in theory, packages customizable by the end-user would remove the productivity problem from the IT developers. [9] Besides these, recent developments affecting the e-business include, e.g., Web Services and Semantic Web. Web Services can be described as modular Internet-based applications that facilitate business interactions within and beyond the organization. As opposed to the traditional business-to-business applications such as EDI, Web Services are typically decentralized, open and unmonitored, shared, and dynamically built, and the user base and scale are not predefined. [10] On the other hand, Semantic Web aims at solving the problem of machines not being able to interpret the meaning and relevance of documents in the web. Semantic Web offers a vision for the future in which information is given explicit meanings, which enables people and computers to co-operate more efficiently. [11]

3. ATTRIBUTES OF MODIFICATION

As described in the previous section, the rapidly changing environment creates new requirements, while simultaneously obsolescing old specifications, at an increasing pace. In order to study the feasibility of different technical and architectural solutions in a changing environment, we first define the concept of flexibility in e-commerce systems. Here we identify the main attributes of e-commerce systems that influence the type of actions required and cost incurred when functionality is altered. We then combine these to form a framework for classifying and evaluating components of systems, as well as entire e-commerce systems.

3.1 Depth of Modification

We first distinguish between two top-level classes of system components: a) core components and b) user components. *Core components* are an integral part of the e-commerce system and are identical in each installation of the system. These components specify the functionality of the system and methods for accessing information in the system. *User components* are related to user requirements and may vary from one installation to another.

The *depth of modification* attribute (hereafter the "depth" attribute) indicates which component classes in the information system are subject to changes. A simple system may allow the end-user to insert, modify and delete database records, while a more elaborate system may also permit changes to the structure of the record. An advanced information system may also allow changes to functionality and internal structures of the system itself. All of these cases are possible *without* reprogramming the system itself; naturally, more elaborate modifications are possible if we allow reprogramming of the system (see Section 3.2 for further discussion on this topic).

We identify four main levels of depth in system components, according to the content and structure that can be modified in these, corresponding to levels 1–4: 1) content in user components only, 2) content and structure in user components, 3) content and structure in user components, as well as content in core components, and 4) content and structure in both user and core components. Level 1 allows modification of *content* in *user* components, typically data related to the application area of the user. At level 2, the *structure* of such information can also be modified, allowing the addition of new information types or the extension of existing types. A level 3 component allows changes in *content* of *core* components, in addition to that of case components. In this case, both functionality and access to information in user components can be altered. At level 4, one is also able to modify the structure of core components.

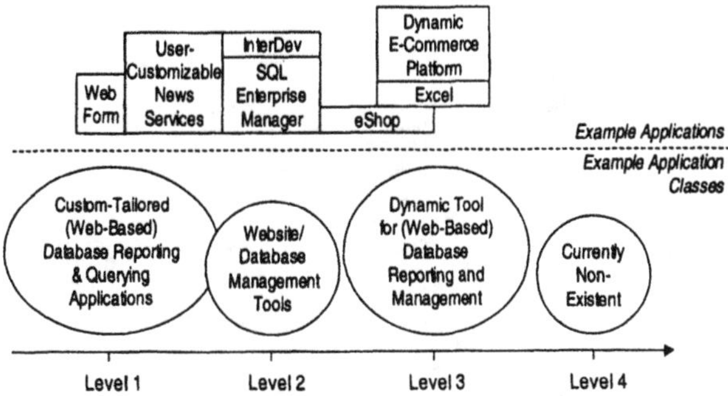

Figure 1. Examples of applications at various depth levels (sophistication level ≥ 3).

In Figure 1 we demonstrate the differences in the depth levels of various generic components (applications in this example) when the *sophistication of the modification method* ≥ 3; i.e., components that can be modified without any reprogramming labor (see Section 3.2). Depth level 1 encompasses stand-alone or web-based e-commerce applications built on a database platform. A typical website/database management tool at level 2 is able to modify both the content and structure of the content; however, automated mechanisms for providing end-user functionality are not included. An integrated, possibly web-based, database management and reporting tool with an automated end-user editor would fulfill requirements at level 3, allowing modification of core components, such as database access mechanisms and some functionality for the end-user. Level 4 applications do not currently exist without reprogramming work.

3.2 Sophistication of the Modification Method

In this section, we categorize the methods available for modifying components into the *sophistication of the modification method* attribute (hereafter the "sophistication" attribute); i.e. the type of action required to modify a system component: 0) non-modifiable, 1) pre-compiled, 2) auto-generated, 3) configurable, and 4) self-configuring. The functionality of a level 0 *non-modifiable* system component is fixed in the design phase and cannot be changed after the manufacturing stage. Hence, modifying a component at this level requires *physical* replacement. A level 1 *pre-compiled* component is also designed to perform a specific function, but can later be manually reprogrammed if modification is required. A level 2 *auto-generated* component can be altered using automated modeling tools, allowing a shorter and more reliable development process. A level 3 *configurable* component can be modified by the end-user at any time without reprogramming. Finally, a level 4 *self-configuring* component will monitor and modify itself autonomously.

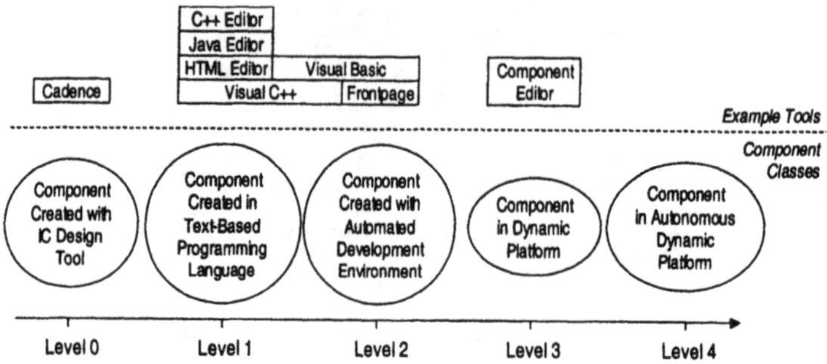

Figure 2. Examples of system component classes and tools at various sophistication levels.

Figure 2 illustrates some examples of system components and development tools at different levels of sophistication. At level 0, the Application-Specific Integrated Circuit (ASIC) is a typical non-modifiable system component. At level 1, the functionality and content of *pre-compiled* components can be created using an editor for textual programming and markup languages, such as C++, Java and HTML. At level 2, an auto-generated component can be re-designed using user-friendly, automated development environments, such as Microsoft FrontPage ™. Between levels 1 and 2 are hybrid components, such as Microsoft Visual Basic ™ and Visual C++ ™, in which some portions are created graphically, whereas others require textual programming work.

Components at levels 3–4 constitute a new class of *dynamic platforms.* At level 3, the end-user can add configurable components, or remove or modify existing ones at any time. The main difference, in comparison to level 2, is that the component *itself* is dynamic, not only the tool that was used to generate it. Level 4 self-configuring components are similar, but are also equipped with mechanisms for autonomously modifying themselves to adopt to circumstances, without end-user intervention. Techniques for implementing components at levels 3–4 are presented in Section 4.

3.3 Operational Continuity

The third attribute, *operational continuity*, refers to the ability to ensure uninterrupted operation in the component subject to modification: 0) interrupted and 1) uninterrupted. At level 0, modifying the component results in interruption of the normal operation of the component and other dependent components. At level 1, no interruption is necessary, and the new functionality of the component is valid from the moment that the modification takes place. Figure 3 illustrates two examples of operational continuity. A typical compiled binary component must be replaced when any modification other than normal data manipulation is required (depth of modification ≥ 2). A dynamically configurable component can be modified without downtime, expect for when modifying the structure of a core component (depth of modification ≤ 3).

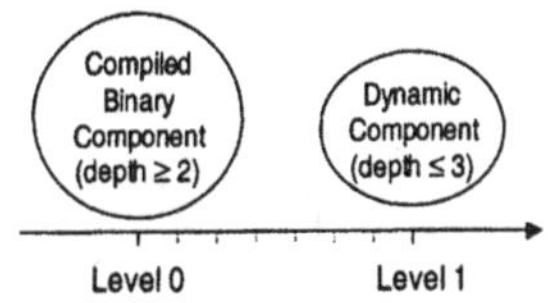

Figure 3. Examples of system components at two levels of operational continuity.

3.4 Freedom from Technical Errors

The *freedom from technical errors* attribute (hereafter the "error-freedom" attribute) signifies the ability to ensure the correct implementation of modifications; i.e., the risk of system instability or data inconsistency due to technical errors is avoided. We can distinguish two primary levels of error-freedom with respect to system operation, when: 0) technical errors possible and 1) technical errors not possible. Figure 4 shows some examples of generic system components at different levels of error-freedom.

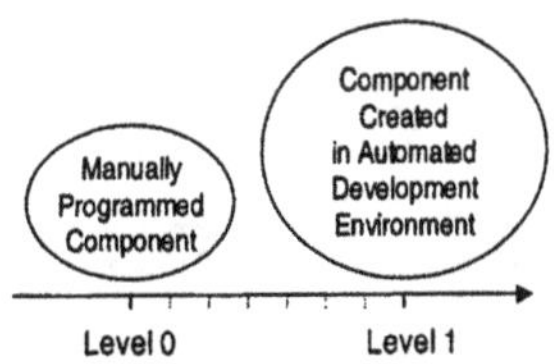

Figure 4. Examples of system components at various levels of error-freedom.

The risk of technical error can be reduced by creating the component in an automated development environment, by using modular and component-based design, or by standardizing interfaces. Here components could be positioned at intermediate levels (between 0 and 1) of error-freedom.

3.5 A Framework for Evaluating Modification

In Figure 5, the attributes of the previous subsections are combined to form a four-dimensional framework for assessing the modification characteristics of an information system. The indices in the axes of the framework correspond to the levels introduced in chapters 3.1–3.4. When depth levels 1–4 or 0–1 for each component are displayed in a single graph, a *modification profile* for the component can be formed. The dotted lines illustrate two imaginary modification profiles.

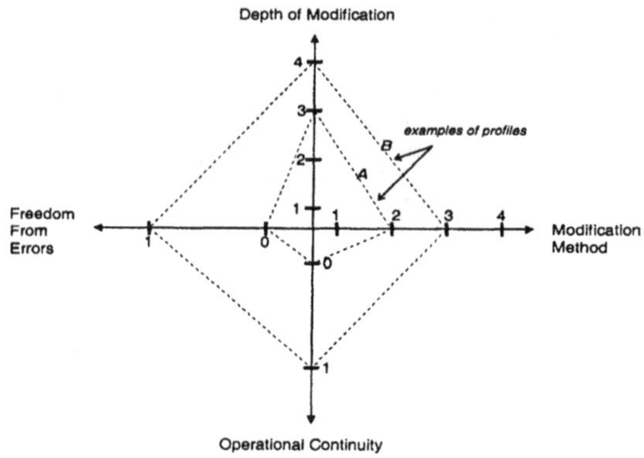

Figure 5. Modification attributes combined.

From the shape of this profile, we can determine the modification characteristics of the component. Narrow flame-shaped profiles similar to example A are a sign of a very static component; modifying such a component would require manual work, could interrupt the operation of the component and cause errors. In contrast, broad and circular profiles similar to example B promise straightforward modification, entailing little manual programming work or negative side effects.

4. THE DYNAMIC E-COMMERCE PLATFORM

Today many e-commerce systems are tailored to match the needs of a particular end-user group (or end-user organization) at a certain time. In this section we outline the design methodology of a next-generation real-time dynamic e-commerce system that is completely configurable by end-users and requires little re-engineering during its life cycle.

When creating an e-commerce system where all components are *configurable* and the content of *core* components is modifiable, a number of design issues must be addressed. Firstly, because the functionality of the system (residing in core components) is dynamic, using standard techniques for implementing functionality – such as programming and compiling code – is not an option. Furthermore, the end-user must be able to modify the internal methods used to access information from the database, as well as all user interface components. The end-user should also have access to all user components, be able to modify the content *and* structure of these, as well as the ability to process information and use this to generate *totally new* information types. The end-user must be allowed to modify components at any time, without shutting down any components in the system or producing technical errors. These requirements are particularly challenging for real-time e-commerce systems, where the flow of information is continuous.

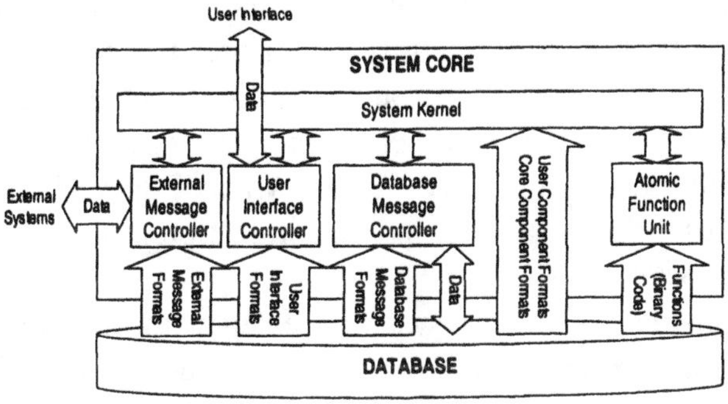

Figure 6. Architecture of the dynamic e-commerce platform.

The dynamic e-commerce system (Figure 6) is divided into two main sections: a) the system core implemented prior to installation and b) database structures that can be modified at any time. The *system core* comprises several control units for the information system. The *user interface controller* is responsible for displaying *user interface atoms*, i.e. components accessible to the end-user, such as windows, buttons, images, etc. The *external message controller* deals with the reception and transmission of messages from and to external information sources and other installations. The *database message controller* manages communications with the database. The *atomic function unit* processes incoming data. Finally, the *system kernel* schedules events in the system core and coordinates communication between other units.

The depth of modification of this system is at level 3, because the *content* of core components, as well as the *content* and *structure* of user components, is stored in the database. However, the *structure* of the core components is fixed and cannot be modified without reprogramming. Hence, the system fails to qualify for depth level 4 (above sophistication level 2).

5. MODIFICATION PROFILE ANALYSES

In this section, we use the framework presented in Section 3.5 to evaluate four common e-commerce applications (1–4) that exhibit various levels of sophistication, a spreadsheet tool (5), and the dynamic e-commerce platform

(6) presented in Section 4. (The spreadsheet tool is included to exemplify a familiar user-configurable system with the possibility of end-user developed applications.) The applications are evaluated against the framework in a situation where a change is required, in order to analyze how the application meets the flexibility requirements of the changing environment.

We first study a generic form in a web site created on a traditional web-server. As seen in Figure 7, a form of this type is very flexible when collecting and changing the data entered in its fields; however, altering the structure of the form requires significant effort. The second example is a user-customizable web site created on a traditional web-server, e.g. a service that allows the user to key in a set of preferences, which then creates a customized web site for the user. As shown in Figure 8, flexibility now extends to the user data structure, but customization options for the user are limited.

Figure 9 illustrates a similar system created with an automated tool. In this case, the improvement is due to the fact that an automated tool reduces the possibility of technical errors. In Figure 10, templates for creating an e-store application exemplify a yet more flexible system, allowing the creation of customized commercial web sites without programming skills. [6]

Although this system is flexible, it nevertheless has the problem of downtime, and modification of core components still requires programming.

An analysis of the spreadsheet application in Figure 11 reveals greater operational continuity than previously presented applications. Figure 12 displays an analysis of the dynamic e-commerce platform presented in Section 4. This approach possesses the characteristics required from a system that is designed to meet the constantly changing requirements; i.e., extensive modifiability, advanced tools for system modification, operational continuity, and error-freedom due to the use of automated tools.

Here the main advantage is that most of the system can be modified without any reprogramming work; the structure and content of user components, as well as the content of core components, can be configured during system operation. Reprogramming and recompilation of the code is required only when the structure of core components is modified.

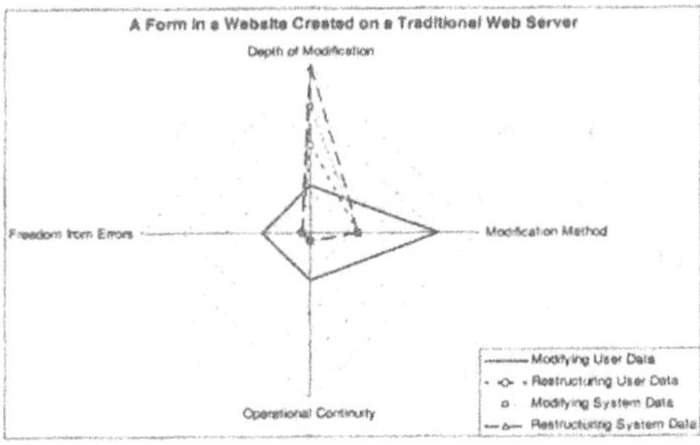

Figure 7. A form in a web site created on a traditional web-server.

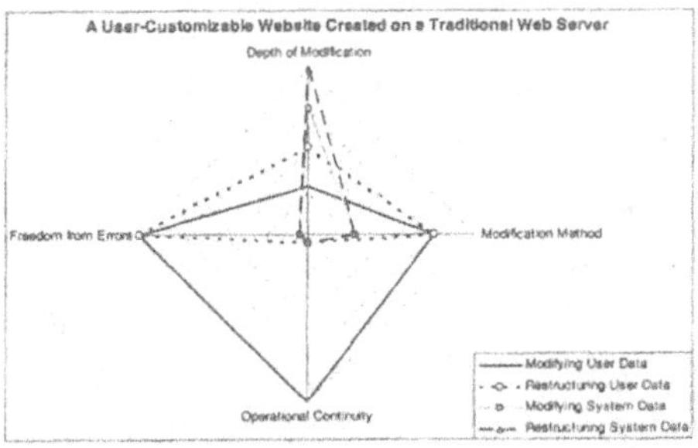

Figure 8. A user-customizable web site created on a traditional web-server.

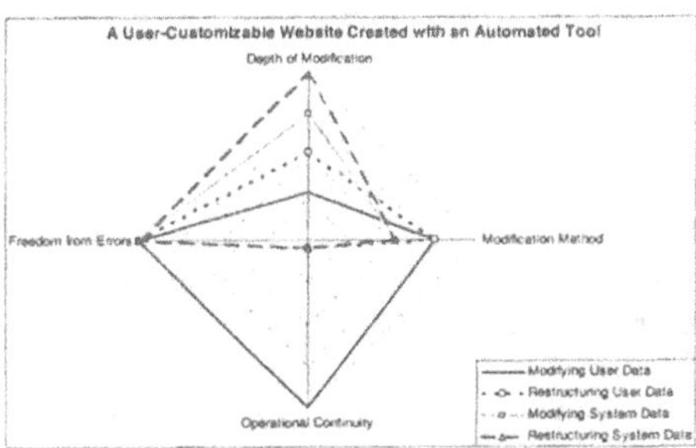

Figure 9. A user-customizable web site created with an automated tool.

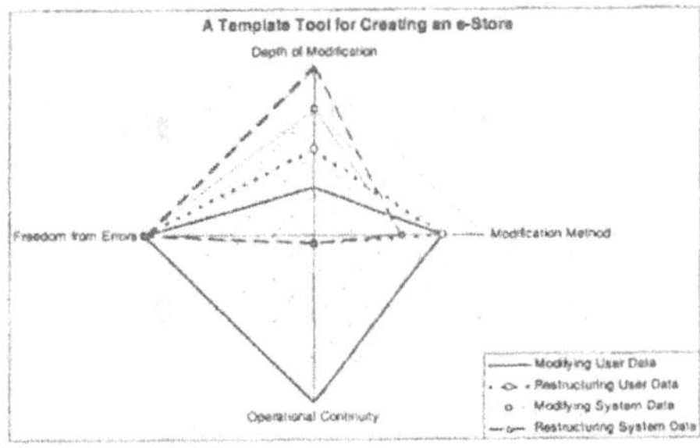

Figure 10. A template tool for creating an e-store.

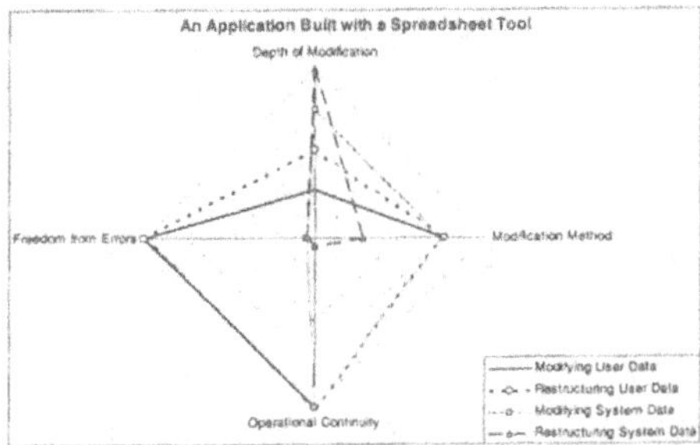

Figure 11. An application built with a spreadsheet tool.

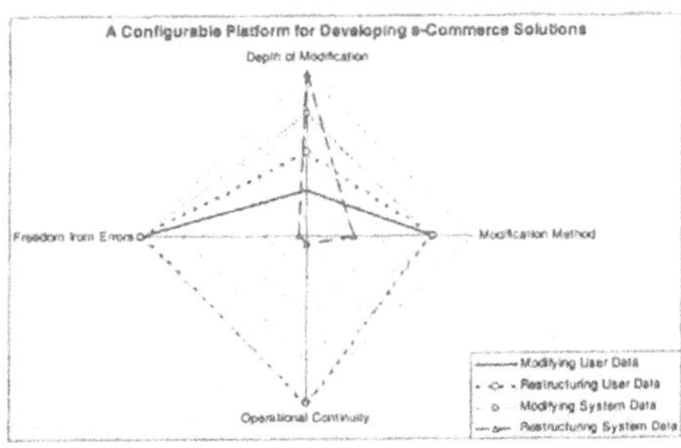

Figure 12. A configurable platform for developing e-commerce solutions.

6. CONCLUSIONS

This article is concerned with the modification of e-commerce systems. We first discuss contemporary solutions for frequently changing system requirements and describe a number of technological and organizational innovations aimed at shortening the product development cycle, whilst maintaining rigor and predictability. We discuss four attributes of modification in e-commerce applications: depth of modification, sophistication of the modification method, operational continuity and freedom from errors, and compose a framework for the evaluation of modification in e-commerce systems. We also introduce a novel configurable e-commerce development platform and assess its modification characteristics against four typical e-business applications and a spreadsheet application. We observe that contemporary e-commerce applications can deal with certain levels of modification with no difficulty, but more fundamental changes in system specification could lead to extensive reprogramming, downtime and the risk of data inconsistency. The configurable e-commerce platform is found advantageous in three specific cases: (1) when system specifications are altered frequently, (2) when changes are of fundamental nature, and (3) when the end-user requires extensive control over the system. Finally, we demonstrate some of the benefits of studying system flexibility using multiple independent attributes; many strengths and weakness of diverse system designs can only be revealed via thorough multi-perspective analysis. In particular, the significance of modifiability is emphasized – the ability to rapidly adapt to a constantly changing environment will be the key to future e-commerce.

REFERENCES

1. Behling, Robert – Behling, Cris – Sousa, Kenneth (1996) Software Re-engineering: Concepts and Methodology in Industrial Management & Data Systems Vol. 96, No. 6, pp. 3–10.
2. Blackburn, Joseph D. – Scudder, Gary D. – Wassenhove, Luk N. – Hill, Graig (1996) Time-based Software Development in Integrated Manufacturing Systems Vol. 7, No. 2, pp. 60–66.
3. Dubé, Line (1998) Teams in Packaged Software Development – the Software Corp. Experience in Information Technology & People. Vol. 11, No. 1, pp. 36–61.
4. Gong, Beilan – Yen, David C. – Chou, David C. (1998) A Manager's Guide to Total Quality Software Design in Industrial Management & Data Systems Vol. 98, No. 3. pp. 100–107.

5. Herzwurm, Georg – Schockert, Sixten (2003) The Leading Edge in QFD for Software and Electronic Business in Int'l Journal of Quality & Reliability Management Vol. 20, No. 1, pp. 36–55.
6. Kotisivut.com (2002) http://www.kotisivut.com/eshop.shtml. (Read: 04/30/03).
7. Paulk, Mark C. – Weber, Charles V. – Garcia, Suzanne M. – Chrissis, Mary Beth – Bush, Marilyn (1993) Key Practices of the Capability Maturity Model, version 1.1. Software Engineering Institute, Carnegio Mellon University, Pittsburgh, USA.
8. Paulk, Mark C. (2001) Extreme Programming from a CMM Perspective in IEEE Software. Nov/Dec, pp. 1–8.
9. Quintas, Paul (1994) Programmed Innovation? Trajectories of Change in Software Development in Information Technology & People. Vol. 7 No.1, pp. 25–47.
10. Ratnasingam, Pauline (2002) The importance of technology trust in Web services security in Information Management & Computer Security. Vol. 10, No.5, pp. 255–260.
11. Sadeh, Tamar – Walker, Jenny (2003) Library Portals: Toward the semantic Web in New Library World. Vol. 104, No. 1184/1185, pp. 11–19.
12. Saksan Pankit Yhdistyvät (2000) in Verkkouutiset http://www.verkkouutiset.fi/arkisto/Arkisto_2000/17.maaliskuu/depa1100.htm (Read: 04/28/03).
13. Siegel, Jon et al. (2001) Developing In OMG's Model-Driven Architecture at http://www.omg.org/mda/mda_files/developing_in_omg.htm (Read: 07/15/03).
14. Supply Chain Management (SCM) Definition (2003) http://www.mariosalexandrou.com/glossary/scm.asp (Read: 04/28/03).
15. Yang, Y Helio (2001) Software Quality Management and ISO 9000 Implementation in Industrial Management & Data Systems. Vol. 101, No. 7, pp. 329–338.

SECTION 4
Auctions and e-Payment

Chapter 10

DYNAMIC ROI CALCULATIONS FOR E-COMMERCE SYSTEMS

Michael Amberg, Markus Hirschmeier

Abstract: The introduction of eCommerce Systems poses a special challenge for estimating value payoffs in the face of uncertain future developments. Practitioners have difficulty to capture all or even some of the benefits of eCommerce Systems with existing traditional capital budgeting models. In this paper, we analyze and evaluate various ways of measuring the business value of eCommerce Systems, considering financial and non-financial, quantitative and qualitative, traditional and innovative models. We intend to identify boundaries of state of the art approaches of economic valuations and present new approaches which integrate the business value of eCommerce Systems into the traditional capital budgeting models. A case example shows an exemplarily approach for a dynamic modification of ROI calculations with the Customer Lifetime Value for CRM Systems.

Key words: ROI Calculations, eCommerce Systems, Customer Lifetime Value, Real Option Pricing Models, Economic (e)Valuations, Capital Budgeting Models

1. MOTIVATION

Amberg and Hirschmeier (2003) showed basic problem fields for the economic valuations of eCommerce projects. The following problem fields summarize the failure of existing economic valuations.

Problematic Anticipation of Benefits: IT practitioners have problems in determining the time series of inflows and outflows of an innovative technology project. Especially benefits are hard to capture in numeric values. Important decisions are more often based on intuition and experience than on quantitative analysis. Simple and static models are often preferred. Similar results can be found in the studies of Tam (1992).

Decisions on Rules of Thumb Basis: Decisions in eCommerce projects are more often based on rules of thumb such as "invest to keep up with the technology" or "invest if the competitors have been successful" than on quantitative analysis. Further investigation of the state of the art methods that were used for determining the value of IT infrastructure in practice can be found in the studies of Weill (1993).

Problematic Application at Early Planning Stages: Practitioners complain that standard evaluation procedures for IT projects are impossible to apply, particularly at early planning stages. Hochstrasser (1994) stated the same problems about 10 years ago.

Negative ROI: The problematic quantification of benefits leads to an asymmetric consideration of cost in the balance of the cost benefit analysis. ECommerce projects often have negative economic valuation results - despite of their strategic importance.

Risk of Incredibility: Many investments in new technologies start with a project status and the goal to create a positive ROI within only a short period of time. To rescue the project, many financial calculations overestimate the benefits in the first years in order to constitute a positive return on investment. Commonly, the credibility and the acceptance of these calculations is endangered.

Risk of False Rejection: Investments in innovative technology is budgeted. Companies have to overcome the problem not to routinely reject truly important investments by using simplistic quantitative techniques (Kester, 1984). This is especially important for companies with independently operating business which act under the objective to maximize the company's equity in short-term. Organizations could loose important opportunities if they strictly rely on traditional criteria to assess innovative technology benefits.

In the following we will give an overview on existing traditional economic valuations and discuss the limitations of the presented state of the art methods.

2. STATE-OF-THE-ART APPROACHES FOR ECONOMIC (E)VALUATIONS AND THEIR LIMITATIONS

Traditional Capital Budgeting Models consider monetary costs and benefits of an investment cumulated for each year over a specified period of time. The traditional methods can be divided in static and dynamic models. Dynamic methods take into account the time value of the money (e.g. net present value (NPV), internal rate of return (IRR) etc.) sometimes even considering the inflation rate. Static methods like the payback method, the accounting rate of return (ARR), the Return On Investment (ROI) etc ignore the time value of money. Traditional Capital Budgeting Models consider the time value of money by comparing the cash outflow at the beginning of a proposed project with the anticipated cash inflows to be generated by the project. Discounted cash flows refer to the present value of a stream of payments to be received or paid in the future.

- **Return on Investment (ROI)** is the most common measure of profitability. Different users define income and investment differently when measuring ROI. Most measures of ROI relate an income statement element such as operating or net income to a balance sheet element such as stockholders' equity. The basic construction of the formula is:

$$ROI = \frac{Income}{Investment} \tag{1}$$

Return on Investment (ROI)

- The **Total Cost of Ownership (TCO)** approach was originally developed by the Gartner Group in 1987. The background was that the state of the art capital budgeting methods at that time only considered the initial hardware costs. The TCO approach takes all operating cost of the investment into account, specifically for IT investments maintenance cost, content development and management costs, training costs, license cost, consulting costs etc. The resulting ratio is a price, the cost of the investment per working place. Unfortunately there is no general systematic for this economic valuation and different TCO methods have been developed for example by Forrester Research and Meta Group.
- **Process Costing Models** (Target Costing, Activity Based Costing) try to overcome the traditional functional oriented view in accounting. With the shift of perspective the economic valuations of products can change significantly. Examples show that the economic valuations can double or halve. The process oriented costing models can contribute to acceptance of the right projects. The method is only useful for repetitive processes with a small variability of decisions.
- **Option Pricing Models:**The basic approach of option pricing is derived from physics, specifically the Brownian motion in thermodynamics. The basic mathematical construct is an exponential function. The formulas describe the distribution of the present value of the project's expected revenues. There are different models, which assume that the option value follows either a geometric, binomial (Cox, Ross, Rubinstein, 1979),

normal (Taudes et al., 1997) or lognormal (Black-Scholes (as cited in Benaroch, Kauffman 1998)) distribution. For example Taudes et al (1997) use the following formula construct:

Value of a software platform = NPV of fixed application portfolio + Option value of implementation opportunities.

$$NPV = e^{-\mu T}(V_0 e^{\alpha T} N(d_1) - IN(d_2)$$

$$d_1 = \frac{1}{\sigma\sqrt{T}}\left(\ln\left(\frac{V_0}{I}\right) + \left(\alpha + \frac{1}{2}\sigma^2\right)T\right) \qquad d_2 = d_1 - \sigma\sqrt{T} \tag{2}$$

Real Options Pricing Model

Limitations of the state-of-the-art approaches: Historically, capital budgeting models come from the time of the industrial age. Heavy machinery produced a predefined output which could easily be quantified and assigned a monetary value. However, the innovative technologies today differ completely in their characteristics from the technology of the industrial age. The strategic impact of the traditional capital budgeting models was to assure the liquidity of the company. Today, the strategic focus has shifted from the financial costs to the strategic benefits of an investment. Under this perspective, even newer approaches like the Total Cost of Ownership (TCO) can be counted to the traditional cost-oriented approaches as the method focuses on the costs, not on the benefits.

The option pricing model is a methodical improvement for economic valuations of CRM Projects as the methodology takes the opportunities into account and provides a better characterization of the investment's true value than would a net present value. The most critical aspect of the option pricing value method is at first sight the estimation of the future opportunities because the developments depend on the economic environment of the company and its industry, world trends and other events beyond the control of the company. Additionally, questions of initiation and pacing of investment have to been determined. The mathematical approach does not provide any help in quantifying the relevant parameters.

Secondly, Real Option quantifications are always involved in questions of organizational change and decision making. Strict real option assumptions do not allow applying the approach for organizational change. Adner & Levinthal (2002) show technical violations and point out the inherent limitations and boundaries of applicability of real options for economic valuations for organizational design and management.

3. PROJECT CLASSES AND METHODICAL REQUIREMENTS FOR ECOMMERCE PROJECTS

The limitations of the state-of-the-art approaches lead to a need of methodical support for the ROI calculations of eCommerce projects. In the following we present a set of project classes according to the problems that arise when calculating ROI of eCommerce projects with traditional capital budgeting models and the solutions that practitioners have found to assess the benefits of an investment.

Projects with Qualitative Benefits (Project Type I): The shift towards intangible benefits leads to a class of projects where benefits are no longer quantified with a monetary value. Qualitative scoring models are used to assess the values of each project so that the projects can be ranked in the order of importance. The ordinal values are often transferred into numerical values on quantitative scales, but the fundamental evaluation rests qualitative. The Balanced Scorecard is one of the best known scoring approaches. In its finance dimension it offers the measurement of benefits through qualitative scoring. Scoring models rely in their calculation of value for eCommerce Projects on the executives' experience and judgements. If experiences with eCommerce projects are rare, the evaluation process could be problematic. A common criticism that is brought to Scoring models is that they leave plenty of room for political influences as the scores and weight of single criteria must be found subjectively. Furthermore, economic valuations based on scoring models are confronted with a number of problematic and implicit assumptions such as full substitutability of criteria and uniform ordinal scales.

Projects with Long-term Benefits (Project Type II): The shift of the time frame for benefits realizations towards a long-term perspective leads to a class of projects, which are characterized through their future potentials and enabling function. Most benefits of these projects are realized in related successive projects. Infrastructure projects are a typical project class with this characteristic. The benefits of these related successive investments depend on variables of the social and technical surrounding like the development and acceptance of electronic standards. These future potentials are highly volatile and the benefits cannot be calculated deterministic.

Projects with Retrospective Success Measurement (Project Type III): The projects of this class are characterized through an economical retrospective on an investment which has already been made. Usually, the economic evaluation is based on an actual/target comparison of the related processes. The main difficulty is based on the fact, that state of the art of project management does not implement any tools which support a future quantification of the efficiency progress and optimization efforts. The development and deployment of appropriate instruments for the measurement of the economical success is an important in the as a controlling tool function for the identification of further potentials for optimization.

Projects with No Accounting Effective (Atomized) Benefits (Project Type IV): This project class is characterized by quantization problems of cost savings. The cost reductions are atomistic distributed on the different cost centers, so that they don't affect the accounting or balance. Like energy particles in quantum physics, the timely savings can only be effective as cost savings, until they reach a certain value. For instance timely savings in employee processes do only have an effect on the accounting if they are high enough to replace a complete position; otherwise the savings are accepted by the controlling. This problem is forced by different perspectives of two different business units, one which is responsible for the Controlling and another, which is responsible for the introduction of IT. Essentially, this is a problem of organizational structure and it can be resolved through appropriate modification of the existing cost centers. An appropriate economic evaluation should be able to equally support both perspectives.

We use a 2x2 matrix grid to point out the need for methodical support. The matrix consists of two dimensions: One dimension shows the state-of-the-art methodologies:This dimension is on the horizontal scale and ranges from deterministic to volatile: The degree of quantification complexity is positively related to higher requirements of the mathematical back-ground and applied formula. The second dimension refers to the typology of benefits in the ROI calculations. The dimension on time perspective is shown on the vertical scale and distinguishes tangible and intangible benefits. Each of the four quadrants can be identified at least with one project class (figure 1).

Methodical Requirements: Projects with a retrospective success measurement and projects with no accounting effective (atomized) benefits can be calculated deterministic. The projects with qualitative benefits and projects with long-term benefits are characterized by a high volatility of their future economic developments. These project classes cannot be calculated in a deterministic way and the state of the art methods of the traditional capital budgeting methods are not suitable for this degree of quantification complexity. There is a need for the development of new methods for the economic evaluation of the projects of this class. The next chapter presents new approaches which contribute to solve the problem classes.

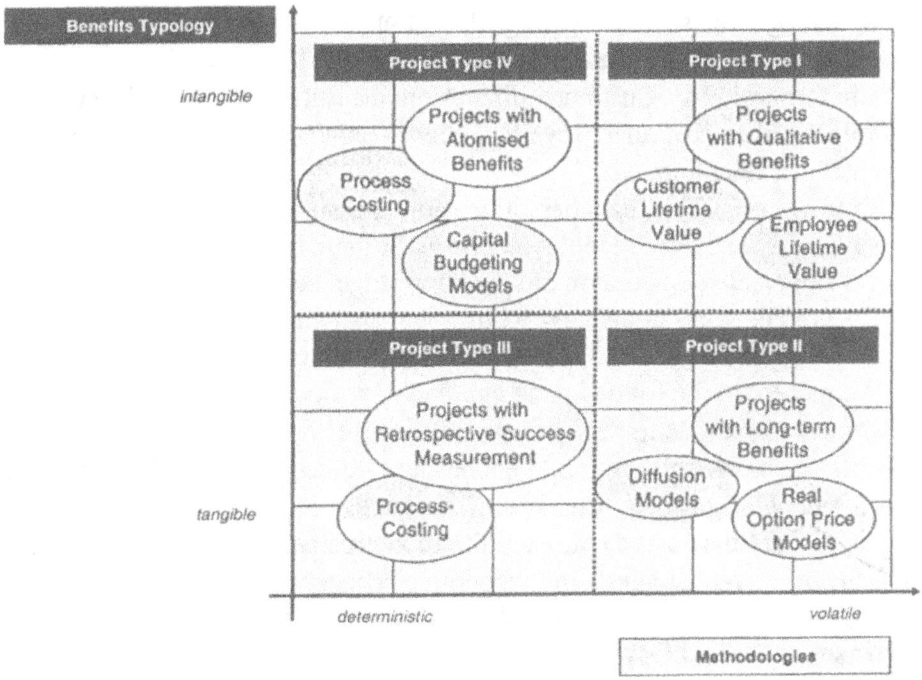

Figure 1. Project Classes and State-of-the-art Methodologies

4. DYNAMIC ROI CALCULATIONS FOR ECOMMERCE SYSTEMS

Amberg and Hirschmeier (2003) presented new approaches known in Management Information Systems literature like Real Option Pricing Models, Process Costing, Scoring Models (Balanced Scorecard), Customer Life Time Models and Business Dynamics Models and discussed them with regard to the practical advantages and their suitability for the economic evaluation of eCommerce projects. In the following we will present an approach that integrates the existing models to an approach for the economic evaluation of eCommerce systems.

The Balanced Scorecard is a state of the art of method for the controlling and evaluation of eCommerce Systems. The Balanced Scorecard identifies four dimensions for an equal and balanced consideration of strategic values of an eCommerce System. The four dimensions cover the processes, the customers, the employees and the costs of an eCommerce System. Except the cost dimension, the dimensions focus on intangible and soft factors which are measured on a qualitative scale:

- **Processes:** The Processes dimension consists of management ratios and operating figures measuring the improvement, effectiveness and efficiency of the business processes.
- **Customers:** The Customer dimension measures qualitatively strategic values like customer loyalty, client satisfaction, corporate image, customer retention, etc.
- **Employees:** The Customer dimension measures qualitatively strategic values like enhanced employee goodwill, increased job satisfaction, skill development and adoption of new knowledge, etc.
- **Costs:** The Cost dimension focuses on financial figures and ratios like revenue mix, service costs, charges, cost structures, etc.

We present in the following an approach for the economic (e)valuation of eCommerce investments which explicitly focuses on the equal and balanced quantification of soft and intangible facts. According to the Balanced Scorecard we consider the business values of processes, customers, employees and the costs to measure the economic benefits of an eCommerce investment. We suggest the economic concepts of Process Costing, Customer Lifetime Value, Employee Lifetime Value, Total Cost of Ownership and Real Option Price Value to quantify the different values and their monetary development over time.

- **Process Efficiency:** Process Costing is the state of the art method for economic evaluations of an investment. The effected business processes are compared between now and a point of time in the past or in the future to quantify the economic improvement of the process efficiency.
- **Customer Lifetime Value (CLTV):** The Customer Lifetime Value is usually defined as the total net income a company can expect from a customer (Novo 01). The exact mathematical definition and its calculation depend on many factors as the industry, etc. The sum of values follows in its development over time usually a Gaussian e-function. This statistical model can be used to anticipate the future benefits development of the customer based benefits of an eCommerce investment.
- **Employee Lifetime Value (ELTV):** To quantify the benefits of the employee related benefits, one could define an Employee Lifetime Value analogous to the Customer Lifetime Value. Accordingly the Employee Lifetime Value can be defined as the total net income a company can expect from an employee. We suggest a similar basic mathematical definition and an equivalent calculation to the Customer Lifetime Value.
- **Total Cost of Ownership (TCO):** The Total Cost of Ownership (TCO) takes all operating cost of the investment into account, specifically for IT investments maintenance cost, content development and management costs, training costs, license cost, consulting costs etc. The resulting ratio

is either a price, the cost of the investment per working place or the total sum of accumulated costs.

- **Future Potential through Real Option Pricing Value (ROPV):** Real Option Pricing Models allow assigning a monetary value to the volatility of related future projects of an investment. There have been several models proposed in Management Information Systems literature. Stickel 1999 shows that innovative projects can be valuated by European call options and that related future projects can be valuated by compound options. The basic mathematical construct of all proposed approaches is an exponential function.

The four categories of benefits namely process efficiency, customer value, employee value and benefits from related future investment can be combined to an integrative ROI calculation, which is schematically shown in equation 7.

$$ROI = \frac{Income}{Investment}$$

$$ROI = \frac{Process\_Efficiency + \sum CLTV + \sum ELTV + Future\_Potential}{Total\_Cost\_of\_Ownership} \quad (3)$$

Dynamic ROI Calculation
CLTV = Customer Lifetime Value
ELTV = Employee Lifetime Value
Future Potential = Benefits of Related Successive Projects

As the concepts of Total Cost of Ownership and Process Costing are state of the art methods, we will focus in the following on the concepts of Customer Lifetime Value, Employee Lifetime Value and Real Option Price Value and show these concepts in more detail.

5. CUSTOMER / EMPLOYEE LIFETIME VALUE MODELS

The Customer Value describes the economic relevance of a customer, meaning his tangible and intangible contributions to business objectives. A recent study conducted by Cambridge Technology Partners Germany used a Customer Lifetime Value Approach for ROI calculations of CRM investments (Graf et al., 2003). We will illustrate in this example an approach for the economic valuation of a CRM Project in the assurance sector with a dynamic ROI calculation. The calculations cannot be shown in full detail due to confidentiality aspects.

In CRM Projects the customer behaviour plays the important role for the development of the benefits. To quantify these benefits in this specific focus of the assurance business the benefits calculation was conducted with the customer lifetime value (CLTV). The following equation shows the formula, which is used for the benefits calculations based on the CLTV (Graf et al, 2003):

$$CLTV = \sum_{i} \sum_{k=0}^{L-A} A_{ik} * \left[D_{ik} * E_{ik} * (1-W_{ik}) * \left(\frac{1}{1+p}\right)^{k} + \sum_{s=k+1}^{L-A} E_{is} * \left(\frac{1}{1+p}\right)^{s} * \prod_{j=k}^{s} (1-W_{ij}) \right] \quad (4)$$

Customer Lifetime Value Method

A = current customer age
L = customer life expectancy
K = current calendar year
Eij = expected customer revenue in branch i in year j after K
Wij = customer's probability of cancelling in branch i in year j
Dij= annual insurer's share of customer in branch i

The precise calculations can be viewed in the study published by Cambridge Technology Partners (Graf et al., 2003).

Suitability of the Lifetime Value Concepts for eCommerce Projects: The integration of dynamic elements like the CLTV into the ROI calculation leads to an s-shaped curve of ROI distribution, which is typically for innovation diffusion processes. The s-shaped ROI development represents a more realistic anticipation of the value development of the CRM investment and provides a much more authentic and credible ROI calculation. The Life Time Value Models are an interesting approach as they capture intangible benefits and important system behaviour like acceptance on a macroeconomic perspective. Problematic is the definition of the function parameters as estimating the future development of the lifetime cycle, qualitative components and changes of retention effects in the future and the complex relationship between building up and skimming the potential value.

6. CONCLUSION

The application of traditional tools beyond their actual boundaries overstresses their internal logic as a tool and framework and undermines their effective application. A helpful approach could be to analyze IT-economics from a macroeconomic point of view. Statistical approaches over adaptive behaviour and system dynamics describe the overall development and can contribute to a new way of determining the costs and benefits of changes processes through eCommerce projects. We showed an exemplary approach how innovation diffusion models can be integrated in economic valuations for customer relationship management systems with the Customer Lifetime Value.

REFERENCES

Amberg, M., Hirschmeier, M. (2003). Reframing of Economic (e)Valuations for Innovative Technologies. International Conference on Synergetics, Cybernetics and Informatics.

Adner, R., Levinthal, D.A. (2002). What is not a Real Option: Identifying Boundaries for the Application of Real Option to Business Strategy.

Barth, W., Kaletsch, U. (2001). Churn Management in Ratail Banking. in: Customer Relationship Management in Banken: Jürgen Moormann and Peter Roßbach, Bankakademie-Verlag Frankfurt, 1.ed. 131-152.

Benaroch M., Kauffman R.J. (1998). A Case for Using Option Pricing Analysis to Evaluate Information Technology Project Investments. Forthcoming in: Information Systems Research.

Buss, M.D.J. (1983, January-February). How to Rank Computer Projects. Harvard Business Review. 118-125.

Graf et al. (2003). Time to ROI, Cost Effectiveness Study of CRM-Systems in the Class of Insurance. Cambridge Technology Partners.

Chalasani P., Jha S., Sullivan K. (1998, February). An Options Approach to Software Prototyping. Technical Report, Carnegie Mellon University, Computer Science Department.

Cox J., Ross S., Rubenstein M. (1979, Sept.). Option Pricing: A Simplified Approach. Journal of Financial Economics. 229-263.

Dixit A., Pindyck, R.S. (1994). Investment under Uncertainty. Princeton University Press.

Dos Santos, B.L. (1991, Spring). Justifying Investments in New Information Technologies. Journal of Management Information Systems. (7:4). 71-90.

Friedlob G.T., Plewa F.J. (1996). Understanding Return on Investment, John Wiley & Sons.

Hochstrasser, B.(1994). Justifying IT Investments. In: The evaluation of information systems investments, ed. by Leslie Willcocks, Chapman&Hall. 151-169

Hull J.C. (1997). Options, Futures and other Derivative Securities. Prentice Hall, Englewood Cliffs, 3. ed.

Laudon, K.C.; Laudon, J.P. (2002). Management Information Systems: Managing the Digital Firm. Prentice Hall, New Jersey, USA.

Mahajan V., Muller E., Bass F. (1993). New Product Diffusion Models. In: Handbook of Operations Research & Management Science, Eliashberg J. (ed.), Elsevier Science Publishers. 349-408.

Margrabe W.: The Value of an Option to Exchange One Asset for Another, Journal of Finance, (33:1), March 1978. 177-186.

Morris H.D. et al. (2003, January). The Financial Impact of Business Analytics. IDC, Massachusetts, USA.

Mukhopadhyay, T., Kekre, S., Kalathur, S. (1995, June). Business Value of Information Technology: A Study of Electronic Data Interchange. MIS Quarterly. 137-156.

Pfeiffer, H. K. C. (1992). The Diffusion of Electronic Data Interchange. Physica, Heidelberg.

Smith, J. E., Nau, R. F. (1995, May). Valuing Risky Projects: Option Pricing Theory and Decision Analysis, Management Science, Vol. 41:5.

Quigg, L. (1993, June). Empirical Testing of Real Option-Pricing Models. Journal of Finance, (48:2). 621-640.

Sassone P.G.. (1987, July). Cost-Benefit Methodology for Office Systems. ACM Transactions on Office Information Systems, (5:3). 273-289.

Tam Kar Yan (1992, December). Capital Budgeting in Information Systems Development. Information & Management, (23:6). 345-357.

Sagner, J. (2002). The Real World of Finance: 12 Lessons for the 21st Century, John Wiley & Sons.

Sternman, J.D. (2000). Business Dynamics: Systems Thinking and Modeling for a Complex World, McGraw-Hill, Boston, USA.

Stickel, E. (1999). Die Bewertung von IV-Entwicklungsprojekten mit Methoden der Optionspreistheorie. In: Scheer, A.W., Nüttgens, M. Electronic Business Engineering, Physica-Verlag, Heidelberg, p. 685-707.

Sullivan K., Chalasani P., Jha S.(1997, June). Software Design Decisions as Real Options. Technical Report, Carnegie Mellon University, Computer Science Department.

Taudes A., Feuerstein, M, Mild, A. (1999). How Option Thinking can improve Software Platform Decisions. Working Paper No. 38, Working Paper Series.

Trigeorgis, L. (1996). Real Options: Managerial Flexibility and Strategy in Resource Allocation. The MIT Press.

Trigeorgis, L. (1995). Real Options: An Overview. in: Real Options in Capital Investment - Models, Strategies and Applications, Trigeorgis L. (ed.), Praeger, Westport, CT. 1-28.

Chapter 11

µP: A MICROPAYMENT SYSTEM

Pedro A. L. Mindlin, Christiane M. Schweitzer, Tereza Cristina M. B. Carvalho, Wilson V. Ruggiero

Abstract: The development of electronic commerce has led to a new trend: the distribution of digital information. Micropayment systems come as an alternative, allowing the implementation of such transactions at low costs. This work introduces µP, a micropayment system based on central generation of electronic coins that are bought by and distributed among customers and easily verified – and thus accepted as payment – by electronic commerce vendors. It differentiates itself by generating a single group of tokens that can be used for shopping in all of these vendors. This process is performed concerning security and scalability requirements.

Key words: Electronic commerce, micropayments, network security.

1. INTRODUCTION

The development of electronic commerce has led to the development of a new category of payment systems. Many of them already exist and are currently being used, such as credit card and electronic money systems. However, these systems usually can't be used for small amount payments and they present high latencies for each transaction.

Micropayment systems were designed to solve both questions to small amount transactions – concerning also security and scalability requirements that are inherent to any payment system to be implemented in open networks.

The possibility of performing small amount transactions opens interesting new paths to electronic commerce: it makes selling information products – with values in the range of cents, such as images or access to information pages or individual news texts – possible.

This paper presents a micropayment scheme adequate for smaller transactions. It is organized as follows. Section 1 is this introduction. Section 2 discusses electronic commerce, including components and functionality of these systems. Section 3 specifies the scheme we developed, explaining its architecture. Section 4 presents the results obtained from implementing such scheme. Finally, section 5 presents our conclusions.

2. ELECTRONIC COMMERCE

Electronic commerce transactions consist of information exchanges such as the product information that the vendor sends to the client and the financial information that the client sends to the vendor, usually through the financial system.

The micropayments is a new category of payment systems, initially proposed circa 1992, which took advantage of new uses for hash functions like SHA [NIST] and MD5 [Rivest-92]. Among the most important proposals of micropayment schemes are the ones by [Rivest-96], [Pedersen] and [Anderson, et. al.]. Micropayment systems were designed for electronic commerce systems and are therefore electronic in their nature. As so, they require no adaptation for the electronic medium.

Micropayment systems typically consist of three entities: the user, who wants to buy something, normally a person using his or her computer connected to the Internet; the vendor, which is the company or person who is selling a product or service through the network; and the agent, which is the company – usually into the financial business – that is responsible for the system, for issuing tokens and making their payment reliable, for maintaining user and vendor accounts.

Micropayments has resurfaced over the last few years, and we believe that it could be successfully deployed, because electronic commerce is now prepared for accepting significant changes.

2.1 Related Work

One of the first important proposals of micropayment was Millicent [Manasse et. al.]. This scheme made use of a script containing the owner identification, a serial number and other information signed by a digital certificate. The user would buy one script from the agent to each vendor he or she would buy products from. At each transaction, he would send the script to the vendor, who would deduct the purchase value from it and digitally sign the new value, maintaining an account history for the user inside the script.

Millicent opened way to lots of other works, like NetCard [Anderson et al], TickPayments [Pedersen], PayWord [Rivest-96] and iKP [Hauser et. Al.], all of them making use of the same general idea of signed certificates used to authenticate digital money.

Among the most important micropayment schemes proposed at that time was PayWord [Rivest-96], which was used as the basis for the system here presented. PayWord is a credit-based system in which the user gets from the agent a digital certificate containing his name, IP address, a public key and other information. The certificate is signed by the agent. The certificate allows the user to create token chains by computing hash functions over a seed and digitally signing the last value in the chain, which is then sent to a vendor. All the subsequent values sent to the vendor during that day (the previous tokens in the hash chain) are verified against the signed token received in the beginning of the day with a simple hash function computing. As hash functions are one-way, it is impossible for the vendor to find out what is the next token to be received, but it is fairly easy to verify it.

Some other micropayment schemes have been proposed since 1997 as improvements to some of the original proposals. Among them are UpayWord [Mu et. al.], and MR1, MR2, MR3 [Micali et. al.], that are improvements over the scheme presented by [Rivest-96].

3. ARCHITECTURE

The first question to be addressed was what underlying micropayment scheme we would use. Some schemes were discarded for imposing the use of dedicated hardware, which would imply high costs. Then, it was decided to use a software-based scheme. The other central question here was the distribution requirement: the system would have to work over the Internet and multiple platforms, especially Windows, Linux and MacOS. This led to the use of Java technology.

Next came the task of creating an independent set of tokens for each vendor. The first idea was to create what we called a **specific micropayment system**: the system would work for only one vendor; the generated token set would only be used to make purchases with that vendor. This scheme would be useful for big companies selling their goods, but would not be useful for small vendors. Thus we expanded the system, developing what we called a **generic micropayment system**, which would create a unique set of tokens to be used with all vendors.

The **generic micropayment system** works as follows. A unique set of tokens is generated when the user registers within the system. When the user visits the registered vendors, he or she sends the tokens to these vendors in order to make purchases. The tokens are then verified locally by each vendor. As the tokens are the output of hash functions, they have to be verified by applying these functions on them and comparing the results with the next value on the chain. Note that the next value on the chain has to be already stored by the vendor, so that it can be compared with the generated one. This leads to the problem of sending the root token to the vendors.

The first solution we devised for this problem was to create a vendor certificate, which is sent to each vendor when the user certificate is created. That way, if the user certificate contains a chain of n tokens, the vendor certificate contains the n+1 token and information about the user to whom that token belongs. When making the first purchase, the user sends to the vendor the token n; the vendor would take this token and feed it to the hash function, obtaining token n+1, which it compares with token n+1 that is in the certificate. If they are the same, the transaction is authorized and the vendor stores token n. At the next transaction with that vendor, the user will send token n-1, which hashes to n, and is stored at the vendor. The vendor authenticates the purchase, and so on.

This scheme seems to function properly, but what happens if token n-1 is spent with some other vendor, which did not receive token n? To solve this problem the vendor will have to hash token n-1 twice, obtaining token n+1, which is in the certificate. Thus, all tokens will have to contain an order number, which represents the number of hash function calculations to be made over that token for it to be verified. For this scheme to work, though, the root token (n+1 in our example) has to be self-verifiable, which can be made through the use of a digital signature by the agent when it generates the vendor certificate. We chose to embed the signed vendor certificate into the user certificate and make the user send this vendor certificate along with each token. The vendor has to verify the agent signature on the certificate and calculate the hash value of the token as many times as its order number.

3.1 Frauds

There are mainly two types of frauds: the vendor and the user frauds.

The vendor fraud happens when the vendor intentionally creates and presents to the agent a valid token, which was not spent by any real user at the vendor's site. This fraud is possible because a user can send different tokens to different vendors. Let's see an example: the user sends token n to vendor A, then sends token n-1 to vendor B and then sends token n-2 to vendor A. If vendor A wants, it can create token n-1 from token n-2 and fool the agent, because obtaining token n-1 from token n-2 is just computing the hash function once over token n-2. In this situation, vendors A and B present the same token to the agent, who, in lack of more information, has no choice but paying both of them. A scheme for protection against these frauds will be presented ahead.

The user fraud happens when the user breaks into the locally stored certificate and changes values, creating false tokens. To circumvent this possibility we opted for storing the user certificate in cryptographic form. Another fraud that can occur is based on the certificate generation algorithm: the user has the right to have his certificate regenerated in the event of accidentally loosing it in his computer; at the time the new certificate is generated, there can be unreported expenses from the user at some vendors, which will lead the agent to generate more tokens than the user actually has.

To prevent these frauds it is necessary for the agent to verify the tokens spent at the vendors. This can be done in four different ways: online, offline, almost online and batch mode. The chosen scheme for μP was almost on line, though we also implemented an optional batch mode. In almost on line mode, at each transaction, the vendor sends the token received to the agent, but does not wait for authorization from it. This scheme allows the agent to keep immediate track of transactions but keeps latency at a minimum. It does not prevent frauds, but makes detection fast and easy to do. The smaller the time interval, the better the fraud protection and the higher the communication and processing cost..

3.2 CRL (Certificate Revocation List)

To keep track of fraudulent users the system will maintain a CRL (Certificate Revocation List). The vendors will access the CRL at the agent's site and maintain it locally updated to use it at every transaction.

3.3 Integration

The μP should be integrated with other systems, such as a web application that delivers the product to the user. The simplest and easiest solution to achieve this is to use a database.

3.4 Components

μP has five elements: the token - a 20 bytes number, coded in Base64, with an associated order number; the certificate - a chain of tokens, each one being the result of the hash function computing on the previous one; the user module; the vendor module and the agent module:

Certificate: The certificate is generated by the agent at user request and is kept both in the agent's database and in the user's local disk. The user's copy is stored in cryptographic form to prevent eavesdropping and frauds. At any given moment there will be only one valid certificate for each user in the generic micropayment system and one certificate for each user-vendor pair in the specific micropayment system.

User: The user interacts with μP when he or she is registering, making a purchase and requesting a new certificate from the agent.

Vendor: The vendor interacts with the system when it registers, sells a product and sends information to the agent. The registering of the vendor is an important moment in which it has to decide if it wants to use the specific or the generic micropayment system.

Agent: The agent is the central element in the system. It generates the tokens, collects money from the users and pays the vendors for the tokens received. It has to maintain accounts for all the users, keeping track of each token they spend, and detecting any possible frauds - in this case it revokes fraudulent users or vendors. The agent has to be always informed of spent tokens so it can be able to generate new certificates in case of loss and so it can be able to detect and prevent frauds. Certificate generation can also occur when the user runs out of unused tokens and tries to make a purchase.

3.5 Processes

μP can be described in terms of its function processes, which are: registering, token generation, purchase transaction and transaction informing and verification by the agent:

Registering: μP was not developed to provide its users with anonymity. Each prospect user must provide the system consistent identification data so he or she can be accepted as a valid user. The user has to provide a username and a password, which will be used for all further operations such as transactions or token purchases to maintain his or her privacy and to positively identify him or her to the system. As any secure registering process, μP will be done over secure SSL connections. The vendors also have to register within the system, and at this time they have to choose whether they want to work with a specific or a generic micropayment system.

Token generation: Token generation occurs when the user is a new one or when he or she does not have enough tokens to make a purchase. It can also happen if the user has lost his or her certificate. In this case the agent invalidates the former certificate and generates a new one with the same amount of unspent tokens. The token generation process can be better understood through Figure 1 and Figure 2, steps A to D:

A – Token request. The user informs the agent his or her name, password and the number of tokens he or she wants to buy.

B – Payment sequence. This phase is comprised of external protocol payments such as SET (Secure Electronic Transaction) for the user to pay for the requested tokens. It will not be part of this study.

C – Token generation. The agent generates the tokens through the following steps: it verifies the password given by the user against its database; it gets from the database the order number k of the next token to be spent by the user; it generates a random number and from this number it generates the token sequence, with order numbers starting from k; it digitally signs the last token (root) with the user's public key in the sequence and generates vendor certificate, and inserts this vendor certificate into the new user certificate; it revokes the user's last valid certificate; it inserts the new certificate into the database and it makes the new certificate available for the user to download.

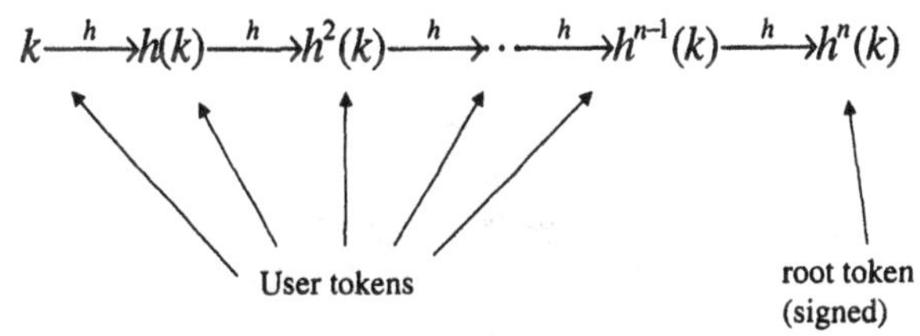

Figure 1. Tokens generation

D – Token delivery. The certificate is delivered through an applet that retrieves it from the agent server and stores it in the user's local disk.

Purchase transaction: This is the most important process in μP . It can be better understood in Figure 2, steps 1 to 8.

The transaction begins when the user clicks on a link at the vendor's page. The vendor then returns an applet to the user. This applet is the most important component of the user module. The applet searches for a user certificate on the local disk, according to the naming rules, and prompts the user for a password. From the password it obtains the cryptographic key, which is used to decrypt the file. The applet then searches for unused tokens and checks if there is the necessary number of them to make the purchase. If the number of tokens is insufficient, the applet redirects the browser to the agent's site, going to step A.

If the certificate is valid and the number of tokens is enough, the applet sends to the vendor the next token – according to the value of the purchase – together with its order number, the signed vendor certificate which contains the root token, the serial number of the certificate, the user name, the product id and the purchase value. The vendor checks if the certificate is valid (i.e. is not in the CRL) and if the received token is authentic by computing its hash value k times and comparing it with the signed root token, showed in the Figure 1.

If the token is authentic, the vendor generates a random number, puts this number in the database and returns the same number to the user. Right after that, the vendor sends information about the purchase to the agent (user name, the token itself, its order, the purchase value and the authorization number issued). If the token is invalid, the vendor cancels the transaction and returns a value of -1 to the user.

When it receives an authorization number different from -1, the applet redirects the browser to the product page, sending the authorization number together with the request for the page. The web application responsible for the product will then check the authorization against the database and deliver the product to the user if the authorization is valid.

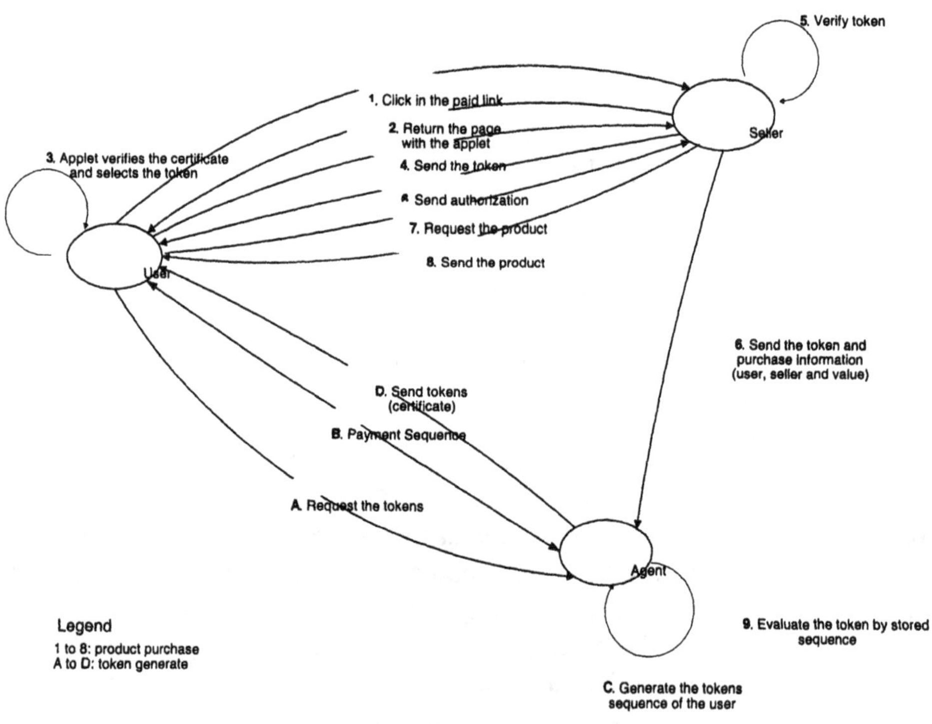

Figure 2. μP processes

Transaction informing and verification: Information about every transaction made at the vendor site can be transmitted to the agent right after the transaction or in batches at defined time intervals. There is a standalone program designed for the vendor that can sweep the database at regular or variable intervals of time and send new transactions info to the agent. On the agent side, a servlet is responsible for parsing the data and storing it in the database, checking immediately for frauds and errors.

4. RESULTS

μP will be evaluated in accordance with the framework proposed by [Schmidt et. al.]. This framework makes qualitative evaluation of micropayment schemes based on nine parameters, clustered in three dimensions: microeconomic, technologic and social. The results of the evaluation can be seen in Table 2. To evaluate micropayment systems with these parameters, the authors attribute to each one of them a grade that ranges from "--" to "-" to "o" to "+" to "++".In the next itens, we will examine each of these evaluations in detail:

Low transaction cost: When evaluating the transaction we can ignore the phase of sending information to the agent and concentrate on the user-vendor interaction. The transaction cost is evaluated in terms of its computational cost, which is the cost of one digital signature verification and k computing of the hash function. As k is a variable number, the cost of each transaction is non-deterministic. Moreover, k ranges can vary from one user to another, depending on the size of the user's certificates. The hash function is fast enough and can be considered low cost, even if computed many times. The digital signature verification, though, can be considered a strong delaying factor. Thus we grade the transaction cost of our system as "+".

Table1. μP evaluation

Dimension	Parameter	μP
Microeconomic	Low transaction cost	+
	Atomic transaction	O
	User base	O
Technologic	Security	-
	Reliability	+
	Scalability	-
	Latency	+
Social	Peer-to-peer payments	Does not apply
	Anonymity	-

Atomic Transactions: μP was designed and implemented so as not to consider a token as spent if the authorization number was not received. If the applet receives a -1 it will not modify the contents of the certificate, keeping the token as new. However, the atomicity of a μP transaction does not consider the final delivery of the product. If the connection is lost at this last stage, for example, the system cannot recover it. A μP transaction ends when the authorization number is received. We consider this a flaw in the system and thus grade atomicity as "o".

User base: The μP user base still does not exist and cannot be evaluated. The grade for this parameter is "o".

Security: μP was implemented to handle secure connections between modules that offer authenticity, integrity and confidentiality services for messages exchanged. This was achieved deploying SSL protocol suites for communication. The fraud detection scheme is still weak and requires human intervention to work properly. One example of a fraud that is difficult to catch is the generation of false tokens by vendors. Although it was designed to be weak against frauds and gain strength from the speed and usefulness, in this particular parameter we have to be strict and grade the system as "-".

Reliability: Due to the fact that the transaction is verified only at the vendor, the system's reliability is distributed among all vendors' sites. There can be some problems concerning the Java applets at the users' computers, because they need access to local disks, but we will consider the situation in which they are authorized to do that by the user and can work normally. There is a central point of failure at the agent when it generates the tokens, but it does not comprise the actual purchase transaction. For these reasons we graded μP as "+" for this parameter.

Scalability: Up to this moment it has not been possible to perform scalability tests on the implemented system. We will consider theoretical scalability, based on the fact that the agent centralizes all operations except the transaction. Although μP is decentralized in the transaction process, it is very centralized when it concerning its, administration. The agent's database has to contain all the tokens generated for all the certificates of all users relating to all vendors (although we must say that we believe the majority of vendors will choose the generic micropayment system), and has also to keep records of each transaction performed by users and vendors. The agent also has to check all transactions for possible frauds. This centralized dependency is a clear scalability flaw in μP and for this reason we graded it "-".

Latency: The transaction latency is the time needed for one digital signature verification and a variable amount of hash function computations, as described in section 4.1. Although it is the most important factor for the latency, the digital signature verification can be considered as a relatively fast operation, if compared to digital signature creation. For this reason we graded μP's latency as "+".

Peer-to-peer payments: μP was not conceived to make peer-to-peer payments. The grade for this parameter is "does not apply".

Anonymity: The current implementation of μP does not allow any level of anonymity. The user has to be registered and for this he or she has to input personal data to the system. However, there is a possible implementation of an anonymous μP: anonymous user-cards would be sold containing hidden passwords that would allow the card holder to access the system anonymously. This is future work, but it raises the grade of this parameter to a "-".

5. CONCLUSIONS

Electronic commerce is steadily growing. This growth leads to new business possibilities, among them the distribution of digital information. It requires a new payment system capable of handling transactions with low cost and latency.

Micropayment systems have the necessary features for small amount transactions through simple and reasonably secure operations. The security model for such systems is a simplified one, as a result of a trade-off between the cost of implementation and the required security level for the transactions.

μP is an experimental but totally functional micropayment system, developed on the grounds of the PayWord system proposed by [Rivest-96]. The main characteristic of μP is the use of an electronic token generated in a central entity but verified as authentic in a distributed way by simple and fast computing operations. It differentiates from PayWord by making use of a unique set of tokens that can be used with any registered vendor. Its web interface makes it easily distributed throughout the various platforms that form the Internet. A more detailed comparison between the two systems can be seen in table 2.

Table 2. PayWord and μP comparison

PayWord	μP
One certificate per vendor per day	A single certificate for all vendors. Also accepts single-vendor certificates
User computes digital signatures	User does not compute digital signatures
Token authentication through one hash function computing	Token authentication via hash function computing and one digital signature
Tokens are generated by user	Tokens are generated by agent
Vendor sends to agent the last token received from each user each day	Vendor sends to agent all tokens received
Good for frequent relationship between user and vendor	Good for occasional relationship between user and vendor
Agent verifies tokens against user certificates	Agent checks tokens against central database
Credit-based	Debit-based (prepaid)

The micropayment system can be used to purchase papers in the Internet. Normally, in the case of a newsletter or magazine, the subscriber have access to use the full services by signature. A non-subscriber could use the micropayment system to buy only some pages of that newsletter or magazine. At this moment, the developed system is in an experimental phase at a banking site. The current purpose is to evaluate the technical and commercial feasibility of this solution to selling images and news.

REFERENCES

1. Anderson, R.; Manifavas, C.; Sutherland, C. Netcard – a practical electronic cash system. http://www.citeseer.nj.nec.com/168207.html.
2. Hauser, R.; Steiner, M.; Waidner, M. Micro-Payments based on iKP. [S.l.], 1996. Technical report. http://www.citeseer.nj.nec.com/hauser96micropayment.html.
3. Manasse, M. et al. The millicent protocol for inexpensive electronic commerce. 1995. http://www.w3.org/Conferences/WWW4/Papers/246/.
4. Micali, S.; Rivest, R. L. Micropayments revisited. In: CT-RSA. [S.l.: s.n.], 2002. p. 149–163. http://www.citeseer.nj.nec.com/502775.html.
5. Mindlin, P.A.L. μP:Uma solução de micropagamentos. Master thesis-PCS/EPUSP. 2002.
6. Mu, Y.; Varadharajan, V.; Lin, Y.-X. New micropayment schemes based on paywords. In: Australasian Conference on Information Security and Privacy. [S.l.: s.n.], 1997. p. 283–293. http://www.citeseer.nj.nec.com/mu97new.html.
7. NIST. Fips publication 180: Secure hash standard (shs). National Institute of Standards and Technology. May 1993.
8. Pedersen, T. P. Electronic payments of small amounts. In: Security Protocols Workshop. [S.l.: s.n.], 1996. p. 59–68. http://www.citeseer.nj.nec.com/pedersen96electronic.html.
9. Rivest., R. The md5 message-digest algorithm.RFC 1321. April 1992.
10. Rivest, R. L. Electronic lottery tickets as micropayments. In: HIRSCHFELD, R. (Ed.). Financial Cryptography. Anguilla, British West Indies: Springer, 1997. p. 307–314. http://www.citeseer.nj.nec.com/rivest98electronic.html.
11. Rivest, R. L.; Shamir, A. Payword and micromint: Two simple micropayment schemes. In: Security Protocols Workshop. [S.l.: s.n.], 1996. p. 69–87. http://www.citeseer.nj.nec.com/rivest96payword.html.
12. Rivest, R. L.; Shamir, A.; Adelman, L. M. A Method for Obtaining Digital Signatures and Public-key Cryptosystem. [S.l.], 1977. 15 p. http://www.citeseer.nj.nec.com/rivest78method.html.
13. Schmidt, C.; Müller, R. A framework for micropayment evaluation. http://citeseer.nj.nec.com/172517.html.
14. Schweitzer, C. M. et al. Sistema para Controle de Distribuição Segura de Imagens. São Paulo, 2001. Technical Report LARC - PCS - EPUSP / Scopus.

Chapter 12

ELECTRONIC AUCTIONS IN FINLAND

Virpi Kristiina Tuunainen, Matti Rossi, Jussi Puhakainen

Abstract: In this paper we explore the electronic auctions in Finland. We have investigated, on the supply side, the major auctions, especially looking at their auction types, business models and popularity. On the demand side, we have conducted a survey of students to seek out the demand for auctions. Our findings show that in a geographically and linguistically isolated market place several auction sites can survive, but there is too little demand to support electronic auctions as separate businesses.

Key words: Electronic auction, auction type, business model

1. INTRODUCTION

Electronic (online) auctions have received considerable publicity during the heyday of "Internet-boom". The online auction industry was born in 1995 with the founding of eBay, now the biggest auction-site in the world, with an 85% market share of the online auction sector (Lansing & Hubbard, 2002). Albert (2002) presents several industry growth related numbers, of which the number of participants perhaps illuminates the situation best: 6.5 million participants was the projection in online auctions in 2000, which was exceeded with considerable margin (35 million participants). Also the number of online auction transactions is considerable; 1.3 million transactions per day in 2001.

Online auctions have several characteristics that make them appealing from the viewpoints of both consumers and businesses. According to Koch and Cebula (2002), electronic auctions can yield to better prices to sellers than via other electronic channels. They point out that while the Internet in general tends to lower prices, product branding, price discrimination and online auctions may raise prices. Lansing and Hubbard (2002) also point out higher prices to sellers, and add low overhead costs. According to them, the online auctions are an example of the ideal online business, where the auctioneers act solely as intermediaries between sellers and buyers, without considerate investments in land, property and inventory (Kambil & van Heck, 2002).

In addition to the famous, big general auctions such as eBay and Amazon's and Yahoo's auctions sites, there has been a surge of more regional or niche-market focused auctions (see for example Lansing and Hubbard, 2002). However, the majority of consumer focused auctions have proved to be short-lived, some due to not reaching critical mass of users and some due to not finding a sustainable business model. In addition, one must note that there may be and probably are differences whether an auction operates in large markets (eBay: USA plus world) or in smaller, more language and location dependant markets such as Finland.

In this study our goal is twofold. On the one hand we have studied the supply side of online auctions by interviewing the management of the major Finnish online auctions. On the demand side we conducted a study of potential users of online auctions. We were especially interested in a few the particular aspects of the Finnish market: people are well wired and mobile, and there are a number of different eAuctions but eAuctions have not gained wide popularity. Our main aim was to try to find reasons for why supply and demand do not seem to be meeting in Finnish market. To explore this we used students as our sample population with the assumption that the younger people, who have grown with internet and mobile networks would be more eager to use the new electronic means of buying and selling goods.

The paper is organized as follows: first, in the next chapter we briefly describe the electronic auctions, and then we look at the auction providers in Finland. In the fourth section we analyze the (potential) user perceptions of online auctions and finally, in the fifth section we discuss the findings and provide some further research.

2. ELECTRONIC AUCTIONS

There are four basic types of auction mechanisms that are widely used and analyzed (see e.g. Klemperer 1999). The main four types are: the ascending-bid auction – also called the open, oral, or English auction; the descending-bid auction – used in the Dutch flower industry and also called the Dutch auction (for more details, see Kambil & van Heck 1998); the first-price sealed-bid auction; and the second-price, sealed-bid auction – also called the Vickrey auction (for more details, see Vickrey 1961 and Kauffman & Wang 2001).

In the ascending auction (English auction), the price is successively raised until only one bidder remains, and that bidder wins the object at the final price. This auction can be run by having the seller announce the prices, or by having the bidders call out prices themselves, or by having bids submitted electronically with the best current bid posted. This third form of the ascending auction was used in this research.

The descending auction (Dutch auction) works in exactly the opposite way: the auctioneer starts at a very high price, and then lowers the price continuously. The first bidder who calls out or submits electronically that she or he will accept the current price wins the object at that price. The auctioneer can announce the prices or a clock can be used that will indicate the price. The clock hands tick downward until a buyer stops them by raising a hand, pushing a button, or by clicking the mouse of his computer. The third form was used in this research.

In the first-price sealed-bid auction each bidder independently submits a single bid, without seeing others' bids, and the object is sold to the bidder who makes the highest bid. The bidder pays her bid – that is the highest price or "first" price bid. This method is used in procurement, that is competing contractors submit prices and the lowest bidder wins and receives her price for fulfilling the contract. This type of auction is also called "reverse" auction, because in this case the seller is the bidder and the buyer the bid-taker.

In the second-price sealed-bid auction each bidder independently submits a single bid, without seeing other's bids, and the object is sold to the bidder who makes the highest bid. However, the price she pays is the second-highest bidder's bid or "second price".

One notable factor of these different auction mechanisms is that according to Vickrey's classical theorem, they all yield to the same final price. In other words, there is no difference from the seller's perspective which mechanism is used. However, with online auctions there have been some notions that this may not be a valid assumption. For example Lucking-Reiley (1999) conducted a field study by selling Magic: the gathering-collectible cards in several online auctions. This research is interesting in the sense, that it avoids the most eminent pitfall of such studies, namely it recreates almost identical starting position to each sale. Normally, it is very hard to make comparisons between different mechanism, since for instance Sotheby's does not run both a second-price auction and an English auction for the same piece of antique furniture. According to Lucking-Reilly's findings, the Dutch auction produced 30-percent higher revenues than the first-price auction format (Kambil & van Heck, 2002).

3. THE FINNISH ELECTRONIC AUCTIONS

In this section we report the findings of the empirical studies conducted in both the supply and demand side of the Finnish auction market. We conducted interviews with more than a half of the major auctioneers and we made several questionnaires for potential users of auctions for the demand side. We used students as our sample for the demand side, because they have the technical means and interest to participate in electronic forms of auctions.

3.1 The Supply Side

The research was focused upon Finnish electronic auctions. 21 such auctions were found, of which 11 participated to the study. The study was conducted through interviews of the management of auctions. In addition, three auctions were studied through Internet only (i.e. no interviews). The interviews were conducted as semi-structured interview, i.e. the basic structure for each interview remained the same, and however, in each interview the respondent had quite a lot of room to go into topics not mentioned in the interview-structure.

The auctions in question were either business-to-consumer or consumer-to-consumer auctions. The semi-structured interview concentrated mainly on the chosen business models of the auctions, their pricing mechanisms, used technology and marketing efforts. The interviewers asked also about the perceptions of the online auctions and their profitability from each auctioneer.

3.2 The Demand Side

Survey data was collected in four sets in Helsinki School of Economics (HSE) and Turku Schools of Economics (TuSE) with students (some also international) as subjects, and in one set in an upper secondary school in Espoo. The university students were gathered on courses, which have wide participation across subjects to avoid bias toward technically oriented students, however, 60% of the students had information systems or ebusiness as their major subject. The upper secondary school students were chosen so that they had a language concentration to avoid the possible bias by very technically oriented students.

Background Information on the Subjects

There were a total of 106 university students, divided into 68 at HSE and 38 TuSE students, and 59 upper secondary school students. All the upper secondary school students were Finns, whereas of the university students 83 were Finnish and 18 from other countries (Russian, Estonian, Ukranian, Canadian, Chinese, etc.). All together there were 79 female and 85 male subjects, with an age range from 17 to 55, with a mean age of 24 years (19 for te upper secondary school students and 28 for the university students). 75 per cent of the upper secondary school students were either employed at the time (on average little less than 7 hours a week) or had worked in past, while 25 per cent had no experience. The working experience of university students varied from ½ year to 33 years, with a mean of 5 years. The university students had studied from ½ year to 8 years, with a mean of 3 years. All the students were well connected with 80% having internet connection at home and full 100% having a mobile phone. The following table lists their internet usage patterns:

Table 1. Internet usage patterns of the research subjects

	N	Minimum	Maximum	Mean	Std. Deviation
Daily use of WWW (avg mins)	162	00	1200,00	95,876	119,9289
daily received email (avg)	160	00	100,00	6,0745	10,4999
daily sent email (avg)	161	00	30,00	2,6412	3,7091
online purchases / last 6 months	155	,00	40,00	0,8323	3,0471

It is noticeable however, that there were very few online purchases performed by the sample group. It is also noteworthy that 16 % of the respondents had bought items from online auctions and 2 % had sold something in them. It can be argued that despite the high degree of wired and wireless connectedness of the participants, very few of them had actually performed online buying and selling and even fewer on auctions.

Table 3. General attitudes towards selling and buying goods in auctions

Web auctions

		Frequency	Percent	Valid Percent	Cumulative Percent
Valid	yes, for buying	17	10,3	10,8	10,8
	yes, for selling	2	1,2	1,3	12,0
	no	133	80,6	84,2	96,2
	no, but used for searching	4	2,4	2,5	98,7
	yes, for both buying & selling	2	1,2	1,3	100,0
	Total	158	95,8	100,0	
Missing	System	7	4,2		
Total		165	100,0		

The table below summarizes the general attitudes towards selling and buying goods in auctions:

Analysis of the results/findings

In the following preliminary results we briefly summarize the findings about the survey. We first look at the supply side by characterizing the auction site by their technical platform, auction type and business model according to the classification in the second section. In all the questions for the supply side we used a Likert scale from 1 (fully agree) to 7 (fully disagree).

3.3 Technical platform

The Internet was the dominant "platform" for auctions, only few auctions offered for example mobile solutions. This is interesting because Finland is one of the foremost countries in mobile-phone usage. The supply side was quite unanimous in their belief that mobile-only auction has at the moment no possibilities, mainly because of the limited screen and user interface of current mobile phones. However, certain companies voiced their belief in Digital Television coupled with mobile phones as a terminal device for making the actual bid, thus serving as the return channel. In the demand side there was little need for this, even among our sample, which is quite advanced in mobile phone use. The following two tables show that most of the sample population were either indifferent or outright rejecting the possibility of SMS bidding and auction updates. This cannot be explained by the unfamiliarity of the medium, because most of the respondents used heavily SMS for communication.

I could bid on known items with SMS-messages (e.g. bidding process started on the Internet).

		Frequency	Percent	Valid Percent	Cumulative Percent
Valid	fully agree	7	4,2	4,4	4,4
	2	14	8,5	8,8	13,1
	3	24	14,5	15,0	28,1
	4	42	25,5	26,3	54,4
	5	30	18,2	18,8	73,1
	6	28	17,0	17,5	90,6
	fully disagree	15	9,1	9,4	100,0
	Total	160	97,0	100,0	
Missing	System	5	3,0		
Total		165	100,0		

I would like to receive online updates on auctions items to my mobile phone or pda.

		Frequency	Percent	Valid Percent	Cumulative Percent
Valid	fully agree	2	1,2	1,2	1,2
	2	6	3,6	3,7	5,0
	3	12	7,3	7,5	12,4
	4	19	11,5	11,8	24,2
	5	20	12,1	12,4	36,6
	6	44	26,7	27,3	64,0
	fully disagree	58	35,2	36,0	100,0
	Total	161	97,6	100,0	
Missing	System	4	2,4		
Total		165	100,0		

(Mean 5,57; Standard deviation 1,536)
(Mean 4,36; Standard deviation 1,584)

3.4 Types of auctions

English auction was the dominant form of auction, a few auctions used Dutch auction mechanism, one stamp and collectibles auction used sealed mechanism. This is understandable, as the English auction is the best known mechanism in Finland. In the stamp and collectible auctions there is a long tradition of auctions from the offline world and thus the more advanced auction mechanisms can be employed.

3.5 Business model

The studied auctions were classified under four categories:

1. general auction, where the auction is either the main business or has high role for the main business
2. marketing magnets, where the role of auction is to draw customers to the main business
3. additional service, where the role of the auction is to support main business
4. "window" to traditional auction, where the eAuction offers an interface to the traditional (physical) auction.

While analyzing the viability of the business models, we encountered several interesting issues. First, the general auctions such as Huuto.net, Keltainen Pörssi and QXL were generally dissatisfied with their results revenue wise. None of the major players was able to include any form of commissions, thus they were dependant upon banner-advertising, strategic partnerships and other means of indirect revenue. It is also noteworthy that none of these companies announced that they were making profits.

Firms using eAuctions as marketing **magnets** were satisfied with results. Online auction seems to be, under the light of this research, a good way to market a site, at least at this moment.

Companies using eAuctions as **windows** to traditional auctions were generally very satisfied with the results. The reason for their satisfaction stems from location dependant reasons. Traditional auction is somewhat hindered by its place-dependency. Internet in general was seen as a vehicle to broaden markets, by making information distribution about sold items both economical and quick.

Generally, eAuctions were seen as good additional services, but as a main business found problematic. After the research, the number one general auction in Finland, Huuto.net, was purchases by local telecom company, Sonera, and added as a additional service to their popular portal. In light of this research is thus seems that independent general auctions in small markets do not have viable business model available. However, it seems that auctions per se are popular, and generate lots of traffic to sites whose main service is something else.

3.6 Products or services for auctions

Suitable products or services for auctions have certain properties. They must be easily describable, and their related information (what, in what condition etc.) must be easy to digitize. Examples of such were multiple, including computer equipment, electronics, collectibles and antiques.

On the demand side the subjects think that they can (or could) buy goods from auctions that they cannot find anywhere else (mean 3,9*), rather than use auctions to purchase things that are bought regularly (e.g. clothes or groceries) (mean 5,0*). They also largely agreed that Web-auctions are best used for making impulse or one-time purchases (e.g. a DVD-player or airline tickets) (mean 3, 39*). Selling goods on eAuctions is seen as a fairly feasible option (mean 3,83*).

It is noteworthy that despite the long relationship with digital medium, most respondents saw product presentation, even to the degree of seeing the physical product, as very important. This poses challenges for mobile auctions, as there are very limited possibilities for transmitting images of the auction items into cellular phones in the near future. However, this will change with the advent of GPRS and 3G phones.

Main criteria for shopping

		Frequency	Percent	Valid Percent	Cumulative Percent
Valid	price	54	32,7	33,1	33,1
	quality	102	61,8	62,6	95,7
	both	7	4,2	4,3	100,0
	Total	163	98,8	100,0	
Missing	System	2	1,2		
Total		165	100,0		

3.7 Price setting value/role of eAuctions

Electronic auctions define theoretically both the correct price from the markets as well as adjust the level of needed information to correct level (example of the latter: a product is put to the eAuction, interactive features of the system allow for feed-back upon more information (more detailed descriptions, digitized pictures etc.). However, it is notable that only the major auctions can achieve the liquidity and popularity needed for the price setting to work. In the table below the main reasons for using eAuctions are listed. It should be noted that the quality is perceived as far more important than price alone.

3.8 Entertainment value of eAuctions

Perceived entertainment value of eAuctions for the subjects is rather modest on average. So it should not be expected to become a major pastime for people.

3.9 Trust issues

Trust in one of the key components in electronic commerce in general and even more so on online auctions. Online auction fraud is an issue, which has grown hand in hand with the growth of the online auction industry. Albert (2002) for example reports that consumer complaints to US National Consumer Leagues Internet Fraud Watch increased by 600% from 1997 to 1998, with online auction related complaints numbering almost 5500 or 68% of all 1998 complaints. The Federal Trade Commission in the US has seen online auction related complaints to grow from 106 in 1997 to a remarkable 10872 in 2000.One must also note that these figures indicate only the reported incidents, the true number of frauds in higher.

I (could) allow my buying habits to be analyzed by the auctioneer.

SCHOOL			Frequency	Percent	Valid Percent	Cumulative Percent
universtity	Valid	fully agree	4	3,8	3,9	3,9
		2	13	12,3	12,7	16,7
		3	22	20,8	21,6	38,2
		4	22	20,8	21,6	59,8
		5	18	17,0	17,6	77,5
		6	16	15,1	15,7	93,1
		fully disagree	7	6,6	6,9	100,0
		Total	102	96,2	100,0	
	Missing	System	4	3,8		
	Total		106	100,0		
upper secondary school	Valid	fully agree	3	5,1	5,2	5,2
		2	3	5,1	5,2	10,3
		3	7	11,9	12,1	22,4
		4	9	15,3	15,5	37,9
		5	10	16,9	17,2	55,2
		6	15	25,4	25,9	81,0
		fully disagree	11	18,6	19,0	100,0
		Total	58	98,3	100,0	
	Missing	System	1	1,7		
	Total		59	100,0		

When asked, the subjects on average, felt that it is important to see the items in real as they are before making a bid (mean 2,78*), rather than trusting textual descriptions of goods being sold in the eAuction (mean 4,88*). This alleviates again the previously mentioned ease of digitization as a key to success of the auctions. It is also important to be able to view the reputation of the seller/buyer (as graded by others at the auction site) (mean 2,78*). Furthermore, most subjects would rather buy from a company than another consumer in an auction (mean 3,11*). Thus it is not surprising that many of the bigger auction companies have taken several incentives in building consumer trust over online auctions.

Most subjects were concerned about security of online payments (mean 3,26*). Also, they do not like the idea of allowing their buying habits to be analyzed by the auctioneer (mean 4,39*). This is in contrast with the ideas of eAuctioneers, who use these as magnets and would like to understand the patterns of behavior of the bidders.

It is especially noteworthy that the younger subjects are noticeably stricter in privacy issues. We could speculate this to be seen as a mark of the people becoming more aware and educated of the need for privacy and the value of their click streams.

I (would) have no problem in trusting textual descriptions of goods being sold in the auction.

SCHOOL			Frequency	Percent	Valid Percent	Cumulative Percent
universtity	Valid	fully agree	3	2,8	2,9	2,9
		2	4	3,8	3,8	6,7
		3	15	14,2	14,4	21,2
		4	22	20,8	21,2	42,3
		5	34	32,1	32,7	75,0
		6	18	17,0	17,3	92,3
		fully disagree	8	7,5	7,7	100,0
		Total	104	98,1	100,0	
	Missing	System	2	1,9		
	Total		106	100,0		
upper secondary school	Valid	3	5	8,5	8,8	8,8
		4	9	15,3	15,8	24,6
		5	13	22,0	22,8	47,4
		6	18	30,5	31,6	78,9
		fully disagree	12	20,3	21,1	100,0
		Total	57	96,6	100,0	
	Missing	System	2	3,4		
	Total		59	100,0		

4. DISCUSSION AND CONCLUSIONS

In order for an eAuction to work, a critical mass of users is needed, which is hard to attain is small markets with strong language barrier and remote location, such as Finland. First mover advantage was the dominant explainer of success for general auctions; however, first mover advantage did not lead into revenue. The existence of close substitutes (competing general auctions) prevented any auctions to charge for their services. Thus, in Finland, even successful general auctions were not able get any revenue from the main business. Revenue had to be found from marketing and affiliated activities. The companies, who used the auctions as marketing magnets were generally quite pleased with their results.

On the auction user side the perceived problems seem to be largely the same as with Internet-based buying and selling in general, that is, related to insufficient quality of on-screen presentation of goods and security issues. Also, the goods seen to be suitable for online auctions (i.e. computer related goods, software, tickets) are the same as those found to be suitable for eCommerce in general. However, there are further issues with the security and trust, because the peer-to-peer nature of auctions forces the buyers to think carefully of not only the price, but also the risks involved. This alleviates the need for repudiation mechanisms, which need a critical mass of users to work.

Our findings with young user that have considerable online experience indicate that there is a slow trend towards online buying and that this way of trading could include auctions as a price setting mechanism. The move, however, is very slow and most people will continue to use fixed price shopping also online. In conclusion, eAuctions are not really trusted as the most efficient mechanism for setting price, and their entertainment value is perceived as rather low, online auctions at the moment are not able to attract a critical mass of users.

The results of this study indicate that the actual user base of Internet auctions is very small and difficult to find in more random sample of respondents. For that reason, in the next phase of the research we will contact the actual eAuction users, that is, those users that are active in one or more auctions operated in Internet. One purpose of this study will be to compare the perceptions on electronic auctions between these groups. In the future it would also be interesting to compare these results with other countries to see, if this is just a cultural phenomenon here or is this a more general trend.

5. REFERENCES

Albert, Miriam R. (2002) E-Buyer Beware: Why Online Auction Fraud Should Be Regulated. American Business Law Journal, 39, pp. 575-643.

Kambi, Ajit and van Heck, Eric (2002) Making Markets, How Firms Can Design and Profit from Online Auctions and Exchanges, Harvard Business School Press, Boston MA.

Kauffman, Robert J. and Wang, Bin (2001) New Buyer's Arrival Under Dynamic Pricing Market Microstructure: The Case of Group-Buying Discounts on the Internet. Journal of Management Information Systems, 18/2, pp. 157-188.

Koch, James V. and Cebula, Richard J. (2002) price, Quality, and Service on the Internet: Sense and Nonsense. Contemporary Economic Policy, 20/1, pp. 25-37.

Lansing, Paul and Hubbard, John (2002) Online Auctions: The Need for Alternative Dispute Resolution. American Business Review, January, pp. 108-116.

SECTION 5
Future Aspects of Communication

Chapter 13

I-CENTRIC COMMUNICATIONS

Radu Popescu-Zeletin, Stefan Arbanowski, Stephan Steglich

Abstract: This chapter introduces the rationales, the framework, and the architecture of I-centric Communications, a new paradigm to design and to develop telecommunication systems. I-centric Communications has been proposed by the Wireless World Research Forum (WWRF) working group 2 based on numerous contributions from academia and industry world wide.

Key words: I-centric communications, 3G beyond service architecture, personalization, ambient-awareness, adaptability

1. INTRODUCTION

In 2000, the IST Programme Advisory Group (ISTAG) published a vision statement [ISTAG] for Framework Programme 5 that laid down a challenge to:

'Start creating an ambient intelligence landscape (for seamless delivery of services and applications) in Europe relying also upon test-beds and open source software, develop user-friendliness, and develop and converge the networking infrastructure in Europe to world-class'.

Ambient Intelligence stems from the convergence of three key technologies: Ubiquitous Computing, Ubiquitous Communication, and Intelligent User Friendly Interfaces. Ambient Intelligence implies a seamless environment of computing, advanced networking technology, and specific interfaces. It is aware of the specific characteristics of human presence and personalities, takes care of needs and is capable of responding intelligently to spoken or gestured indications of desire, and even can engage in intelligent dialogue. Ambient Intelligence should also be unobtrusive, often invisible: everywhere and yet in our consciousness – nowhere unless we need it. Interaction should be relaxing and enjoyable for the citizen, and not involve a steep learning curve. [ISTAG]

The prerequisite for service integration is global connectivity, based on a fast developing web of interconnected communication networks, comprising both fixed and wireless networks. In addition, the provision of a global service infrastructure, based on network-independent open service platforms is the other fundamental prerequisite, hiding the complexity of network diversity, and allowing the fast and efficient creation, provision, and management of future services.

Ambient intelligence requires appropriate communication services and information systems that adapt automatically to the human communication behavior. Users have to be supported in order to cope with the huge amount of information available and the increasing complexity of (communication) services. In addition, the continuous reachability of people due to advanced mobile communication requires adequate means for information filtering and communication control.

2. MOTIVATION

It is common understanding that future services will have to adapt to individual requirements of human beings [BoV01]. The communication system has to provide the intelligence required for modeling the communication space of each individual adapting to its interests, environment, and life stage.

I-centric communications considers the human behavior as a starting point to adapt the activities of communication systems to it. Human beings do not want to employ technology. Humans rather want to communicate interacting in their individual communication space.

Introducing communication systems that follow the I-centric vision can only be an evolutionary process. A transition phase between 3rd generation telecommunication networks towards 3G beyond (3Gb) systems is needed. 3Gb is anticipated as enabling platform for user-driven service provisioning incorporating a variety of access mechanisms, mobile service execution environments, and a huge amount of mobile services [Arb02b].

IP-based technologies that allow the integration of available heterogeneous networks into a single platform capable of supporting user to roam between them, while not interrupting active communications have already been introduced [FSAN]. This development will be assisted by the rise of new mobile devices maintaining various access interfaces that will allow connectivity over a range of providers and technologies. Finally, the emergence of a variety of access devices will dictate the liberation of users from a single device and allow service mobility between devices and networks [Mohr02, Arb03a].

2.1 Trends towards 3Gb Telecommunication Systems

The driving forces for today's telecommunication world have been the growth of Internet (especially broadband internet access at home) and 2G mobile telecommunication. Both are now converging around the 2.5 and 3G networks and services.

From a service perspective, the main difference between 2G/2.5G and 3G telecommunication systems is the new air interface enabling higher transmission data rates. For 3G beyond systems a global consensus exists that a new system architecture needs to be developed. This system architecture has to support a number of new features that have been identified already:

- blurring business roles
- personalized, ambient-aware, adaptive end user services
- augmented environments as part of the ubiquitous communication system
- new networking services: ad-hoc, p2p
- all IP services: always best connected, packet switched, broadband
- diverse access technologies, global coverage, global roaming
- further convergence of voice, data, and mobile communications
- new wireless links (high/low data rate, long/short range)

Beside all these technical trends, a harmonization between different application domains is expected. The integration between services for office and home environments, which has been started already, will continue towards service environments covering the complete communication space of individual users.

3G beyond systems will support mobile service usage irrespective where users are, what kind of terminal they are using, what kind of bearer technology is underneath, and what kind of information has to be delivered.

2.1.1 Telecommunication and Wireless Networks

Started in the mid 90th, telecom markets are still characterized by the convergence of traditional telecommunication services. Many of these services are available today without a harmonized appearance towards the user.

Service provisioning and especially personalization of services is perceived as an upcoming and important success factor. Although, some services offer the possibility of personalization, the support of individual user needs is proprietary, at best. Market analysis has already required future services to be user-centric. They should adapt to user needs, and to the current situation, the user is in. Traditional communication services, designed for large user groups, are not able to address individual user needs.

Provision of ambient-aware services is the next frontier towards the realization of such concept. Location based techniques are the first to be exploited in this direction, and some operators are offering location-based services in GSM and UMTS already.

Another trend is the introduction of service and content adaptation. In the area of content adaptation, technologies have been developed that allow description and transformation of content in such a way that it can be presented by devices with different characteristics. Research has also been performed on coding algorithms that allow information to be trans-coded in relation to e.g. the current network status. In addition, media conversion is an available feature today and it has been introduced in commercial Unified Messaging Systems already.

Some of the main architectural breakthroughs in respect to a general service architecture have been developed by the TINA consortium [TINA]. The main results are the definition of a telecommunication related business model, an associated session model, and the specification of related reference points. The TINA business model implies a centralized architecture, where components are assigned to fixed domains. In a mobile environment as envisioned for 3G beyond, this approach and the business model itself have to be extended.

2.1.2 Internet Protocol Everywhere

The era of monolithic telecommunication networks with centralized intelligence is developing towards decentralized structures where the borderline between the traditional roles of network provider, content provider, service provider, and user vanishes. This is due to the penetration of network technologies everywhere and by everyone. The integration of these networks pertaining to even different administrative domains is solved on OSI layer 3 by using the Internet protocol (IP).

Another major trend is the proliferation of IP-based devices. End-systems will not only be desktop/laptop computers but also Smart-IP devices. All these devices will be addressable in a global IP communication infrastructure connected via various wired and wireless networks.

Considering these facts, it is obvious that any microcontroller-based device has the potential to be part of the communication space in the future. Wireless networks like Bluetooth, WLAN, or ZigBee enable these devices to be connected to each other without new wires, and IPv6 supports their interconnection by accommodating a massive number of devices. Available computer power and the networks lead to the ubiquitous access to services –

anytime, anywhere, and for anyone. Devising according service architecture is fundamental for future communication systems.

Present communication systems are designed and developed for specific end-systems and for a specific communication service. A vertical design from network technology up to the user interface and device capabilities takes place. Services are kept presentation oriented and each of them has its own way to handle it. Every service is developed for a certain network technology, dedicated to a user community in which individuals are reduced to the common denominator defined by the service designer. The hope is that everybody will buy the service and the associated device. This implies the communication infrastructure engineered to offer the broadest solution. Scalability, performance, and controllability of the network infrastructure are resulting problems.

2.1.3 Service Platforms

Taking emerging 3G business models and value networks into account, services are expected to work in concert of many actors. The 3GPP promotes the possibility to create services on standardized tool sets, e.g. OSA. OSA defines an architecture that enables operator and third party applications to make use of network functionality through an open standardized API. In this way, network complexity is transparent to the applications that are presented independent of the underlying network technology.

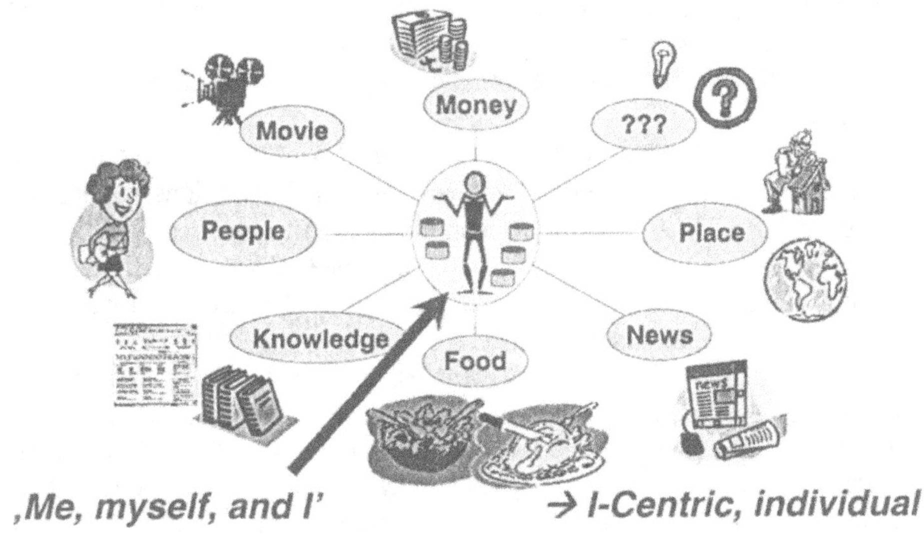

Figure 1: Individual communication space

2.2 Vision for I-centric Communications

The communication behavior of human beings is characterized by frequent interactions with a set of objects in their environment. Humans solve the problems of their daily life, e.g. money need to be managed, food has to be bought and to be prepared for eating, movies are watched for entertainment, places are visited and news are consumed, and other people are met. The set of objects, controlled by each individual human, define its individual communication space as shown in Figure 1.

A communication space of an individual is limited: 'I do not know everybody in the world, I am not interested in everything, and I do not have all necessary devices required by all communication services everywhere at all times'.

Furthermore, individuals are interested in semantic and not necessarily in the kind of presentation of a specific service. Services in an individual communication space have to support the quality of the human senses, and since quality of senses is individual, they have to adapt their presentation to each individual automatically. Services have to adapt to the life stage and the environment of each individual.

Following this view, a new approach is to build communication systems not based on specific technologies, but on the analysis of the individual communication space. The result is a communication system that adapts to the specific demands of each individual (I-centric). Such a communication system acts on behalf of human's demands, reflecting recent actions to enable profiling, and self-adaptation. I-centric Services adapt to individual communication spaces and situations. In this context 'I' means I, or individual, 'Centric' means adaptable to I demands and a certain environment.

The rationales above require intelligence in service provisioning in order to personalize, adapt to situational and environmental conditions, to monitor and to control the individual communication space. An I-centric communications system provides the intelligence, which is required for modeling the communication space of each individual by adapting to its interests, environment, and preferences.

The multitude of devices, wearables, different telecommunication technologies, positioning and sensing systems are considered as enabling technologies for I-centric communications. Universal information access (including service interworking, media conversion), flexible management of equipment and facilities, and personal communications [Eck96] form the basis of such systems.

2.3 I-centric Communications – Basic Terminology

I-centric means to take a bottomless look at human behavior and to adapt the activities of communication systems to it. This abstract description of humans' communication activities requires a set of definitions that allow the mapping of abstract requirements (or wishes) to the physical communication environment later on.

All entities, humans are interacting with, will be called objects further on. They can be activated or deactivated by an individual, or environmental conditions, to perform an action according to specific needs of an individual. Objects can represent one or more physical entities performing a certain service.

An object is a logical representation of hardware or software entity, or even a representation of a certain individual, and provides well-defined services from the perspective of an (other) individual.

An I-centric system should support massive numbers of objects and should be tolerant against object failures. The population of objects is always changing because they spontaneously enter/leave/roam the environment and hence the communication space. Already standardized mechanisms for naming, lifecycle, monitoring, fault tolerance etc. have to be taken into account to determine whether they suit the requirements of I-centric communications.

Due to changes in human being's daily live, the amount or the concrete instances of objects are changing over time. Nevertheless, the sum of objects, an individual might interact with form his individual communication space. Objects may pertain to different communication spaces. They can be controlled by individuals, other objects, or services. Individuals can directly ask for a service to be performed by an object, whereas environmental condition may influence the status of objects indirectly. Communication between different individuals takes place by sharing objects of their communication spaces. In this case, objects representing communication facilities in the different communication spaces are connected to establish a physical connection between two individuals. What kind of physical resources are used for the communication is decided dynamically and depends on individual preferences of involved parties, their available communication facilities, and additional ambient information. The process of how to select and activate objects and physical resources underneath is one of the main activities of an I-centric communications system.

Individual communication spaces are growing and shrinking in the time axes based on the individual life stage, personal interests, working and living environments, and the availability of new kinds of telecommunication services and devices. The size of the individual communication space varies over time due to the appearance or disappearance of objects.

Each individual has only one individual communication space. It contains all objects this individual might want to perform requests on. Objects that pertain to individual communication spaces of different individuals must handle concurrent access from different individuals or must delegate the concurrency control to the I-centric communications system.

2.3.1 Context and Active Context

The context provides the definition of relationships and causalities between different objects of an individual communication space and the individual. A context represents a 'universe of discourse' in an individual communication space. Individuals communicate with objects in their environment in a certain context.

Objects may pertain to different contexts (even to contexts of different individuals), because individuals might want to have a certain object involved in different activities.

Contexts are independent from any concrete environment. If an individual wants to act in a certain context this context has to be activated. An active context defines the relationship of an individual to and between particular numbers of physical resources at a certain moment in time, in a certain environment. The activation and deactivation of a context should usually be done automatically just by analyzing individuals' activities, but an individual should also have the possibility to do this explicitly.

Activating a context means:

- the identification of objects that are required by the context,
- the evaluation of the relationships and causalities between objects defined by the context,
- the discovery of the actual vicinity of the individual to identify physical resources that provide the functionality required by the identified objects,
- the activation/configuration of these physical resources to perform the task required by the context.

Context only refers to objects as an abstract model of what kind of objects have to be taken into account in a certain context, whereas an active context refers to physical resources that have been identified during the activation process. Active contexts are of dynamic nature reflecting the current environment an individual resides in.

The activation and deactivation of contexts is one task of I-centric Services. To activate a context the I-centric Service performs the activities described above. In addition, the I-centric communications system has to manage concurrent access to objects and conflicts caused by contrary wishes, expressed by individual(s).

A context is active when it is adapted to a certain environment at a certain moment in time. It defines the relationships and causalities of an individual to a particular number of physical resources at certain a moment in time, in a certain environment.

Acting in a context means to use only services that are provided by objects, which are part of that very context. Starting to interact with objects that are not part of an active context causes the activation of another context. That means, on one hand individuals are allowed to act in several contexts in parallel, and on the other hand, the I-centric communications system must handle conflicts that might occur due to contrary causalities defined in the different contexts.

To handle each individual communication space and associated contexts, a general model of domain information and relationships to objects and physical resources is needed. This model must be flexible to be enhanced due to the introduction of new locations, devices, etc.

2.3.2 Preferences and Ambient Information

Individuals have different preferences in different situations. With preferences, individuals express their choices of services characteristics in certain contexts. Therefore, preferences provide a powerful mechanism to influence the behavior of I-centric Services by giving them explicit instructions.

Preferences are conditional choices of service characteristics of an object depending on context and ambient information. Preferences are applied to objects during the activation of a context.

I-centric Services evaluate preferences to adapt their behavior to what is 'really wanted' by an individual in a certain environment at certain moment in time. Therefore, preferences have to be either gathered from individuals interactively or automated by monitoring, and they have to be expressed in a machine computable form.

The description of preferences, which can be processed automatically, is another challenging task. Preferences can capture many aspects like mood, interests, live stage etc. that are even hard to describe in words. Furthermore, the kind of preferences that are relevant to different individuals may differ completely. A model for describing preferences must be as generic as possible to avoid restrictions that might prevent the expression of a certain preference. On the other hand, the model has to provide some structuring or categories to allow the assignment of preferences to a certain I-centric Service.

In general, ambient information is information that can be collected, gathered, or sensed from the environment. Ambient information comprises temporal and spatial characteristics like any user input, temperature, noise level, light intensity, and presence of other people just to give a few examples. Ambient information is sensed by sensing facilities, like motion detectors or microphones, and transmitted through sensor networks. Ambient information may also include geographical information, environmental information, and life conditions.

Ambient information is information that can be collected, gathered, or sensed from the physical environment using the objects of the individual communication space of a certain individual.

A semantic model is needed to describe preferences and ambient information. Such kind of model incorporates knowledge representation to qualify available information and ontology languages to relate syntax and semantic to each other. The focus for I-centric communications here is to define a harmonized semantic model that includes human aspects as well as the process to gather, store, evaluate, and exchange preferences as well as ambient information.

2.3.3 I-centric Service

I-centric Services define, manage, and (de)activate contexts in an individual communication space taking the preferences of individuals and ambient information into account. They support an individual (I-centric), adaptive, personalized, and ambient-aware, way to interact with objects in individual communication spaces.

I-centric Services need ambient information in order to adapt to the environment. Temporal and spatial characteristics are only two examples of information, which may affect the service behavior. Note, that a certain environment can restrict the functionality requested in a certain context. Interacting in a 'TV context' while driving a car may reduce the available functionality to 'record the movie for later viewing' or to listen just to the audio part.

I-centric Services activate contexts by choosing the equipment to be controlled, their quality of service to be finally connected via heterogeneous networks to create an I-virtual private network.

The process of choosing and controlling the equipment of the physical environment is supervised by the service logic of I-centric Services. The service logic controls the activation of contexts by combining multiple objects dynamically. It parameterizes objects by defining what, when, and how one or more objects behave in a given condition. The service logic decides based on profiles and on the status of the objects how those objects should behave in a certain situation. This enables sensitive services that adapt to the environment dynamically.

Nowadays, service logic is in most cases 'hard-coded'. Once implemented, it cannot be changed afterwards. The basic idea of I-centric communications is to provide individuals with their own services that might change over time.

The process of creating or modifying I-centric Services has to be accompanied by ontology definitions that describe what services an object is providing. Interactive applications are envisaged that allow individuals to assemble their service by simple 'drag and drop' mechanism. Like a LEGO™ toolbox, the individual should be able to create and to deploy its I-centric Services.

2.4 Reference Model for I-centric Communications

Figure 2 shows the reference model for I-centric communications. It follows a top-down approach starting with the introduction of individual communication spaces, related contexts, and objects. In general, the topmost layer recalls the I-centric vision that human beings interact with objects of their communication in a certain context. It is common understanding that I-centric Services have at least to support three different features, namely ambient-awareness, personalization, and adaptability. To emphasis that these features are needed for I-centric communications they have been assigned to the individual communication space.

The service platform for I-centric communications is responsible for shaping the communication system, based on individual communication spaces, contexts, preferences, and ambient information. Preferences are provided by the personalization feature, whereas ambient information is provided by ambient-awareness feature.

The IP based communication subsystem provides the linkage between different objects in the communication spaces. These links have to be maintained and managed even when they are subject to change because of roaming between different network topologies or access networks. IP communication is seen as the common denominator to harmonize heterogeneous network infrastructures.

The Wired or wireless Networks layer implements all aspects of the physical connection(s) between different objects. Due to the hierarchical structure of the reference model, a connection in the IP based communication subsystem might use multiple connections in underlying network.

The main features of I-centric communications (ambient-awareness, personalization, and adaptability) affect all layers. Therefore, supporting functions have to be provided as a vertical solution. The reference model introduces the concept of Generic Service Elements that implements common functionalities on all layers. Generic Service Element can be seen as a toolbox from which complex services can be assembled and executed dynamically.

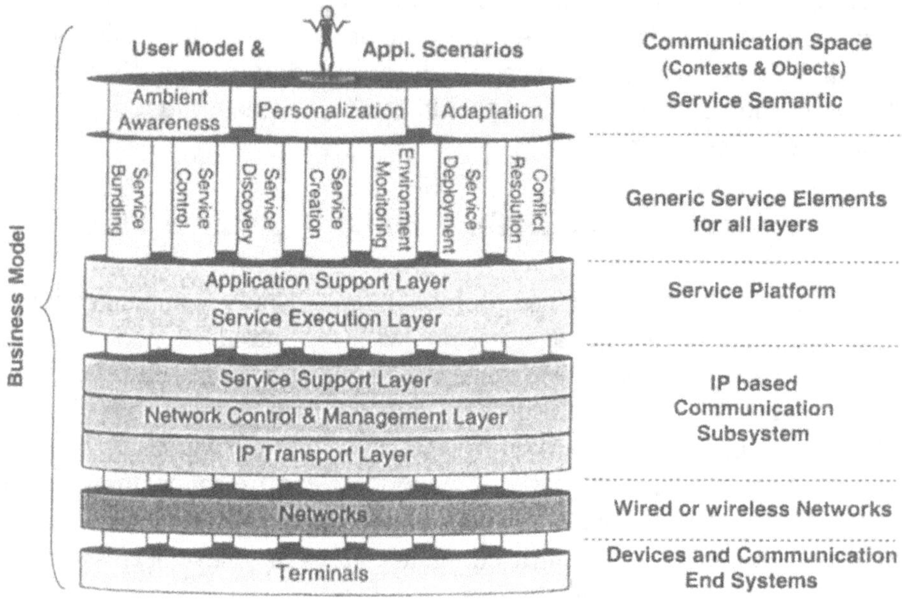

Figure 2: Reference Model for I-centric Communications

Accompanying to all these technical issues, the Business Model for I-centric communication identifies the relationships and information flows between all active roles within an I-centric system.

2.4.1 Business Model

The borders between traditional roles and administrative domains: network provider, content provider, service provider, and retailer are blurring. An individual may become service provider, or content provider, or retailer. Additionally, roles may change dependant on context, which implies a very flexible business model.

The business model for I-centric communication has to cover:

- roles, relationships, and reference points
- business topologies
- service lifecycle (creation, deployment, management, and billing)
- benefits for parties involved in the market value network

The first objective of a business model for I-centric communication is a model for describing relationships between involved parties in a global business community. Based on these relationships, roles and reference points are defined. This allows the participation of each business partner on a global business on one side and provides the freedom in development and integration on the other side. Reference points provide standardized points of contact and information exchange between business partners.

2.4.2 Personalization

Information and services must become increasingly tailored to individual preferences to make the usage of services easier and the perception of the individual communication space richer.

Objects available in an individual communications space have to adapt to the preferences of individuals. Personalization models each individual in the I-centric Service platform by managing its preferences and providing these preferences to I-centric Services.

To reach this goal, personalization federates profile information (containing preferences). Personalization incorporates dynamic behavior to enrich stored and federated information to enable pro-active I-centric Services. This leads to an overall profiling infrastructure managing the individual preferences.

The main research issues behind personalization are:

- how to gain preferences from individuals (interactively or automated)
- how to store preferences in profiles (profile format & categories),
- secure privacy sensitive parts of profiles.

2.4.3 Ambient-Awareness

In I-centric systems, services are tailored to contexts of the individual communication spaces. The services automatically adapt themselves to changes in the environment of nomadic individuals.

The federation of ambient information from various sources, according to individual's mobility and roaming is an integral part of ambient-awareness. Intelligent inference systems for missing information are needed in order to incorporate as much information as possible to provide an automatic, ambient-aware environment to the individual.

Ambient-awareness is the functionality provided by an I-centric system to sense and exchange information about the current environment, an individual is in at a certain moment in time.

Sensors networks will play a major role in providing ambient information. Sensor technologies will be embedded in mobile equipment, communication networks, living and working environments to sense who the user is, where he is, what he is doing, what the environmental conditions are, to provide this ambient information to I-centric Services.

2.4.4 Adaptability

Adaptability is mainly based on information provided by personalization and ambient-awareness. It provides the functionality to adapt I-centric Services to personal preferences and environmental conditions. Therefore, adaptability can be seen as a function that activates a context based on whatever information is provided by ambient-awareness and personalization.

In general, I-centric adaptability translates the wishes of individuals, which are usually inaccurate, incomplete and sometimes even contradictory, into a set of rules precise enough for processing to be automated with sufficient reliability. It has implications in the structure of the services to allow adaptability and is the engine, which activates a context at a certain moment in time in a certain environment.

Typical situations when adaptation takes place include a substantial change in characteristics of connectivity, entering into a new service domain, or changing terminal device in service session.

By technical means, adaptability requires the adaptation of media, content, and service behavior. During the last years, a variety of concepts for adaptation has been developed [Pfe99]:

- communication streams can be altered during transmission (e.g. bit rate adaptation),
- media types can be changed (e.g. text-to-speech conversion),
- type of presentation can be adapted (e.g. downscaling an image),
- altering the content of a message (e.g. adding or stripping off information), or
- modifying the service behavior (e.g. by customer service control).

Adaptability cannot be only reactive. When the battery of a mobile device dies or the connectivity breaks, many actions become impossible. Therefore, adaptation must also be proactive, which in turn requires predictability of the near future.

2.4.5 Service Platform for I-centric Communications

A Service Platform for I-centric communications is responsible for shaping the communication system, based on individual communication spaces, contexts, preferences, and ambient information. Finally, it (de)activates objects (advised by I-centric Service), identifies causalities between them based on sensed environmental data, controls the services offered by these objects, and converts data structures and operations for interworking between services. The equipment is configured dynamically, its state is profiled, distributed objects are controlled, service creation and deployment are supervised, and the interworking among domains is enabled by the platform.

To fulfill the functionalities requested by I-centric communications, I-centric Service platforms have requirements on the underlying communication subsystem. This is caused mainly by the empowerment of any individual to act as a service provider or network provider in a paradigm shift from a provider centric paradigm to a decentralized I-centric paradigm.

On the other hand, the requirements are based on information that has to be provided by lower layers to the service platform. Traditional platform approaches (e.g. object-oriented middleware platforms) try to hide as much as possible technical parameter between the different layers. I-centric Services have to be provided with ambient information.

However, the paradigm shift addressed here does not only concern the individual as a provider of network related services. A service platform allowing global mobility and transparent access to any kind of service over a common IP platform is the basis for allowing everyone to provide a wide range of services.

2.4.6 Generic Service Elements

I-centric communications systems have to cope with issues like numerous service providers, always-connected individuals, automatic service adaptation, and ambient-awareness. Aspects like dynamic service discovery and service provisioning in (for individuals and services) unknown environments and personalized services usage requires new mechanisms to support I-centric communications systems.

To simplify the definition and realization of I-centric Services and applications, a set of reusable software components support functionalities common for different services and applications. These components are called Generic Service Elements to emphasis their general applicability for all kind of services.

A GSE is a functional software component that can be used by other GSEs, services, or applications and it is hosted by the I-centric Service platform. GSEs provide functionalities common to different services and applications to ease and shorten their development process.

Because I-centric Services should work under changing environmental conditions, serving changing individual preferences, the most promising candidates for common functions are: Service Discovery, Service Management, Service Deployment, Service Composition, Service Logic, Service Control, and Environment Monitoring.

Consequently, well-defined collections of interface specifications designed for certain business domains are needed. Such interfaces must provide framework functionality, like hot plugging of services, dynamic (re-)binding, service mobility, AAA-services, support for automated SLA negotiation, contracting, and so on.

3. SUMMARY

This chapter has introduced the vision and the reference model for I-centric communications. A service architecture compliant to the Reference Model for I-centric communications is needed, to implement I-centric communications systems. The Service Architecture has to define building blocks and their interworking to provide the functionality requested by the reference model.

The service scenarios, expected to become possible with I-centric communications, require the design of interaction mechanisms for distributed objects. Furthermore, object and service discovery mechanisms, and object repositories are needed to enable intelligent service brokerage. Service and object ontologies facilitate these discovery and brokerage mechanisms.

Profiling, decision making, and intelligent device control has to be embedded in the business logic of I-centric Services. A user/context model for I-centric communications is wanted to feed the business logic of personalized, ambient-aware, and adaptive services.

Service Composition and Service Bundling are the elements to combine the services provided by the objects of individual communication spaces. Due to the variety of objects and services to be combined in I-centric Service, a Service creation environment is needed that suits the requirements of individual users.

To come up with a coherent service architecture for I-centric communications systems these aspects have to be integrated into a single framework. This work is currently carried out by Wireless World Research Forum WG2.

REFERENCES

[Arb02b] Popescu-Zeletin, R.; Arbanowski, St.: I-centric Service Architectures for B3G. Proc. of the International Forum on Future Mobile Telecommunications & China-EU Post Conference on 3Gb, Beijing, China, November 20-22, 2002

[Arb03a] Popescu-Zeletin et al.: Service Architectures for the Wireless World. Computer Communications, Vol. 26, No. 1, January 2003, pp. 19 - 25, invited special issues, ISSN 0140-3664

[Eck96] Eckardt, T.; Magedanz, T.; Ulbricht, C.; Popescu-Zeletin, R.: Generic Personal Communications Support for Open Service Environments. Proc. IFIP World Conference on Mobile Communications, Canberra, Australia, Sep. 1996

[ISTAG] IST Programme Advisory Group: Scenarios for Ambient Intelligence in 2010, Final Report. IPTS-Seville, February 2001 (see www.cordis.lu/ist/istag.htm)

[Mohr02] Mohr, W.: The Wireless World Research Forum (WWRF) Towards Systems Beyond 3G. The Proceedings of the Korean Institute of Communication Sciences, vol. 19, No. 7, July 2002, pp. 56, invited paper.

[Pfe99] Pfeifer, T.: Automatic Conversion of Communication Media. - Dissertation, Technical University of Berlin, Institute for Open Communications Systems (OKS), 1999

[BoV01] WWRF: Book of Visions, Edition December 2001 http://www.wireless-world-research.org/

[FSAN] Full Service Access Networks Initiative http://fsan.mblast.com/initiative.asp

[TINA] TINA-C: Baseline Service Architecture http://www.tinac.com/specifications/documents/sa50-main.pdf

Chapter 14

A COMMUNICATION FRAMEWORK TOWARDS FLEXIBLE ASSOCIATIONS OF BUSINESSES IN EVOLVING ENVIRONMENTS*

Hendrik Ludolph
Technologies de l'information, HEC Montreal
Montreal, Canada H3T 2A7
Hendrik.Ludolph@hec.ca

Gilbert Babin
Technologies de l'information, HEC Montreal
Montreal, Canada H3T 2A7
Gilbert.Babin@hec.ca

Peter Kropf
Departement d'informatique et de recherche operationnelle, Universite de Montreal
Montreal, Canada H3C 3J7
kropf@iro.umontreal.ca

Abstract The Internet and electronic commerce have become indispensable for many of us. To adequately use the increasing amount of data available, attempts are made to extend data processing from a lexical view towards a multi-level view, including meaning and/or context (e.g., DAML, Web Services). The goal of this paper is to introduce a formal framework, which models communications from such a multi-level perspective. Therein, we discuss fundamental ideas of communication, such as agents involved and their respective structure. We integrate the concept of an agent's adaptive behaviour in order to assure a high degree of understanding. The framework is illustrated using a practical example to depict its usefulness and how it may be further developed.

*The completion of this research was made possible thanks to funding provided by NSERC (grants DG-155899-03 and CRD-224950-99), Bell Canada's support through its Bell University Laboratories R&D program, and support by the CIRANO

1. Introduction

Information technology has become the corner stone in today's society. Businesses, organizations, governments and individuals rely on IT systems for their prosperous functioning, behavior, and development. These IT systems are composed of many different units that interact and work in concert to satisfy some goal. Furthermore, they do not stand alone but interact with their environment, be it other IT systems or humans. In the context of e-business or e-commerce, examples of such systems include ERP systems or transactional Web sites. Clearly, as far as the interaction between systems is concerned, the (public) Internet plays a predominant role. However, it must be noted that in many cases, such as in the banking sector, private networks are often used in place. Communication of information, knowledge or in general any cognitive structure between different systems and with a system's environment, which may include humans, is therefore a central element. In general, such systems are called *Communication and Information Systems* (CIS).

Organizations or systems of this kind are set up, designed, and implemented by humans, and are therefore subject to human rationality. Such a necessarily bounded rationality results in a limited view, which leads to *satisficing*, as it is called by Herbert A. Simon [Simon, 1996], which renders a system and its environment static, making a system to appear as acting and existing in empty space [Plaice and Kropf, 2000]. Indeed, current IT systems are limited to fixed, pre-defined ontologies which do not allow for a system's evolution or adaptation as a result of interaction in a space that may be described in a holistic way such as Aristotle's *aether*. A system is transformed into a new evolved system by the knowledge transfered by communication from one system to the other, from a system to its environment or *vice versa*. In the event of a desired change in a systems behavior and functionnality, the standard procedure today is to replace the existing system with a new release or a completely new system. A first attempt to allow for greater system flexibility and evolution at a technical level stems from agent technology [Luck et al., 2003] and to some extent from Web Services where different (new) onthologies may be dynamically integrated. From an economic point of view, adaptation is necessary for economic survival and sustained competitiveness [Heylighen and Campbell, 1995] and fitness [Kauffman, 1995].

Interaction through communcation is the driving force for change and evolution. We therefore propose in this paper a communication framework as the basis for adaptive or coevolutive behavior of communication and information systems. CIS are defined and characterized at many different abstraction levels, from technical specifications of data transmission or data structures and methods up to the communication of facts, knowledge or the sharing and adaptation of entire cognitive sturctures. This leads to a recursively defined structure of

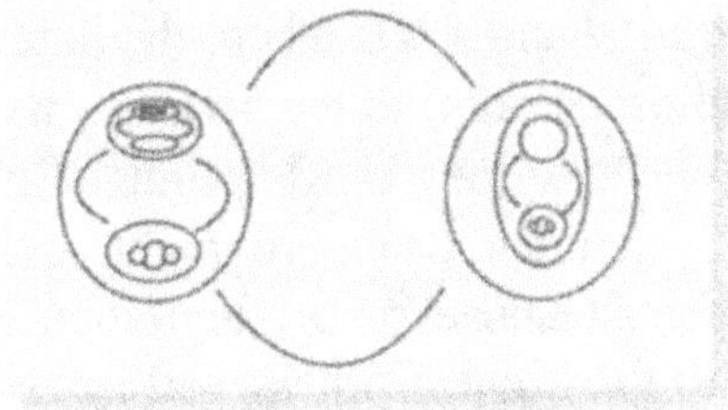

Figure 1. Recursiveness of communication

a system, which we call an *agent*, and all possible communications. The next section discusses evolutionary aspects of communication followed by the formal definition of the proposed communication framework. Using the proposed framework, section 1.3 analyzes how agents may evolve through interaction. Before concluding the paper, we present in Section 1.4 a first attempt of identifying relevant levels of abstractions of the framework.

1.1 (Co)evolutionary aspects of communication

Communication is any kind of interaction between systems that happens at any conceivable abstraction level. If we consider human communication, we could decide not to include communication above the human mind–based cognitive level as we are *a priori* not able to conceive such kind of interaction, albeit it might exist. Nevertheless, in order for us to set up a complete model including all abstraction levels, we follow a generic approach of recursiveness within the communication event to assure the coverage of all necessary elements to install and maintain high levels of mutual comprehension. This means that every system as well as every communication level serves as a sublevel embedded into a higher structure and as a superlevel concerning a related lower structure (Fig. 1). In short, recursiveness may be applied to the grouping of the involved systems and the grouping of possible communication levels as stated above.

Communication is the relation between two systems. From a system's perspective, it is perceived as the relations of that system to its environment, i.e., the rest of the universe. The environment, hence other systems, is by definition beyond the direct influence of the system; it nevertheless influences the functioning of that system. More precisely, the environment "...is considered as the system of surrounding things, conditions or influences, affecting somehow the existence or development of someone, something...," [Krippendorf, 1986] hence another system, all part of communication as we see it.

We further infer that, in a bidirectional manner, a system is not only influenced by its environment but it also influences other systems as a part of their environment. A familiar example may be that of competing companies, such as the "rat race" between *Intel* and *AMD* where the latter has to adapt (e.g., by producing more powerful processors) to the first, i.e., its environment and *vice versa.* To stay "competitive," a system must optimize fitness, where fitness is a complex function of the system and its environment, an index of the likelihood that the system would persist and evolve [Heylighen and Campbell, 1995]. Those configurations with the highest fitness will be selected to contribute at best to a system's survivability, which by the way doesn't mean replacement. This fitness function concerning system's mutual influence emphasizes evolution to a changing environment and is called coevolution [Kauffman, 1995].

Aligning the above considerations to our context, we state that quasi-continuous CIS ought to obey to the same principles. We base this assumption on the fact that CIS, as they support business processes, have a *coordination* or *controlling* function. They serve to distribute data and information aiming the control of processes, operations, employees, teams, etc. In order to adequately fulfill this function, a control system must mimic or map the organizational structure for which it is installed [Conant and Ashby, 1970].

2. A communication framework

In what follows, we propose a formal framework describing interactions between systems, which takes into consideration the recursive nature of both systems and communications, as well as the coevolution principles stated above. The building blocks of the model are the following:

- *agents* are systems that may interact with each other. An agent may be hierarchically structured. Note that we use the term "agent" here in a broad sense, not limited to the agent paradigm. For us, an agent is any system (computer module, computer program, human, organizations, etc.) that is *actively involved* in the exchange of data;

- a *communication signal* is a single transmission of data from one agent to another agent. This corresponds to a single message transmitted, without any feedback;

- a *communication event* is some non-empty arbitrary sequence of communication signals. This corresponds to an interaction between agents and will therefore imply many communication signals;

- a *cognitive structure* is a structured representation of data. It is used to describe an agent's knowledge as well as the data transmitted in a communication signal.

2.1 The cognitive structure

We first must consider how data, stored by agents and transmitted by communication signals, should be structured. We distinguish here between data, which are mere facts and values, from information which is data that leads to a reaction. As stated earlier (sect. 1.1), data is represented at different abstraction levels. Therefore, any representation of data must consider these different levels. Consequently, we have that:

- Λ is some multidimensional space of abstraction levels on which a partial order is defined;
- $\lambda \in \Lambda$ is some abstraction level to represent data;
- $\check{\lambda}$ is the lowest abstraction level recognized by an agent or transmitted by a communication signal;
- $\hat{\lambda}$ is the highest abstraction level recognized by an agent or transmitted by a communication signal;
- ψ^{λ}, called a *partial cognitive structure*, is some representation of the structure of data at abstraction level λ, such that $\psi^{\lambda} = f(\psi^{\check{\lambda}}, \ldots, \psi^{\lambda-1})$. It therefore represents the emergent data obtained by combining data at lower abstraction levels. Furthermore, $\forall\lambda \notin [\check{\lambda}, \hat{\lambda}], \psi^{\lambda} = \emptyset$; and
- Ψ is a *total cognitive structure.* Given an agent or a communication signal, we have that $\Psi = (\psi^{\check{\lambda}}, \ldots, \psi^{\hat{\lambda}})$.

2.2 Agents and the agent hierarchy

We now depict in greater details the agents' hierarchical structure and thereafter its relation to the cognitive structure. We distinguish between *atomic agents*, which are the smallest possible agents that may be involved in communications, and *complex agents*, which represent hierarchical groupings of agents. Hence, complex agents represent recursive structures. Agents may therefore be characterized as follows:

- $l \in \mathbf{N}$ is some hierarchical level of agent composition;
- A^0 is the set of atomic agents; and
- $A^l = \{a | a \in \mathcal{P}(A^{l-1}) \wedge \text{Card}(a) \geq 1\}$ is the set of complex agents at level $l > 0$.

This definition implies that an agent $a \in A^l$ is either atomic ($l = 0$) or some arbitrary grouping of agents, such that any member of that group is an agent of level $l-1$ ($\forall a' \in a, a \in A^l \wedge l > 0 \wedge a' \in A^{l-1}$). When the number of member agents is 1 (Card(a) = 1), we say that agent a is a virtual group. This is useful, for instance, to represent merging of organizations with different hierarchical levels.

Given two agents $a \in A^l$ and $a' \in A^{l'}$, we say that agent a is a member of a', noted a **in** a', if and only if

$$a \textbf{ in } a' \equiv (a \subseteq a') \vee (\exists a'' \in a' | a \textbf{ in } a'').$$

Cognitive structure of agents. Every agent has its own cognitive structure, which emerges from those of its composing agents. Consequently, we have:

- ψ_a^λ is the cognitive structure of agent a at level λ;
- $\check{\lambda}_a$ is the lowest abstraction level at which agent a is able to manipulate data. It is therefore the lowest level λ_a at which a cognitive structure ψ_a^λ is available for agent a. For $a \in A^l$ we have that

$$\check{\lambda}_a \leq \min_{a' \in a} \check{\lambda}_{a'} \leq \min_{a' \in a} \left[\min_{a'' \in a'} \check{\lambda}_{a''} \right] \leq \ldots;$$

- $\hat{\lambda}_a$ is the highest abstraction level at which agent a is able to manipulate data. It is therefore the highest level λ_a at which a cognitive structure ψ_a^λ is available for agent a. For $a \in A^l$ we have that

$$\hat{\lambda}_a \geq \max_{a' \in a} \hat{\lambda}_{a'} \geq \max_{a' \in a} \left[\max_{a'' \in a'} \hat{\lambda}_{a''} \right] \geq \ldots; \text{ and}$$

- Ψ_a is the *total cognitive structure* of agent a. We have that $\Psi_a = (\psi_a^{\check{\lambda}_a}, \ldots, \psi_a^{\hat{\lambda}_a})$.

2.3 Communication signals

In our framework, a communication signal is formally defined as tuple $\omega = \langle a, \Psi_\omega, a' \rangle$, where:

- a is the emitting agent with global cognitive structure Ψ_a and a' is the receiving agent with global cognitive structure $\Psi_{a'}$;
- ψ_ω^λ is the cognitive structure of the data transmitted by ω at level λ;
- $\check{\lambda}_\omega$ is the lowest abstraction level of data transmitted by ω;
- $\hat{\lambda}_\omega$ is the highest abstraction level of data transmitted by ω;
- Ψ_ω is the *total cognitive structure* of communication signal ω. We have that $\Psi_\omega = (\psi_\omega^{\check{\lambda}_\omega}, \ldots, \psi_\omega^{\hat{\lambda}_\omega})$; and

- $\omega^{\circlearrowleft} = \langle a, \Psi_\omega, a \rangle$ is an implicitly induced *loopback signal*, which corresponds to the emitting agent being concious of (i.e., "listening" on) ω.

There is no restriction on the relationship between a and a'. For instance, we may have that $a = a'$, in which case, an agent is communicating with itself. We may also have that a **in** a' or that a' **in** a, in which cases an agent is communicating with a super group or with a subgroup, respectively.

We define $\delta(\psi_a^\lambda, \psi_\omega^\lambda) \in [0,1]$ as the *cognitive difference* between agent a and communication signal ω at abstraction level λ, such that:

- $\delta(\psi_a^\lambda, \psi_\omega^\lambda) = 0$ if and only if $\psi_a^\lambda \supseteq \psi_\omega^\lambda$,
- $\delta(\psi_a^\lambda, \psi_\omega^\lambda) = 1$ if and only if $\psi_a^\lambda \cap \psi_\omega^\lambda = \emptyset \wedge \psi_\omega^\lambda \neq \emptyset$,
- $\delta(\psi_a^\lambda, \psi_\omega^\lambda) \in]0,1[$ otherwise.

By extension,

$$\Delta(\Psi_a, \Psi_\omega) = \sum_{\lambda=\min(\check{\lambda}_a, \check{\lambda}_\omega)}^{\max(\hat{\lambda}_a, \hat{\lambda}_\omega)} \delta(\psi_a^\lambda, \psi_\omega^\lambda)$$

Similarly, we define $\delta(\psi_a^\lambda, \psi_{a'}^\lambda) \in [0,1]$ as the *cognitive difference* from agent a to agent a' at abstraction level λ, such that:

- $\delta(\psi_a^\lambda, \psi_{a'}^\lambda) = 0$ if and only if $\psi_a^\lambda \supseteq \psi_{a'}^\lambda$, hence $\delta()$ is clearly non-commutative,
- $\delta(\psi_a^\lambda, \psi_{a'}^\lambda) = 1$ if and only if $\psi_a^\lambda \cap \psi_{a'}^\lambda = \emptyset \wedge \psi_{a'}^\lambda \neq \emptyset$,
- $\delta(\psi_a^\lambda, \psi_{a'}^\lambda) \in]0,1[$ otherwise.

By extension,

$$\Delta(\Psi_a, \Psi_{a'}) = \sum_{\lambda=\min(\check{\lambda}_a, \check{\lambda}_{a'})}^{\max(\hat{\lambda}_a, \hat{\lambda}_{a'})} \delta(\psi_a^\lambda, \psi_{a'}^\lambda)$$

A communication signal ω can, in principle, be exercised between agents a and a' at any two levels l and l' within the agent hierarchy (i.e, $a \in A^l$ and $a' \in A^{l'}$). Nevertheless, the probability that a and a' understand each other decreases as the distance between l and l' increases, since this may also increase the cognitive difference from a to a' (i.e., $\Delta(\Psi_a, \Psi_{a'})$ increases) or from a' to a (i.e., $\Delta(\Psi_{a'}, \Psi_a)$ increases). For instance, consider two humans within the same society and the same educational background, compared to two humans within the same society, compared to two humans, compared to two creatures from different species, etc. [Jin et al., 2001].

A true meaningful communication signal must imply some change (however infinitesimal it may be) in the cognitive structure of either the emitter or the receiver, or both. Changes in the emitter's cognitive structure are not a direct result of a communication signal itself, but rather of the loopback signal that follows from that communication signal ($\omega^{\circlearrowleft}$).

2.4 Communication events

In reality, it seems awkward to consider single communication signals; interactions between agents usually imply a sequence of communication signals being transmitted between them, minimally to provide feedback on an original communication signal. Consequently, we introduce the notion of communication events, which represents an ordered sequence of communication signals. Formally, a communication event Ω is an ordered list of communication signals $\langle\!\langle \omega_1, \ldots, \omega_i, \ldots, \omega_j, \ldots \omega_n \rangle\!\rangle$, where ω_i occurred before ω_j when $i < j$.

3. Explaining how an agent evolves

We already pointed out that an agent's evolution is a consequence of its interactions with other agents. A basic motivation for evolution is what we consider to be an intrinsic feature of agents, namely minimizing the energy they use to emit/receive a communication signal $w = \langle a, \Psi_w, a' \rangle$. Energy is used at two distinct points: by agent a in constructing the message to emit (Ψ_w) and by agent a' in interpreting the message received. In the following, we explain how agents evolve using the above definitions (Sect. 1.2).

Let us first consider a communication event $\Omega = \langle\!\langle \omega_1, \ldots, \omega_n \rangle\!\rangle$ that involves only two agents, a and a', such that a is not a member of a' ($\neg(a \textbf{ in } a')$) and *vice versa* ($\neg(a' \textbf{ in } a)$). For such a communication event, we have that

$$\forall \omega_i, i \in [1, n], \omega_i = \langle a, \Psi_{\omega_i}, a' \rangle \vee \omega_i = \langle a', \Psi_{\omega_i}, a \rangle.$$

Agents a and a' aim at maximizing what we call their *internal* and *external coherence*. We define internal coherence as the adequation between an agent's cognitive structure and the cognitive structure of messages it emits. An agent a maximizes internal coherence by minimizing the cognitive difference between a and all messages it emits. Formally, we have

$$\min \sum_{\substack{\omega_i \in \Omega \wedge \\ \omega_i = \langle a, \Psi_{\omega_i}, a' \rangle}} \Delta(\Psi_a, \Psi_{\omega_i}).$$

Similarly, external coherence is the adequation between an agent's cognitive structure and the cognitive structure of messages it receives. Hence, maximal external coherence for agent a is acheived by minimizing the cognitive difference between a and all messages it receives. In formal terms, we have

$$\min \sum_{\substack{\omega_i \in \Omega \wedge \\ \omega_i = \langle a', \Psi_{\omega_i}, a \rangle}} \Delta(\Psi_a, \Psi_{\omega_i}).$$

In this particular context, both agents a and d can optimize their cognitive structure in order to minimize the energy deployed. The only factor that may empede that reduction of energy deployment is the nature of these agents, or more concretely their capacity to modify their cognitive structures.

To illustrate this case, consider the following B2C situation, where an enterprise (agent a) is interacting with a single consumer (agent d). In this context, agent a is most likely composed of many agents, acting as a whole rather than as individuals, which in turn, may be organized in teams. Hence, $a \in A^l, l > 0$. Similarly, we can easily assume that d is an atomic agent (i.e., it is not decomposable), and therefore that $d \in A^0$. In order to complete a sale, many communication signals ω_i may be exchanged between a and d, composing a communication event Ω. The communication event therefore corresponds to the negotiation occuring between a and d in order to understand what the needs of a' are and what a may supply to fulfill those needs.

There is coevolution, since at the end of the communication event Ω, d knows more about products available at the enterprise a, while a learned about the needs of its single customer. Depending on how the communication event was concluded, it may in turn bring a to change its sales methods and even its product line. In the framework, this means that the enterprise conginitive structure (Ψ_a) will be modified to take these changes into consideration.

In this limited context, both agents could evolve to the point that only minimal interactions are required since:

- enterprise a knows perfectly what its customer d buys. In fact, a may adjust its product list to meet all requirements of d to the point that only products required by d are sold by a,
- customer a' only needs to indicate the quantity to deliver, since a has only a' as client and it already knows the name, the billing address, the shipping address, and the product characteristics for that unique client.

When an arbitrary number of agents are involved, the situation may also be explained as a maximization of internal and external coherence. Consider a communication event $\Omega = \langle\langle \omega_1, \ldots, \omega_n \rangle\rangle$ involving an agent a_0 interacting with agents $a_1, \ldots, a_m$, such that $\neg(a_j \textbf{ in } a_k)$ with $j, k \in [0, m] \wedge j \neq k$. In this case, we have that

$$\forall \omega_i, i \in [1, n], \omega_i = \langle a_j, \Psi_{\omega_i}, a_k \rangle \text{ with } j, k \in [0, m] \wedge j \neq k.$$

Here, agent a_0 maximizes internal coherence by minimizing the cognitive difference with all the messages it emits,

$$\min \sum_{j=1}^{m} \sum_{\substack{\omega_i \in \Omega \wedge \\ \omega_i = \langle a_0, \Psi_{\omega_i}, a_j \rangle}} \Delta(\Psi_{a_0}, \Psi_{\omega_i}),$$

while it achieves maximal external coherence by minimizing the cognitive difference with all the messages it receives,

$$\min \sum_{j=1}^{m} \sum_{\substack{\omega_i \in \Omega \wedge \\ \omega_i = \langle a_j, \Psi_{\omega_i}, a_0 \rangle}} \Delta(\Psi_{a_0}, \Psi_{\omega_i}).$$

The optimization of internal and external coherence is more difficult to achieve in this context since agent a_0 must not only consider its capacity to change but also the impact of change on its energy deployment when interacting with all the agents in its environment (i.e., agents $a_1, \ldots, a_m$). This in turn leads to satisficing.

In order to illustrate this situation, we extend the example presented above. Here, we assume that the selling enterprise is agent a_0. This enterprise will interact with many customers (agents $a_1, \ldots, a_m$). Here, the evolution of the enterprise is constrained by the requirements of all its customers, which may be contradictory, since different customers may need different products or features. In this case, the enterprise cognitive structure (Ψ_{a_0}) is adapted to consider these different requirements and possible contradictions. However, this adaptation may not occur as fast as the market requires it.

To keep these customers, the enterprise must adjust its products to fulfill as much of these requirements as possible, while minimizing production effort (and hence costs). Furthermore, in addition to the quantity and product ordered, each customer a_i must identify himself to enterprise a_0 whenever he orders a product, since many customers interact with enterprise a_0.

4. A preliminary identification of abstraction levels

In defining Λ (Sect. 1.2.1), we stated that it was "some multidimensional space of abstraction levels." Originally, we were considering a one-dimensional space, such that we could determine the ordering of all possible abstraction levels. It did not take long before we realized that abstraction levels cannot be structured in such a linear space.

What we offer here is a preliminary identification of two of many potential dimensions, and of their respective abstraction levels. The first dimension relates to *modeling* of data. At the lowest abstraction level, we find facts (or simply data). The next level along that dimension is concerned with models (or metadata). Then follows metamodels (or meta-metadata), etc.

A second dimension relates to the *representation* of data. We base this dimension on [Habermas, 1984][Kropf et al., 1998][Shannon and Weaver, 1964] [Ulrich, 2001]. At the lowest level, we have symbols, which are the building blocks of representations. Then, we have the lexical level, which describes rules for assembling symbols into words. This level is followed by syntactic, then semantics. At this point, we limit the levels along this dimension to pragmatics (i.e., contextual information).

Clearly, any abstraction level within the modeling dimension may be refined by levels of the representation dimension, and *vice versa*. This simple observation is what lead us to a multidimensional Λ. For instance, a model (metadata) is represented using symbols (boxes, arrows, letters, etc.) which are connected together to form a diagram following construction (lexical and syntactic) rules. The diagram may be interpreted by analysts (semantics). And so on.

5. Conclusion and Future Work

In this paper, we presented a formal communication framework, which may be used to describe and explain these interactions and relationships, and others as well. We mainly focussed on the identification of fundamental concepts pertaining to interactions among agents, and how these agents evolve as a consequence of these interactions.

We feel however that the real impact of the framework does not lie in its expressiveness, but rather in the way it helps us reason about communications and evolution. Furthermore, we envision information systems, developed by using the framework, that may "understand" their environment and adapt to it. For instance, by better understanding the cognitive structure of communication events, we could dynamically determine what minimal data is required in electronic transactions between two agents, and hence modify dynamically the forms that customers must fill out when ordering products on a B2C Web site.

Such future development may not be forseen without considering the hurdles that lie ahead. Many questions come to mind. How should the abstraction levels space Λ be decomposed to adequatly account for specific business contexts? How do we create software artefacts that have intrinsic understanding of the cognitive structure received (Ψ_ω), referred to as the "symbol grounding problem" [Harnad, 1990]? How do we create software artefacts (i.e., agents) that have adaptable cognitive structures (Ψ_a)? How do we create software artefacts (i.e., agents) that can decide when and how to adapt?

In the short term, as the number of Web Services and the number of XML dialects grow, it will become increasingly important to understand how interactions between enterprises occur. Clearly, Web Services do not solve anything unless we have some way to describe what the service is providing, and not only the how. A service name is not sufficient. The same name may have different meanings in different contexts. Furthermore, there must be mechanisms to simplify the deployment and the use of all these remote services.

The framework presented herein will be used to provide a better understanding of interactions between enterprises, not only at the lexical and syntactic levels (format of data exchanged), but also from semantical and pragmatic perspectives (meaning of data exchanged). As such, it will bring about solutions to the problems enterprises face when deploying Web Services. In the long term, the framework will also provide a basis for the development of truly adaptable CIS, which will "understand" their environment [Pfeiffer and Scheier, 1999], and will coevolve with that environment.

References

Conant, R. C. and Ashby, W. (1970). Every good regulator of a system must be a model of that system. *International Journal of Systems Science*, 1(2):89–97.

Habermas, J. (1984). *The Theory of Communicative Action*. Beacon Press, Boston.

Harnad, S. (1990). The symbol grounding problem. *Physica D*, 42:335–346.

Heylighen, F. and Campbell, D. (1995). Selection of organization at the social level : obstacles and facilitators of metasystem transition. *World Futures – The Journal of General Evolution*, 45:33–99.

Jin, E. M., Girvan, M., and Newman, M. E. J. (2001). The structure of growing social networks. *Working paper at Sant Fe Institute*.

Kauffman, S. (1995). *At home in the universe - The Search for the Laws of Self-Organization and Complexity*. Oxford University Press.

Krippendorf, K. (1986). A dictionary of cybernetics. unpublished report, Principia Cybernetica Web.

Kropf, P., Babin, G., and Hulot, A. (1998). Reduction des besoins en communication de corba. In *Colloque International sur les Nouvelles Technologies de la Repartition (NOTERE'98)*, pages 73–84, Montreal, Canada.

Luck, M., McRumey, P., and Preist, C. (2003). *Agent Technology : Enabling Next Generation Computing*. AgenttLink.org.

Pfeiffer, R. and Scheier, C. (1999). *Understanding Intelligence*. MIT Press, Cambridge, Massachusetts.

Plaice, J. and Kropf, P. (2000). *Intensional Programming II*, chapter Intensional Communities, pages 292–296. World Scientific Press, Singapore.

Shannon, C. E. and Weaver, W. (1964). *The Mathematical Theory of Commmunication*. The University of Illinois Press, Urbana.

Simon, H. A. (1996). *The Science of the Artificial*. MIT Press, Cambridge, Mass.

Ulrich, W. (2001). A philosophical staircase for information systems definition, design, and development. *Journal of Information Technology Theory and Application*, 3(3).

Chapter 15

INTRODUCING NEW BUSINESS MODELS IN PROVISION OF QOS NETWORKS

Borka Džonova-Jerman-Blažič

Abstract: This paper presents a novel approach to setting services required inside and across IP networks known as Quality of Service provision. The first decade of 21st century is a turning point for the Internet provision business as the end users are the customers that will select a service with given or chosen Quality of Service. In the case of service providers that means facilitating dynamic creation of both the service and of the customization. This requires new business approach in form of business model known as mediation. Mediation enables smooth transition and satisfaction of the service requirement condition set up by the end user and the involved service providers. This paper provides insight in the business layer and the relevant business model that is introduced in IP networks enabling QoS provision. It describes briefly the developed model and its implementation within the re-engineered provider network.

Key words: re-engineering, service provider B2B, B2C, SOAP, QoS, SLA, SLS, XML, ebXML, Access Mediator, Service Mediator, Resource Mediator,.

1. INTRODUCTION

Regulation, globalisation, the growth of the Internet, and the services supplied over it, are driving the disaggregating process of traditional vertically integrated tele-operators in the past decade. The telecom activities have been reorganised into many more or less independent domains and business entities inside or outside the tele-operator companies. This allows the creation of a wide range of new communications service providers. As a consequence of this diversity of services and technologies tele-operators are facing massive dislocation in the networking platforms used to support new multimedia services for both wire-less and wire-line realisations. Part of this technology shift is the move towards 'ALL IP' networking. However the currently deployed IP networks are based upon a 'best effort delivery' principle that supports mostly data services where delay can be tolerated. Next Generation 'All IP' networks, both wire-line and 3G wire-less, have to support Service Level Agreements (SLA) and Quality Of Services (QoS) Guarantees for combinations of services, some of them delay tolerant, and others such as voice and video, that are delay intolerant. The delivery of these guaranteed QoS New Generation Networks depends on network control mechanisms and management capabilities working together across interconnected networks operated by different operators. This new picture of service provision away of the classical telecom subscriber scheme requires new approach regarding management of the business processes where different parties end users are involved. This paper provides insight in the business layer and the relevant business model that where developed for IP networks with QoS provision. It presents the basic components of the model, the technology used for implementation of the model and illustrates the approach with one service example.

The business model was developed and implemented within the European project CADENUS "Configuration and Provisioning of End-User Services in Premium IP networks" [1] from the 5 Framework Programme [2]. The model is designed in a way that enable automation of a number of business processes with pure technical background, collectively defined as SLA-based service creation for end-user services by introducing new component in the provider network known as mediation. The paper is structured in three parts, starting with the outline of the model followed by its architecture and implementation presentation. The paper at the end summarise the achievements and the prospects in further deployment of the scenarios and models developed in the CADENUS project.

2. OUTLINE OF THE MODEL

The trend toward using e-commerce based on B2B integration of trading partners is not unique to the telecoms industry as the opportunity offered by the B2B technology presents itself way to lower the cost of operation and doing business in many different fields. However, the telecoms activities in connection with the delivering advanced end-user services based on QoS and SLA, together with adequate service level guarantees, requires coordination between providers in the service value chain. Such value chain includes various partners e.g. content and service providers, retailers, third-party service providers, and network connectivity providers. Realization of end-to -end network connectivity from end-user to content provider, which is required in order to deliver the service to the end-user, presents in itself a difficult task regarding coordination and management. The players 'buyers' and 'sellers' involved in the value chain of service provision/consumption with QoS such as: video on demand (VoD), voice over IP (VoIP), video-conferencing or virtual private network (VPN) are expecting the service to run with full automation and speed. Though it is recognised that the value chain of this process may be complex, a simplified baseline configuration could be used for the design of the model, consisting of the following business entities:

User/Customer. A user or a customer could be a residential users or smaller enterprises using dial-up access to the Internet. Larger enterprises are also users but they are usually connected on a permanent basis utilizing leased

Retailers/3rd party broker. This entity is either reselling the services of a single Service Provider (SP), or is brokering between a numbers of SPs. A 'retail SLA' is associated with customer subscription.

Service Provider(s). These entities are delivering content, application services or simply service management. Services considered include managed multi-domain connectivity services and content delivery services such as video-on-demand, VPN, e-learning services. The SP does not own network facilities, but contracts these from Network Providers (NPs), through SLAs for IP connectivity (with Service Level Specifications - SLSs that describes the technical details of the SLAs). The SP may make such contracts with a number of Network Providers (NPs). Connectivity associated with an end-user contract for particular service will typically span from the end-user site to the SP's site, or between a number of end-user sites (e.g. in the case of a managed VPN service).

Network Provider(s). These entities are delivering IP QoS connectivity. Such connectivity might be limited to a **single domain**, in the case in which the NP sells only its own connectivity services, or it could span **multiple domains**. Typically, NPs have agreements with other NPs and the first in the chain can sell to the SP a service integrating the other NPs' services with its own (in this case, such an NP "resells" the services of the other NPs). Network Providers can be further classified with respect to the roles in the value-chains related to the provision of Internet-based services.

Internet Service Providers:

- *Backbone Service Provider*: Offering broadband IP network service. They operate at the root level of interconnection hierarchy and handle aggregated IP traffic;
- *Regional/National Service Provider*: Operating a larger network of Point of Presence offering Internet access for a certain region or even on a national and trans-national level. They may offer value-added services

like VPN and intranet services for enterprises, differentiated service classes, etc;

- *Internet Access Provider*: Offering dial-up access or other access methods to the Internet. Additional services typically include e-mail accounts and web homepage hosting.

Network Operators:

- *Access Network Operators*: Access connectivity is typically provided by the local telephone company through POTS or N-ISDN services;
- *Backbone Network Operators*: Operators offering network services to larger Regional and Backbone Service Providers for linking their routing nodes and interconnecting to Internet exchange points;
- *CIX/NAP Operators*: Commercial Internet eXchange points that provide
- interconnection between Service Providers.

The underlying business framework in that value chain is the Service Level Agreement (SLA) between different members of the value chain and the related networks which can be named in this context a SLA Networks or SLAN. SLA Network is a network in which all traffic classes, including traditional best-effort traffic the most usual today in the Internet are contracted between users and providers in a form of Service Level Specifications (SLS) [9, 10].

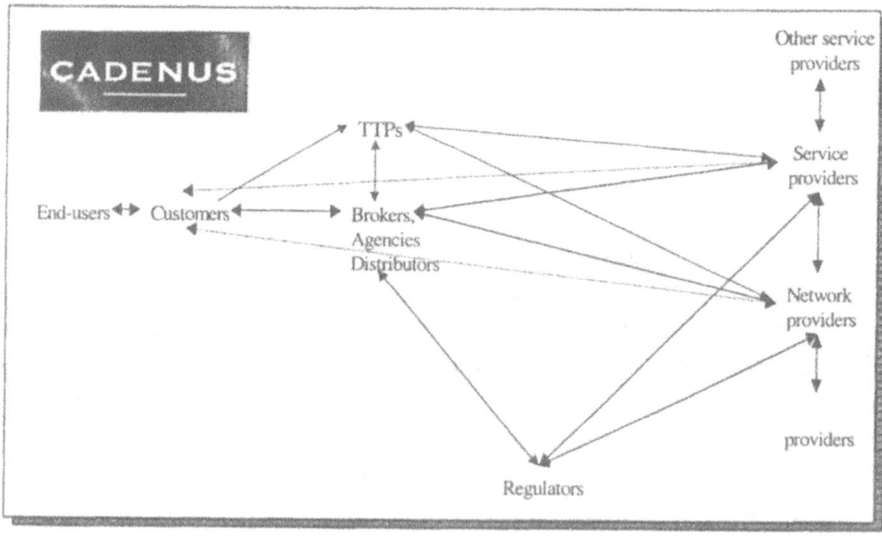

Figure 1. *Entities of the model and their interactions*

SLS is a sort of technical mapping of the elements agreed and specified in the SLA. The most simple example is the translation of the site with the service such as Video on Demand to its IP network address within the SLS.

SLAN is considered as a network commitment: provision of a default quality for default services. User part SLA is a User commitment: to use only services which are SLA conformant. SLA between SPs is also a product of an agreement process or a result of a negotiation process between two (or more) providers on functions, parameters and virtual paths (addresses) at both sides of a SLA. Service creation which is the main goal of these processes is a creation of service logic, service data with associated data management and a creation of needed virtual paths that assist the virtual SLA protocol. The entities of the model and their interactions are presented on Fig.1.

3. THE MODEL ARCHITECTURE

3.1 Components of the model

The approach taken for the architectural design taken in the CADENUS project has a strong focus on enhancing the ability to deliver value-added services to end-users, *by enabling a coordinated behaviour among the actors in the value chain* [5]. Though it is recognized that the value chain may be arbitrarily complex, a simplified baseline configuration in CADENUS has been chosen, consisting of the three main business entities: a retailer, either reselling the services of a single service provider, or brokering among a number of service providers, service provider (SP) and network provider (NP).

To implement this business models a new component known as *Mediation* was set up. The architecture of the CADENUS Mediation Component separates the functionalities of the mediator into 3 major blocks, termed Access Mediator (AM), Service Mediator (SM) and Resource Mediator (RM) (see Fig. 2). By defining these three types of mediators two strategic goals were achieved: first, the business model addresses all points of user/provider/domains interactions, including negotiations, selecting, profiling, etc. and second, it clearly separate not only the service from the resource control and management, but also the service from the service creation machinery. This enables the functionality of Mediation Component to be mapped in a straightforward way to software systems run by the three types of business entities as follows:

- the AM belongs to a 3rd party broker;
- the SM belongs to the SP;
- the RM belongs to the NP.

The **access mediator main** role is to grant access to the services provided by Services Provider. It is responsible for the contract (that is included in the SLA), the compliance of the services supplied to customers, it provides to customers a menu of available services, it authenticates users at the usage process, proceeds their demands and might compose services.

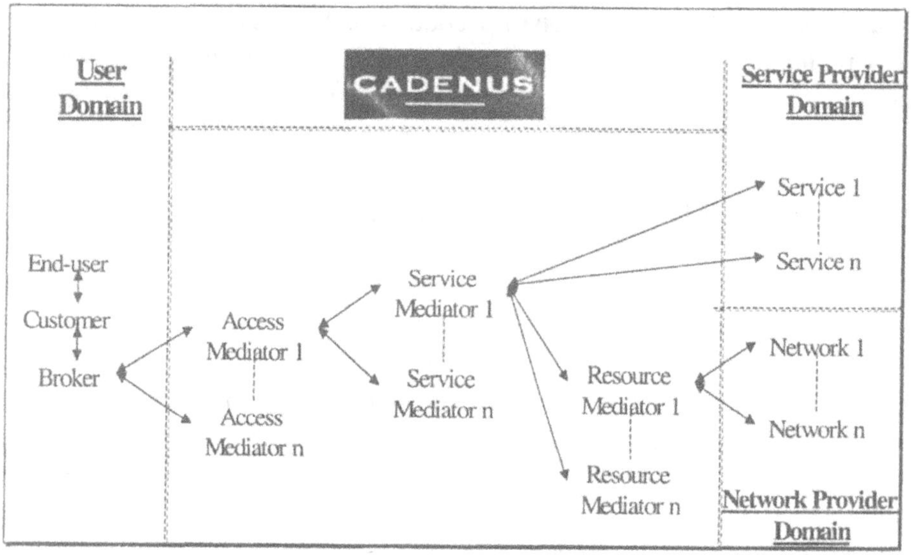

Figure 2. *The mediator component in the model*

It communicates with several Service Providers in order to processes the access mediation function that grants access to a given Service Provider. The access mediator enable other actors involved in the service value chain to be hidden from his customers as it « screens » the subsequent entities involved in the services delivery. The Access mediator knows both end-users access network link and terminal type they use but it does not perform the accounting normally as it is carried out by the service provider itself

The **Service mediator** provides services that are put in bouquets by the access mediator. It deals with services requests but does not establish a direct contact with end-users (for SLA determination, subscription) nor perform accounting. It deals with other Service Providers to compose its services and with Network Providers to have its services supported. The Service mediator is enclosed within the generic services provider roles as it performs the service treatment.

The **Resource Mediator** is responsible for the network performance asked by Service Providers. It translates services demands into specific resources demands and is granting access to the most appropriate resources when the resources provider is given. Otherwise, it chooses the most appropriate IP resources provider thus performing a broker role. The Resource Mediator is granting access to the IP QoS edge of the backbone. It communicates with other Resource mediator(s) to ensure the IP network connectivity. The Resource Mediator performs the so-called "access to the resources function" but it does not carry out the network resource control function. This is left to co called **Service Authority and directory** (see Fig.3) that is ensuring traditional "Gold, Silver, Bronze" classification of the service common and guaranteed by all actors involved in the value add chain. It registers all the actors and all the services, which could be offer to the end users. It should be noted here that QoS agreements are contracted between:

- End-Users and the Service (via the access mediator and/or service mediator),
- Service and Resources Providers (via the resources mediator).

For better understanding this model is compared with other traditional business models. The most known and familiar is the travel agency model. In the CADENUS model the AM acts as the travel agency, which proposes many providers for the same holidays destination. The SM corresponds to the tour operator, which offers complete or partial bundles (i.e. flight, hotel, car rental). The service logic corresponds to the individual offer (i.e. hotel, flight). Like the service logic may be resold by many SMs, the same room location can be purchased from different tour operators. Finally, the RM acts as the central reservation place for the different parts of the trip and the NC performs the real job.

3.2 Message flows between components

Interfaces and message flows between CADENUS mediation architecture components are defined in the context of the different scenarios implementing different services but they have the following common components (see Fig.3):

- Propose New Service - between SM and SA (Service Authority),
- Publish Service Profile - between SM and SD (Service Directory),
- Set up a new SLA - between user, AM, SM, SD and RM,
- *Renegotiate SLA - between user, AM and SM,*
- *Delete SLA - between user, AM and SM*

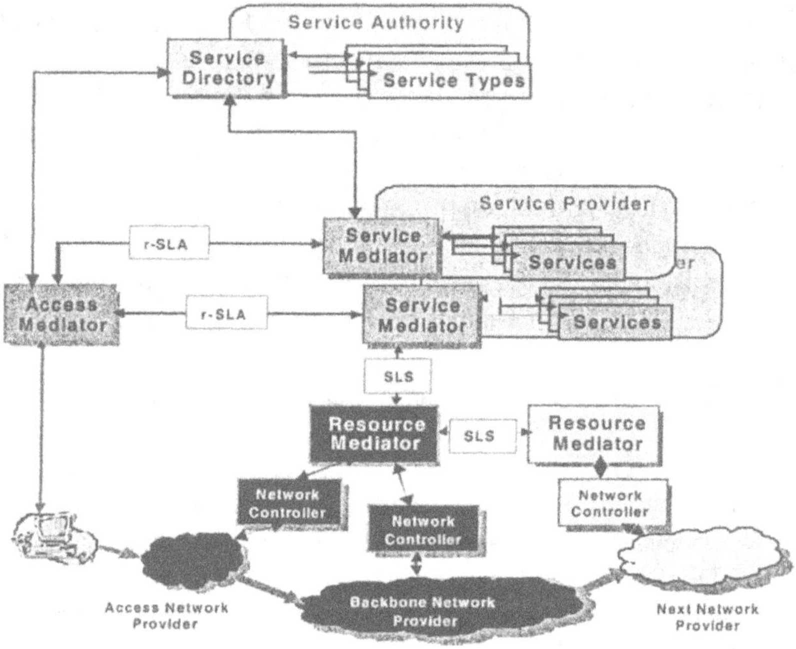

Figure 3. The CADENUS Architecture

The creation and exchange of these messages in creation of real service can be best understood with a description of an example service scenario. The CADENUS model was implemented to three different scenarios for a service: IP Virtual Private Network, Video Distribution Service and Voice over IP. Here the creation and delivery of a *Video Distribution* service, or, more in general, of Distributed Multimedia Applications is used as an example. Such services include audio/video transmissions where a user connects to a video-server archive containing a number of movies that can be sent, in a streaming fashion, to a client host. In the same category are also classified applications such as Videoconference and Tele-medicine, where video and audio data are generated from live sessions. There are four main steps involved in the service creation process:

- define and put in a standard format a description of the business process associated with service trading – in this case Video distribution;
- define a standard Graphical User Interface needed to allow user's customization of the service parameters, as an a example see Fig.4;
- define a standard template for the Service Level Agreement, as an example see Fig.4;

- define the rules to be applied in the translation from the SLA to the corresponding SLS(s) at negotiation time.

When these tasks have been completed, the only remaining operation is the publication of the business process specification, together with the newly defined components (the service GUI and SLA template) into an on-line registry/directory.

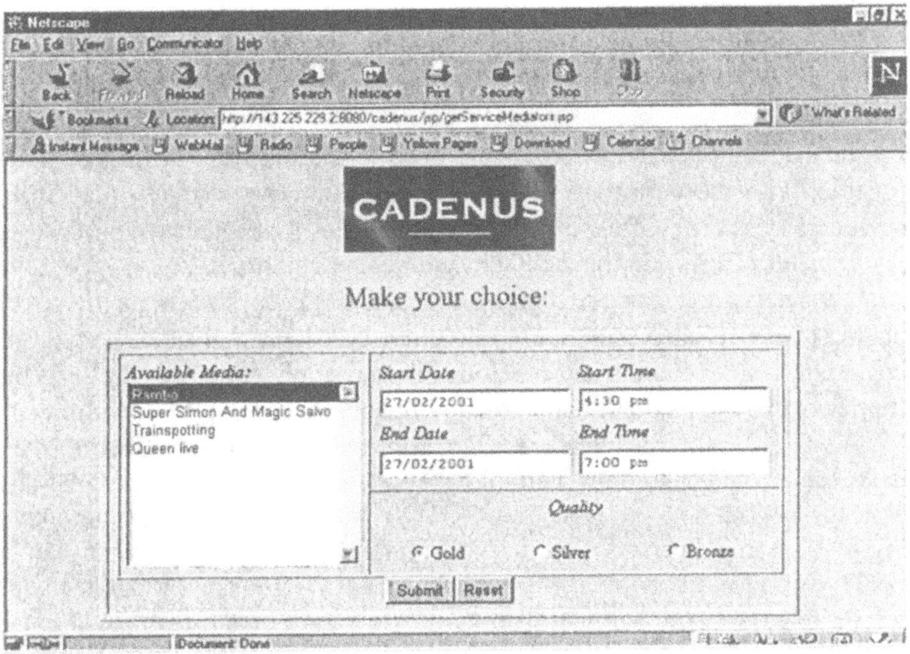

Figure 4. GUI for Video Distribution service

4. IMPLEMENTATION OF THE MODEL

The architecture defined in CADENUS (AM, SM and RM) was implemented in 6 blocks: the access mediation (AM), the service mediation (SM), logic of service, the service authority, the resources mediator(RM) and the resource control (network controller)[7]. The two more services e.g. logic of services and resource control help in translation of the services requests into resources that belong to the service plane. The resource control controls the resources and the QoS inside its sub-domain

CADENUS model implementation embraces the latest standard proposals coming from the electronic business research community, with respect to both the modelling methodology and the actual design for implementation. The business process specification selected is fully compliant with the ebXML (electronic business XML) framework, and the distributed, on-line registry has been implemented via UDDI (Universal Description, Discovery and Integration).

The ebXML framework aims at creating a single global electronic marketplace where enterprises of any size and in any geographical location can meet and conduct business with each other through the exchange of XML based messages.

The specification of a business process is the main required activity while creating a new service; then, in order to enable effective negotiation, it is needed that any interested party defines and publishes a Collaboration Protocol Profile (CPP), where a reference to the business process is made, together with the definition of the role that the party wants to play inside such a process. The CPPs, in turn, form the basis for Collaboration Protocol Agreements (CPAs) established between business parties.

Ultimately, the business processes specified in the CPAs drive the business service interfaces to execute those processes and send the required documents. A detailed description of the application of the ebXML framework to the mediation architecture may be found in [10], [11].

The Graphical User Interface (GUI) for the service and the SLA template have been implemented as customisable web components, whose goal is to ease the negotiation process. The GUI for Video Distribution is shown in Fig.4. It is implemented as a friendly web-based interface, which can be easily exploited even by users who are totally unaware of the technical details related to the service. The GUI is obtained and shown to the user whilst negotiating the service, as the outcome of a series of interactions between the various components of the CADENUS architecture.

The template SLA is an XML file that contains all the information necessary to uniquely identify the two parties user and service provider), together with the service instance that has been negotiated; XML style sheets are applied to this file in order to customise the way the SLA is presented to the final user.

The user must first indicate the service to which he wants to subscribe, specifying the QoS level and, optionally, the service lifetime (seeFig.4).

As stated in the previous section, the negotiation process is implemented in ebXML. The following steps formalise the sequence:

- the user subscribes to (or is authenticated from) the proper AM;
- the user asks for the negotiation of a new service instance;
- the AM allows the user to choose one of the available services (in this case Video Distribution is the user's choice);
- the AM contacts a centralised repository in order to retrieve the service GUI associated with the selected service: in the simplified case as

presented on Fig.4 a GUI contains the list of the available movies, the time schedule for the service and the possible levels of QoS. Every movie title which appears in the GUI is available at least from one SM;
- the service GUI and the data are sent to the end-user.

At this stage, the AM does not make any semantic interpretation of the service under negotiation, but simply acts as a broker between the end-user and the Service Mediator. This has the advantage of relieving the AM from the responsibility of being aware of any specific service definition: the only entity involved in the definition process is, as one would expect, the SM.

Having received the service GUI, the user fills in the required fields and submits his request to the AM. This event triggers the following actions on the AM's side:

- the AM contacts all the SMs which registered as sellers of the specified service (in our case, Video Distribution). The list of such SMs may either have been obtained with the previous access to the repository (when the service GUI has been fetched), or be retrieved through a further access. A document containing the service parameters specified by the user is sent to the SMs in the list, in order to let them become aware of the service the AM (on behalf of the end user) is willing to receive;
- starting from the document just received, each SM creates one or more associated Service Level Specifications (SLSs) which are delivered to RMs;
- the RMs, based on the received SLSs, make an evaluation of the impact that the service is going to have on the network and translate it in the form of a 'cost' to be paid for service enforcement: such a cost is then returned to the SM;
- the SM is now capable to formulate an offer, which is sent back to the AM: the offer comprises a contribution coming from the cost information provided by the RMs and an additional fee related to its own value-added service (e.g. content provisioning, brokerage activity with respect to network configuration, management of service options, etc.);
- once all of the quotations coming from the SMs have arrived, the AM sorts them according to the user's preferences, which may be derived from the user's profile. The sorted list of available offers is presented to the user: each single offer is built on the basis of the standard SLA template defined during the service creation phase;
- the user selects the offer which he deems most suitable. This operation, which has a legal value, is in all respects equivalent to the signature of a formal contract (Service Level Agreement).

For the actions described above, depending upon the application requirements, it might be requested the SLA to be translated into more than one SLS. This can happen, for example, when the application needs a duplex channel to work properly: one way to reserve resources (for the streamed multiplexed audio/video content) on the path from the video server to the end-user's system and the other to cope with streaming control data flowing in the opposite direction. This is true for the most common streaming protocols (RTP/RTSP, modified UDP versions, etc.) available nowadays. The situation obviously changes in those cases where the application needs to reserve completely independent audio and video channels (thus requiring one SLS each) or, stated in more general terms, whenever it is desirable to make a reservation for multiple, separate flows. It could be necessary to create more than one SLS also when different guarantees are to be assured over different time intervals

The traffic characterisations in the example presented on Fig.4 are expressed in the form of a sequence of time slots and related QoS parameters, generally in the form of a token bucket. Such characterisations are usually represented as *metadata* (i.e. "data about data"): in this way the information is linked closely to the media and becomes easily accessible to the Service Mediator in an automatic fashion. This approach of integrating metadata with multimedia content for the guaranteed delivery of digital resources looks to be extremely useful, in so far as users don't need to know anything about the communication requirements for the delivery of a certain multimedia document. All the work related to the negotiation of QoS guarantees with the network infrastructure can be managed and performed transparently by the mediation entities.

The Service Directory assumes a role which is of paramount importance for the CADENUS framework:

- it contains the business processes of the standardised services, together with the associated components (GUI and SLA template);
- it gives a SM the possibility to publish its own profile, together with the services it offers;
- it gives the AM the possibility to retrieve information about the portfolio of services and about the SMs that are offering them;
- it acts both as a registry and as a repository.

The directory is implemented by exploiting the UDDI (Universal Description, Discovery and Integration) technology, a framework for the description and discovery of services based on the creation of a world-wide registry aimed at facilitating integration. UDDI uses XML to represent data and SOAP (Simple Object Access Protocol) to exchange messages, thus solving the integration and interoperability problem via a layered approach. XML provides a cross-platform approach to data encoding and formatting, whereas SOAP makes it simple to package information that has to be exchanged across system boundaries. The Publisher's API enable companies to register information about the Web Services they offer; such information can then be retrieved by other companies via the Inquiry API. The data provided in a business registration are built of three different components:

- white pages, including name, address, phone number and other contact data related to a business entity which is providing services;
- yellow pages, basically a categorization of the companies/services based on taxonomies and/or standard identification mechanisms;
- green pages, containing technical information about the Web Services offered by a company (e.g. endpoint URL, names and arguments of the methods that can be invoked, etc.).

The problems related to scalability and reliability of the UDDI business registry are coped through an implementation that is logically centralized but physically distributed, with multiple root nodes (also called site operators) that replicate each other's data on a regular basis. Once a registration is made at a specified root node, the data are automatically shared with the other site operators, thus becoming freely available to anyone who is interested in discovering the Web Services that have been exposed by a given company.

5. CONCLUSION

The next generation IP networks are gaining more and more proselytes. Its appeal is due to the given opportunity of a standard and consistent way for network configuration, independently of the underlying architecture and QoS provisioning model assumptions. While this technology is powerful and alluring, it's also generally untested and unproven. The IST project from the 5th Framework Program of EU, CADENUS–Configuration and Provisioning of End User Services in Premium IP Networks has developed an architecture that aims to test and validate the policy and business based approach in a real network providing QoS. For that the consortium has developed a business layer in the service provider network and the underlying business model introduced to enable an automated QoS provision.

The successful deployment of an automated Service Level Agreement management system based on the CADENUS architectural framework intuitively introduce efficiencies in areas such as sales, order capture, order management etc., as well as provide an important mechanism through which companies can expand their business and market reach. The CADENUS architecture has been shown to be flexible and adaptable to likely future market developments. The functional separation inherent in the design allows for various mediator configurations, which permit the business roles of a commonplace in marketplaces to be provided for all services. The architecture has been shown to potentially deliver significant benefits to the user as well. While the potential viability of the CADENUS architecture has been presented and demonstrated, there are however a number of developments and refinements necessary before these benefits can be realised. Principally network operators need to complete their automated service activation programmes.

ACKNOWLEDGMENT

Part of this work has been funded under the European Commission 5th framework IST program. The author would like to acknowledge all colleagues in the CADENUS project for their work, contributions and input.

REFERENCES

[1] Project CADENUS URL:http://www.cadenus.or

[2] M. Smirnov (Ed.), "SLA Networks in premium IP", CADENUS Deliverable D1.1, http://www.cadenus.org/deliverables/d11-final.pdf, March'01.

[3] S.P.Romano, M.Esposito, G.Ventre, SLA for Premium IP Networks, RFC draft, May 2001, draft-cadenus-sla-00.txt

[4] M.Smirnov, Service Creation in SLA Networks, RFC draft, draft-cadenus-slan-creation-00.txt, November 2000

[5] O. Dugeon (Ed.), "Service Provisioning in Premium IP: Recommendations to Telecom Operators and ISPs", CADENUS Deliverable D6.2,http://www.cadenus.org/|| deliverables/new/D62final.pdf, March'01

[6] G. Cortese (Ed.), "Mediation Component Release 1 –Requirements and Architecture", CADENUS Deliverable D3.1, http://www.cadenus.org/deliverables/d31-final.pdf, March'01

[7] G. Cortese (Ed.), "Service Configuration and Provisioning Framework: Requirements, Architecture, Design", CADENUS Deliverable D3.2,http://www.cadenus.org/ deliverables/d32final.pdf, November'01

[8] M. Smirnow (Ed.), "Group Event Notification in CADENUS Architecture", CADENUS http://www.cadenus.org/deliverables/new/gen_deliverable_v02.pdf, May'02.

[9] UN/CEFAT and OASIS ebXML Technical Architecture Team: "ebXML Technical Architecture SpecificationV1.04", February 2001, http://www.ebxml.org/specs/ebTA_print.pdf

[10] S.D.Antonio, B.Fadini, S.P. Romano, G.Ventre, "Designing Service Negotiation Entities for the Electronic Market-place", Proceeding of the International Conference on Software Engineering and Knowledge Engineering, Ischia, Italy, July 2002, ftp://ftp.grid.unina.it/pub/pubs/CADENUS/TR-COMICS-10-2002.pdf

[11] S.D.Antonio, S.P. Romano and G.Ventre, "An E-commerce Framework for Service Assurance over the Internet", Proceedings of the 5th International Conference on Electronic Commerce, Montreal, Canada, October 2002

SECTION 6
Internet and the WEB

Chapter 16

THE SEMANTIC WEB

Rudi Studer, Sudhir Agarwal, Raphael Volz

Abstract: The paper presents the vision of the Semantic Web and describes ontologies and associated metadata as the building blocks of the Semantic Web. Current research topics and promising application areas are discussed as well.

Key words: Semantic Web, ontologies, metadata.

1. INTRODUCTION

The development of the World Wide Web is a great success story with respect to the number of users and the amount of information that is nowadays offered by the WWW. However, most of the information that is available has to be interpreted by humans; machine support is rather limited. In order to get rid of that limitation, Tim Berners-Lee, the inventor of the WWW, coined the vision of the Semantic Web: to make the contents of the WWW accessible and interpretable by machines [Berners-Lee et al., 2001; Fensel et al., 2002; Davies et al., 2002]. Today it is almost impossible to integrate information that is spread over several Web or intranet pages. Consider, e. g., the query for a data mining expert in a company intranet, where the only explicit information stored are the relationships between people and the projects they work in on the one hand, and between projects and the topics they address on the other hand. In that case, a skills management system should be able to combine the information on the employees' home pages with the information on the projects' home pages in order to find the respective expert. To realize such scenarios, metadata have to be interpreted and appropriately combined by machines [Lau and Sure, 2002].

The process of building the Semantic Web is still in genesis, but first standards, e.g. for the underlying data model and an ontology language already appeared. However, those structures are now to be filled with life in applications. In order to make this task feasible, one should start with the simpler tasks first. The following steps show the direction where the Semantic Web is heading:

1. 1 providing a common syntax for machine understandable statements,
2. 2 establishing common vocabularies,
3. 3 agreeing on a logical language,
4. 4 using the language for exchanging proofs.

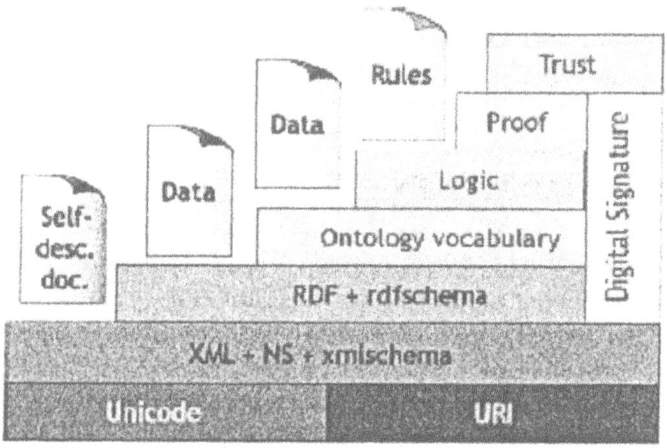

Figure 1. The layers of the Semantic Web.

Berners-Lee suggested a layer structure for the Semantic Web (cf. Figure 1). This structure reects the steps listed above. It follows the understanding that each step alone will already provide added value, so that the Semantic Web can be realized in an incremental way.

2. LAYERS OF THE SEMANTIC WEB

Unicode/Unified Resource Identifiers, XML, RDF, ontologies, logic, proof, and trust are suggested by Berners-Lee[1]: and discussed in detail for instance in [Patel-Schneider and Fensel, 2002] which also addresses open issues.

1 see http://www.w3.org/DesignIssues/Semantic.html

On the first two layers, a common syntax is provided. Uniform resource identifiers (URIs) provide a standard way to refer to entities[2], while Unicode is a standard for exchanging symbols. The Extensible Markup Language (XML) fixes a notation for describing labeled trees, and XML Schema allows to define grammars for valid XML documents. XML documents may refer to different namespaces to disambiguate between equally named tags. The formalizations on these two layers are nowadays widely accepted, and the number of XML documents is increasing rapidly.

The Resource Description Framework (RDF) can be seen as the first layer which is part of the Semantic Web. According to the W3C recommendation [W3C, 1999], RDF "is a foundation for processing metadata; it provides interoperability between applications that exchange machine-understandable information on the Web." RDF descriptions consist of three types of entities: resources, properties, and statements. Resources may be web pages, parts or collections of web pages, or any (real-world) objects which are not directly part of the WWW.

In RDF, resources are always addressed by URIs. Properties are specific attributes, characteristics, or relations describing resources. A resource together with a property having a value for that resource forms an RDF statement. A value is either a literal, a resource, or another statement. Statements can thus be considered as object-attribute-value triples.

The middle part of Figure 2 shows an example of RDF statements. Two persons (i.e., their Web pages) are represented as resources `URI-GST' and `URI-AHO'. The statement on the lower right consists of the resource `URI-AHO' and the property `cooperates with' with the value `URI-GST' (which again is a resource). The resource `URI-SWMining' has as value for the property `title' the literal `Semantic Web Mining'. Such statements may be attached to web pages by annotation tools, as e.g. OntoMat Annotizer [Handschuh et al., 2002].

The data model underlying RDF is basically a directed, labelled pseudograph. RDF Schema defines a simple modeling language on top of RDF which includes classes, is-a relationships between classes and between properties, and domain/range restrictions for properties. RDF and RDF Schema are encoded in XML syntax, but they do not employ the tree semantics of XML.

The next layer is the ontology vocabulary. Following [Gruber, 1993], an ontology is "an explicit formalization of a shared understanding of a conceptualization". This high-level definition is realized differently by various research communities and thereby in ontology representation

2 URL (uniform resource locator) refers to a locatable URI, e.g. an http://... address. It is often used as a synonym, although strictly speaking URLs are a subclass of URIs, see http://www.w3.org/Addressing.

languages. However, most of these languages have a certain understanding in common, as most of them include a set of concepts, a hierarchy on them, and relations between concepts. Some of them also include axioms in some specific logic. We will discuss the most prominent approaches in more detail in the next section.

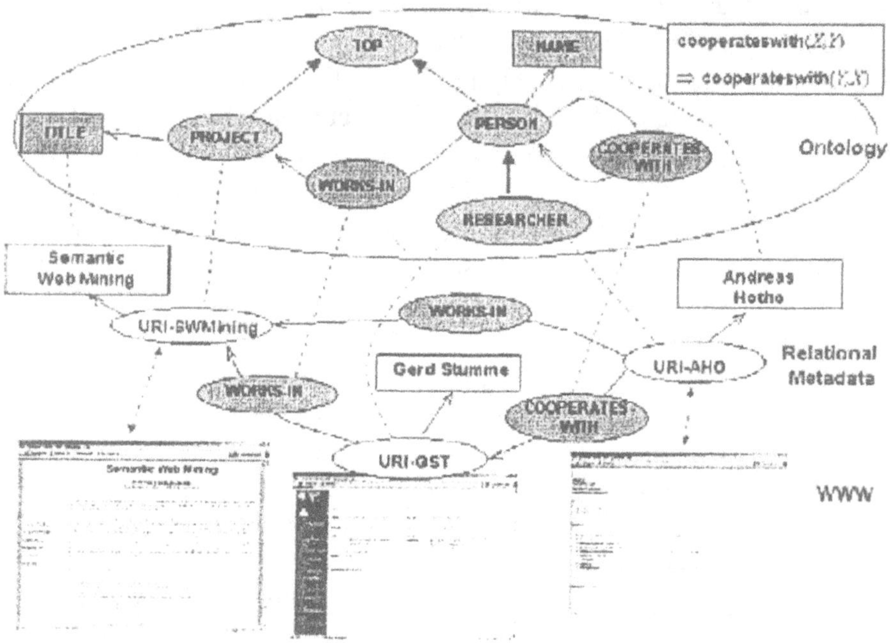

Figure 2. The relation between the WWW, relational metadata, and ontologies.

Logic is the next layer according to Berners-Lee, cf. Figure 1. However, nowadays research considers usually the ontology and the logic levels together, as ontologies are already based on logic and should allow for logical axioms [Kifer et al., 1995]. By applying logical deduction, one can infer new knowledge from the information which is stated explicitly. For instance, the axiom saying that the 'cooperates with'-relation is symmetric (cf. Figure 2) allows to logically infer that the person addressed by `URI-HO' is cooperating with the person addressed by `URI-GST' although only the person "GST" specifies his cooperation with the person "AHO". The kind of inference that is possible depends heavily on the logics chosen.

Proof and trust are the remaining layers. They follow the under standing that it is important to be able to check the validity of statements made in the (Semantic) Web. Therefore the creators of statements should be able to provide a proof which is verifiable by a machine. At this level, it is not required that the machine of the reader of the statements finds the proof itself, it `just' has to check the proof provided by the creator. These two layers are rarely tackled in today's research. Therefore we will focus our interest on the XML, RDF, ontology and logic layers in the remainder of this paper.

3. ONTOLOGIES: LANGUAGES AND TOOLS

A priori, many formal knowledge representation mechanisms[3] may play the role of a Semantic Web language.

Description Logics (DL) are currently the most popular framework for defining a Web ontology language. DLs are subsets of first order logic which aim at being as expressive as possible while still being decidable. The description logic SHOQ(D) provides the basis for the W3C proposal for a Web ontology language, i.e., OWL [Dean and Schreiber, 2003]. OWL is the result of joining the efforts of two research programs: The DARPA Agent Markup Language DAML+OIL[4] was created as part of the DAML research programme, OIL (Ontology Inference Layer) [Fensel et al., 2000] was the result of the Semantic Web research programme of the European Union.

Several tools are in use for the creation and maintenance of ontologies and metadata, as well as for reasoning within them. Ontoedit [Sure et al., 2002] is an ontology editor which is connected to Ontobroker [Decker et al., 1999], an inference engine for F{Logic [Kifer et al., 1995]. Ontobroker provides means for semantic based query handling over distributed resources. F{Logic has also inuenced the development of Triple [Sintek and Decker, 2002], an inference engine which allows to model features of UML, Topic Maps, or RDF Schema.

FaCT[5] provides inference services for the Description Logic SHIQ. The reasoning implemented by the FaCT inference engine may also be used in the OilEd [Bechhofer et al., 2001] ontology editor.

The Karlsruhe Ontology Framework KAON [Bozsak et al., 2002], [Maedche et al., 2003] is a novel open-source infrastructe that takes a holistic approach to ontology management and is targeted for business applications. It includes a comprehensive tool suite allowing easy ontology creation and management, as well as advanced components for building ontology-based applications supporting latest Web standards.

3 See [Studer et al., 1998] for a general discussion.

4 http://www.daml.org

5 http://www.cs.man.ac.uk/_horrocks/FaCT

4. RESEARCH AREAS RELATED TO THE SEMANTIC WEB

There are many research areas related to the Semantic Web, like multi-agent systems or human-computer interfaces. Because of space restrictions we only discuss databases and Web Mining.

Databases The database community has a rather long history in addressing semantic aspects, most notably in the context of semantic data models that were a prominent research topic in the 1980s [Hull and King, 1987]. Semantics played and still play an important role in topics like view management, schema transformation and integration, and query processing, just to mention a few of them. However, the Semantic Web raises new challenges that were not present in the more classical database scenarios [Sheth and Meersman, 2002]: the collection of metadata that will be created as part of the Semantic Web can be seen as a massive new distributed database whose size can be of the same order of magnitude as the data itself and whose complexity may be even higher. Therefore, methods for achieving scalability and robustness in the Semantic Web have to be developed.

The specific characteristics of the data models used in the Semantic Web pose also new research questions, e.g. for view management. In [Volz et al., 2003] a first approach has been developed to provide a view mechanism for an RDF-based data model. Handling properties as first class citizens and being able to cope with non-strict typing are some of the specific characteristics that have to be addressed by a view mechanism for the Semantic Web.

Semantic Web Mining The novel research area of Semantic Web Mining aims at combining the two fast-developing research areas Semantic Web and Web Mining. Web mining is the application of data mining techniques to the content, structure, and usage of Web resources. This can help to discover global as well as local structure within and between Web pages. This means that Web mining is an invaluable help in the transformation from human understandable content to machine under standable semantics. Three areas of Web mining are commonly distinguished: content mining, structure mining, and usage mining [Berendt et al., 2002].

As the Semantic Web enhances the first generation of the WWW with formal semantics, it o_ers a good basis to enrich Web Mining: The types of (hyper)links are now described explicitly, allowing the knowledge engineer to gain deeper insights in Web structure mining; and the contents of the pages come along with a formal semantics, allowing her to apply mining techniques which require more structured input. On the other hand, Web

Mining can help setting up the Semantic Web. It can help to learn structures for knowledge organization (e. g., ontologies) from the Web and to provide the population of such knowledge structures.

5. APPLICATION AREAS

Different application areas benefit from the Semantic Web. We briefly present some areas which are currently under development.

Web Services Semantic Web enabled Web Services are one of the heavily discussed topics within the Semantic Web community. Fensel et al. define in [Fensel and Bussler, 2002] a Web Service Modeling Framework which provides a conceptual model for describing and developing web services. There, Semantic Web technology is used for service description and service discovery.

An application scenario for the configuration of Web services is presented in [Felfernig et al., 2002]. This approach is ontology based and aims at overcoming the huge effort needed for enabling B2B solutions such as eProcurement. Paolucci et al. [Paolucci et al., 2002] and Burstein et al. [Burstein et al., 2002] use DAML-S for defining an ontology for Web Services. The formers show the advantage of a declarative description for supporting the matching process by exploiting ontologies. The latter presents different aspects of web service ontologies and focus on the service grounding. The objective the authors aim at in [Agarwal et al., 2003] is that a human user can seamlessly surf the existing World Wide Web and the emerging web services and that he can easily compose and invoke web services on the y without being a software engineer. They present a framework, which trades o_ between having a reasonably easy to use interface for web services and the complexity of web service workflows.

E-Learning Sharing knowledge is the main idea of education. With the growing amount of educational material in the WWW, this idea gets a new dimension and generates new technical challenges. Metadata schemes for the exchange of educational Web resources have been in use for a number of years. These metadata schemes, for example LOM (Learning Objects Metadata)[6], usually extend the Dublin Core standard[7]. However, these standards lack a precise machine-interpretable semantics to describe the content of the learning objects. In [Schmitz et al., 2002] an approach for

6 see http://ltsc.ieee.org/wg12/

7 http://dublincore.org

accessing and browsing distributed learning repositories is described that exploits ontologies and associated relational metadata.

Peer-To-Peer networks Peer-To-Peer networks can be seen as distributed repositories. In order to retrieve data from another peer, it is of paramount importance to know what the other peer provides. While some Peer-to-Peer scenarios (e. g., music networks such as Gnutella) work well with limited amount of metadata, most applications need more semantic information in order to find relevant peers efficiently. The Edutella framework, for instance, was established to support the exchange of RDF based repositories between peers. [Nejdl et al., 2002] describe use case scenarios for annotation and replication and propose the Modification Exchange Language MEL for such distributed RDF repositories. [Nejdl et al., 2003] discuss the RDF-based P2P networks as a specific example of a new type of P2P networks, schema-based P2P network, and describe the use of super-peer based topologies for these networks. In [Hatala and Richards, 2002], requirements and corresponding solutions for easy interaction between content providers and consumers are described.

Knowledge Management and Knowledge Portals New trends in knowledge management pave the way from the more document oriented view on knowledge management to a more knowledge item oriented view [Staab et al., 2001]. Such approaches again rely on Semantic Web methods, especially ontologies and metadata. In [Davies et al., 2002] a collection of methods and tools is described that exploit Semantic Web techniques for applications like skills management or for supporting virtual enterprises.

Knowledge portals provide views onto domain-specific information on the World Wide Web for facilitating their users to find relevant information. The extensive maintenance needed for keeping a portal up to date can be simplified by using an ontology as conceptual backbone for acquiring, maintaining, and providing information. SEAL [Maedche et al.2002] is a comprehensive architecture for a semantic portal offering a broad range of tools for improving the cost/benefit ratio of semantic portals. SEAL-II [Hotho et al., 2001] tackles the soft spot between unstructured knowledge and richly structured knowledge. To achieve that objective, SEAL-II includes a broad collection of techniques to instantiate knowledge and to browse through collections of documents and knowledge elements.

6. CONCLUSION

The development of the Semantic Web is a fast moving process that raises a lot of research challenges that have to be addressed in an interdisciplinary way. Furthermore, it paves the way to new types of applications that exploit the semantic basis of this new generation of the WWW. These aspects make the Semantic Web so fascinating.

ACKNOWLEDGMENTS

We thank all our colleagues at the Institute AIFB and at the FZI Research Center for Information Technologies for a lot of fruitful discussions. Part of this research is funded by EU in the IST project WonderWeb (project no. 2001-33052), by bmb+f in the SemIPort project (project no. 08C5939) and by bmb+f in the SESAM project.

7. REFERENCES

Agarwal, Sudhir, Handschuh, Siegfried, and Staab, Ste_en (2003). Surfing the service web. In ISWC2003: 2nd International Semantic Web Conference 2003, Sanibel Island, Florida, USA.

Bechhofer, S., Horrocks, I., Goble, C., and Stevens, R. (2001). OilEd: A reasonable ontology editor for the Semantic Web. In Proceedings of KI2001, Joint German/Austrian conference on Artificial Intelligence, volume 2174 of LNAI, pages 396-408. Springer.

Berendt, B., Hotho, A., and Stumme, G. (2002). Towards semantic web mining. In ISWC2002: Ist International Semantic Web Conference, Sardinia, Italy, Lecture Notes in Computer Science, pages 264-278. Springer.

Berners-Lee, T., Hendler, J., and Lassila, O. (2001). The semantic web. Scientific American, Online at http://www.sciam.com/article.cfm?articleID=00048144-10D2-1C70-84A9809EC588EF21.

Bozsak, E., Ehrig, M., Handschuh, S., Hotho, A., Maedche, A., Motik, B., Oberle, D., Schmitz, C., Staab, S., Stojanovic, L., Stojanovic, N., Studer, R., Stumme, G., Sure, Y., Tane, J., Volz, R., and Zacharias, V. (2002). KAON - Towards a Large Scale Semantic Web. In Bauknecht, K., Tjoa, A. Min, and Quirchmayr, G., editors, E-Commerce and Web Technologies, Third International Conference, EC-Web 2002, Aix-en-Provence, France, September 2-6, 2002, Proceedings, volume 2455 of Lecture Notes in Computer Science, pages 304{313. Springer.

Burstein, M., Hobbs, J., Lassila, O., Martin, D., McDermott, D., McIlraith, S., Narayanan, S., Paolucci, M., Payne, T., and Sycara, K. (2002). DAML-S: Web Service Description for the Semantic Web. In ISWC2002: Ist International Semantic Web Conference, Sardinia, Italy, Lecture Notes in Computer Science, pages 348{363. Springer.

Davies, John, Fensel, Dieter, and van Harmelen, Frank, editors (2002). Towards The Semantic Web: Ontology-Driven Knowledge Management. Wiley. Dean, M. and Schreiber, G. (2003). Web ontology language (OWL) reference version 1.0. Internet: http://www.w3.org/TR/owl-ref/.

Decker, S., Erdmann, M., Fensel, D., and Studer, R. (1999). Ontobroker: Ontology based Access to Distributed and Semi-structured Information. In et al., R. Meersman, editor, Database Semantics: Semantic Issues in Multimedia Systems., pages 351-369. Kluwer Academic.

Felfernig, A., Friedrich, G., Jannach, D., and Zanker, M. (2002). Semantic Configuration of Web Services in the CAWICOMS Project. In ISWC2002: Ist International Semantic Web Conference, Sardinia, Italy, Lecture Notes in Computer Science, pages 192{205. Springer.

Fensel, D. and Bussler, C. (2002). The Web Service Modeling Framework WSMF. Electronic Commerce Research and Applications, 1(2):113-137.

Fensel, D., Horrocks, I., van Harmelen, F., Decker, S., Erdmann, M., and Klein, M. (2000). Oil in a nutshell. In Knowledge Acquisition, Modeling, and Management, Proceedings of the European Knowledge Acquisition Conference (EKAW-2000), volume 1937 of LNAI, page 1 ff. Springer.

Fensel, Dieter, Hendler, James, Lieberman, Henry, and Wahlster, Wolfgang, editors (2002). Spinning the Semantic Web. MIT Press.

Gruber, T. R. (1993). Towards Principles for the Design of Ontologies Used for Knowledge Sharing. In Guarino, N. and Poli, R., editors, Formal Ontology in Conceptual Analysis and Knowledge Representation, Deventer, The Netherlands. Kluwer Academic Publishers.

Handschuh, S., Staab, S., and Ciravegna, F. (2002). S-CREAM | semi-automatic CREAtion of Metadata. In 13th International Conference on Knowledge Engineering and Knowledge Management (EKAW02), volume 2473 of LNAI, page 358 ff.

Hatala, M. and Richards, G. (2002). Global vs. Community Metadata Standards: Empowering Users for Knowledge Exchange. In ISWC2002: Ist International Semantic Web Conference, Sardinia, Italy, Lecture Notes in Computer Science, pages 292-306. Springer.

Hotho, A., Maedche, A., Staab, S., and Studer, R. (2001). SEAL-II | the soft spot between richly structured and unstructured knowledge. Journal of Universal Computer Science (J.UCS), 7(7):566-590.

Hull, R. and King, R. (1987). Semantic database modeling: survey, applications, and research issues. ACM Computing Surveys (CSUR), 19(3):201{260.

Kifer, M., Lausen, G., and Wu, J. (1995). Logical foundations of object-oriented and frame-based languages. Journal of the ACM, 42:741-843.

Lau, T. and Sure, Y. (2002). Introducing ontology-based skills management at a large insurance company. In Modellierung 2002: Modellierung in der Praxis - Modellierung fuer die Praxis, Tutzing.

Maedche, A., Motik, B., Stojanovic, L., Studer, R., and Volz, R. (2003). An infrastructure for searching, reusing and evolving distributed ontologies. In WWW2003: The Twelfth International World Wide Web Conference, Budapest, Hungary.

Maedche, A., Staab, S., Studer, R., Sure, Y., and Volz, R. (2002). SEAL - Tying up information integration and web site management by ontologies. IEEE-CS Data Engineering Bulletin, Special Issue on Organizing and Discovering the Semantic Web.

Nejdl, W., Siberski, W., Simon, B., and Tane, J. (2002). Towards a Modification Exchange Language for Distributed RDF Repositories. In ISWC2002: Ist International Semantic Web Conference, Sardinia, Italy, Lecture Notes in Computer Science, pages 236{249. Springer.

Nejdl, W., Wolpers, M., Siberski, W., Schmitz, C., Schlosser, M., and Brunkhorst, I. (2003). Super-peer-based routing and clustering strategies for rdf-based peer-to-peer networks. In WWW2003: The Twelfth International World Wide Web Conference, Budapest, Hungary.

Paolucci, M., Kawamura, T., Payne, T.., and Sycara, K. (2002). Semantic matching of web services capabilities. In ISWC2002: Ist International Semantic Web Conference, Sardinia, Italy, Lecture Notes in Computer Science, pages 333{347. Springer.

Patel-Schneider, P. and Fensel, D. (2002). Layering the semantic web: Problems and directions. In ISWC2002: Ist International Semantic Web Conference, Sardinia, Italy, Lecture Notes in Computer Science, pages 16{29. Springer.

Schmitz, Ch., Staab, S., Stumme, G., Studer, R., and Tane, J. (2002). Accessing distributed learning repositories through a courseware watchdog. In Proceedings of E-Learn 2002: World Conference on E-Learning in Corporate, Government, Healthcare and Higher Education, Montreal, Canada.

Sheth, A. and Meersman, R. (2002). Amicalola report: Databases and information systems research challlenges and opportunities in semantic web and enterprises. SIGMOD Record, 31(4).

Sintek, M. and Decker, S. (2002). Triple | a query, inference, and transformation language for the semantic web. In ISWC2002: Ist International Semantic Web Conference, Sardinia, Italy, Lecture Notes in Computer Science, pages 364{378. Springer.

Staab, S., Studer, R., Schnurr, H.-P., and Sure, Y. (2001). Knowledge processes and ontologies. IEEE Intelligent Systems, 16(1):26-34.

Studer, R., Benjamins, V. Richard, and Fensel, D. (1998). Knowledge engineering: Principles and methods. Data + Knowledge Engineering, 25(1-2):161-197.

Sure, Y., Erdmann, M., Angele, J., Staab, S., Studer, R., and Wenke, D. (2002). OntoEdit: Collaborative ontology development for the semantic web. In ISWC2002: Ist International Semantic Web Conference, Sardinia, Italy, Lecture Notes in Computer Science, pages 221{235. Springer.

Volz, R., Oberle, D., and Studer, R. (2003). Views for light-weight web ontologies. In Proceedings of ACM Symposium of Applied Computing (SAC), Melbourne, Florida, USA.

W3C (1999). Resource Description Framework (RDF): Model and Syntax specification. W3C recommendation, 22 February 1999, www.w3.org/TR/REC-rdf-syntax-19990222/.

Chapter 17

WEB PERSONALIZATION BASED ON USER´S TRADE-OFFS

Maria-Cleci Martins, Iria Garaffa, Maximiliano Kling

Abstract. Current e-commerce personalization schemes do not take into account e-shoppers' decision-making processes. Such schemes are typically driven by information from customers' previous preferences or those of their peers. They do not employ any technique to understand how customers value (or perceive the value of) different product characteristics and use it in the personalization scheme. This paper discusses the use of economic theory on utility choices to unveil the e-shopper's decision-making process and employ them to search for products and services with greater chances of being purchased. A methodology is suggested as well as its potential benefits to users and companies. Mainly, the approach has the potential to increase customer satisfaction, site revisits and sales. Additionally, the methodology provides the company with information on customers' choice behaviour that can be used to improve businesses systems (e.g., product design, marketing).

Key words: Personalization, decision-making, discrete choice modeling, knowledge modeling

1. INTRODUCTION

Companies and individuals can benefit from the Internet as a market place. On one side companies can reach and target each customer individually, on the other the easy access to information allows e-shoppers to compare product characteristics from different brands in different online stores without too much effort.

Typically, information about products and suppliers can be accessed from database servers using either list browsing or key word search. However, the amount of information available on the Internet has substantially increased the cognitive effort for e-shoppers to choose the best option. Browsing large lists, arranged in tree-like structures can be time consuming, while key word search often results in too many items being returned (or none). Thus, instead of facilitating the choice (and the sale), the Internet makes the customer's choice decision-making process more difficult. This might even cause users to abandon the website without buying anything. This is a crucial aspect for e-commerce companies where the customer's needs should be addressed with the best online option, in order to increase revenues, service level and customer's satisfaction.

Personalization is an approach that aims at reducing the amount of information shown to the user in an e-session by taking into account individual characteristics. E-commerce recommender systems (such as Ardissono and Godoy, 2000; Domingue et al., 2002, Kanjas, 2002) address the personalization issue by filtering the amount of non-requested products shown to the e-shopper in a given session. The filtering can be performed using statistical, machine learning, and knowledge-based methods.

Recommender systems are useful when customers do not know exactly what product or service they need, or when the company wants to introduce different products to the user. However, when customers roughly know their needs and the type of product or service to address those needs, the problem is to *find the best available online option*. In this case, recommender systems appear to be of no help. As a comparison, in a physical store, the shopper would be able to use other senses (vision, touch, smell) to recognize available products and compare them before choosing one, or even ask a sales person for advice. On the Internet however, they have to rely on their decision-making skills and the available information to choose the best option.

Given the challenge faced by personalization methods, incorporating an approach that helps e-shoppers to quickly find what they want presents potential benefits to the e-tailor and its customer. For instance, users will tend to revisit the site more often, increasing the chances of sales, as they are satisfied with the level of the online service.

This paper proposes the use of the economic theory on discrete choices (Ben-Akiva and Lerman, 1985) to help shoppers with their choice decision-making process when buying online. Our approach is to refine the search for products and services using the parameters the e-shopper would use in his or her choice decision-making process. We will use the information on what customers want and how they decide between available options to search for the best option. This methodology has been used in the transportation field to forecast travel demand (De Carvalho, 1998; Fowkes, 2001).

This paper is organized as follows. Section 1 presents the motivation for the research; Section 2 describes methods for personalization of web sites.

Then, in Section 3 the proposed methodology is presented. Section 4 discusses the proposed approach's benefits and drawbacks, followed by a conclusion.

2. RELATED WORK

Personalization on web sites is concerned with schemes that select the type and quantity of content that is shown to the e-user, according to the user's profile (Miyahara & Okamoto, 1997). Mainly, they show products and services the user did not ask for, hoping that some of them will catch the users' attention. The main filtering techniques are content-based and collaborative.

Content-based filtering makes recommendations based on comparisons between resources and the user's profile. Results retrieved are based on their similarity to what the e-shopper has previously shown interest.

Collaborative filtering works by showing products or services that are recommended or used by the e-shoppers' peers. One way of accomplishing collaborative filtering is through clustering techniques that group users with similar characteristics and interests.

2.1 Knowledge-based recommendation

Kanjas (2002) describes a range of filtering approaches in intelligent filtering and recommender systems, and also a comparison between the different methodologies. In this paper we will be interested in knowledge-based recommender systems, namely ontology and case based, because they seem to be the perfect host for our approach.

Case-based recommender systems use case-based reasoning (CBR) to make recommendations. CBR uses a knowledge base composed of cases and the methodology consists of finding a previous case that is similar to the new one.

A special application of a knowledge-based recommender system that uses CBR is Dublet (Hurley and Wilson, 2001). It recommends apartments for rental in Dublin, Ireland, based on a description of the user's preferences. Dublet employs information extraction to create cases dynamically by extracting information from the web of apartments for rent and retrieves units that match the user's preference. Dublet performs knowledge synthesis (creation) and extends the power of knowledge distribution of the CBR system by being operational in cell phones.

Ontology-based recommendation systems use ontologies for representing knowledge related to online shopping. An ontology (Gruber, 1993) is an explicit representation of a view of a domain of discourse (a conceptualisation) usually composed of a set of concepts and relationships. An ontology is usually composed of human-readable text describing a set of terms and formal axioms that constrain the interpretation and well-formed use of these terms[i]. Alice (Domingue et al., 2002) is an example of an ontology-based recommender system.

Ontologies describing customers, products, typical shopping tasks and the external context form the basis for the Alice architecture. Ontologies are instantiated as they are linked to the company's databases. The approach in this paper can be used to define slots in the customer ontology (about the customer's evaluation of the product attributes), and in the product ontology (to indicate the range of attributes users might be interested in evaluating, as well as, the way they evaluate them).

An advantage of using ontology is that the coded semantic information can be reused across a variety of applications. For example, those ontologies can be reused to develop decision support systems for companies adopting the personalization methodology suggested in this paper.

3. PROPOSED APPROACH

We propose a personalization approach that is based on the economic utility theory on consumer behavior. The theory provides a means to map the likely implicit choice decision process performed by the consumer when buying a product or service and is based on the following assumptions.

- Products or services are viewed by customers as a vector of characteristics, such as cost, brand, also defined as attributes.
- Customers are optimizers and they compare options based on the value of their attributes.
- Customers make trade-offs between attributes of a product/service to reach their decision. For example, in a transport service, a less comfort can be accepted if the fare is reduced.
- Customers are *maximizers* and they always choose the best-perceived option within a knowledge domain.

The term choice refers to the cognitive process by a consumer who, after evaluating the alternatives in a choice set, decides to select one of them (Louviere, 1988). Discrete choice data reports the values of the attributes characterizing alternative options in a choice set, and the individuals' choices. The main feature of discrete choice data is that the observed response (i.e. the dependent variable) is discrete: we only observe whether or not the customers choose one alternative option.

3.1 Discrete Choice Data Survey

Stated Preference (SP) is a technique used to collect discrete choice data (Pearmain and Kroes, 1990). This can be understood as a simulation game where individuals are asked to state their preferences for a set of possible options (i.e. choice set). A choice set is composed of at least two alternatives (represented by the values of their characteristics, or attributes). For instance, a trip can be characterized by the attributes: cost and in-vehicle travel time. A choice set would consider the transportation modes car and train, each mode being represented by its cost and the travel time.

The design of an SP experiment must consider that the respondents will have to make trade-offs between attributes of the product or service. For example, less travel comfort is acceptable if there is price compensation. Discrete Choice Modeling is the methodology to find those trade-offs for each individual.

Note that to develop a SP experiment we do not need to know exactly which options are available and their exact attribute values. However, it is considered that respondents should be given choice sets with possible options, and use attribute values as close to reality as possible. The knowledge base will support the development of the SP design on the fly. Customers' stated preference data are used to calibrate a mathematical model to unveil the parameters that e-shoppers would use to evaluate and choose one online option.

3.2 Modeling Approach

Discrete choice modeling (Ben-Akiva & Lerman, 1985) allows one to examine the influence that various factors (or attributes) have on the choices people make. Given the modeling assumptions from the economic theory, discrete choice data can be analysed using Logit modeling approach.

Logit models assume that there is an underlying unobserved variable (β_j) which determines the individual's choice. This then, is modeled as a function (U_i) of observed characteristics (X_{ij}) and a random term (ε) to account for the unknown influences on the customer choice (See Equation 1). Logit is a mathematical model that employs exponential behavior to explain predicted probabilities (P_i) of a particular response ("buy" or "not buy") regarding an alternative "i" (See Equation 2) belonging to the choice set with "J" options. Thus, the choice likelihood of an alternative is expressed as a function of its attributes and the other options available in the choice set. An optimization procedure such as Newton-Raphson (Ben-Akiva & Lerman, 1985) is employed to estimate the coefficients (β) on the observed characteristics (X_{ij}) in the Utility function (U_i).

$$U_i = f(\beta_j X_{ij}, \varepsilon) \tag{1}$$

$$P_i = \frac{e^{Ui}}{\sum_j e^{U_j}} \tag{2}$$

3.2.1 Analysing the results from Discrete Choice Modeling

An analyst usually evaluates results from the calibration. However, in case of e-commerce, there is the need to perform this task every time an e-shopper is online. To overcome this, case-based reasoning methodology can be employed.

A case-based reasoning system will be used to classify calibration processes, determining whether the results from the SP data are adequate. CBR implements classification via the k-nearest neighbour case-based classifier. It determines the class of a new case as a function of its k most similar cases from its case base. Aha (1998) describes CBR systems as implementing lazy problem-solving, a demand-driven method that performs inexpensive and accurate classification tasks that involve a large number of feature subsets, by omitting the training that decreases the speed of alternate methods. Accordingly, with a memory of calibration processes and their respective classifications, the system can immediately determine whether the process is complete or not (in which case, more data is required and the calibration re-run).

3.3 Framework

We have set the theory to build a refined personalization system that links customers hidden preferences for a product or service and the options available on the Internet. The aim of the system is to reduce the cognitive effort posed to the customer in his or her choice decision-making process given the amount of information available on the web and use the collected data to better satisfy this customer.

The implementation of the proposed approach requires specific data on the customer's choice decision-making process. The data will be obtained from an interactive SP game. The discrete choice data is then used to calibrate a Logit model that will reveal the trade-offs the user employed in his or her choice decision-making process. Calibration results (i.e. weights) are evaluated using an expert system (rule or case based) while the user is online.

Afterwards, accepted weights are employed to rebuild the customer's utility function for the particular product or service. Then, the task of the web service provider is to search the Internet for the available products or services and calculate their utility values. Those with highest utility are then shown to the e-shopper. Besides helping to identify suitable options for the e-shopper, the personalization based on discrete choice modeling also give companies clues about how to better characterize their products or services according to their customers' information needs.

This approach can be used not only as a stand-alone tool to help anonymous customers with their shopping tasks, but also as a refinement for a personalization system the company uses. A suggested architecture for a knowledge-based application is featured in Figure 1.

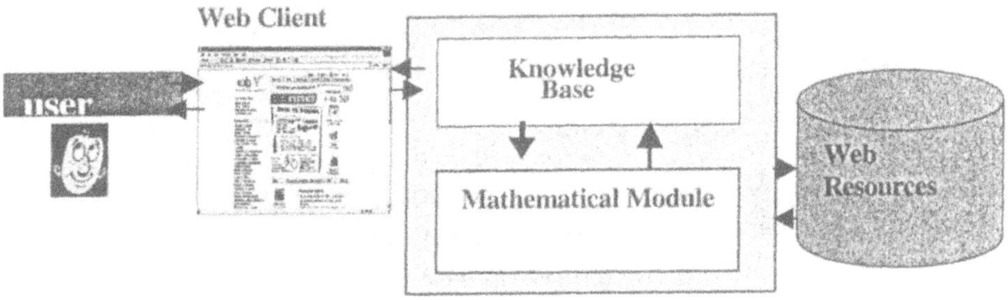

Figure 1 - Personalization system suggested architecture

The suggested approach is a knowledge-based personalization system. The system is based on ontologies containing the company's knowledge about its customers, products, etc. E-shoppers' preferences obtained from the SP game are uncovered by using the mathematical module, which contains the discrete choice modeling approach. The outputs from a Mathematical Module are used to improve the knowledge base. A product ontology would contain the knowledge about products such as price, brand as well as the likely customer's decision process when buying it (type of reasoning, utility function, main attributes considered in the decision-making). Current e-business standards (Dörr et. al., 2001) can be used as the base for this ontology, but additions should be made to incorporate attributes relating to the customer decision-making process.

We are illustrating our idea with an online car rental service pictured on a *shopbot* webpage (Figure 2). *Shopbot* is an e-commerce portal where users have access to different web service providers (i.e. companies) and can compare their offers as well as buy them. In the web site the e-shopper begins the search process filling a form with some parameters (e.g., car size, pick-up day, pick-up location). Those parameters are used to search the server database for available options. Quite often the search is unsuccessful at the first time. There are different reasons for that; the specific supplier may not have branches on the pick-up location, or the requested car size is not available. Eventually, the user has to change the search parameters a couple of times in order to find an offer.

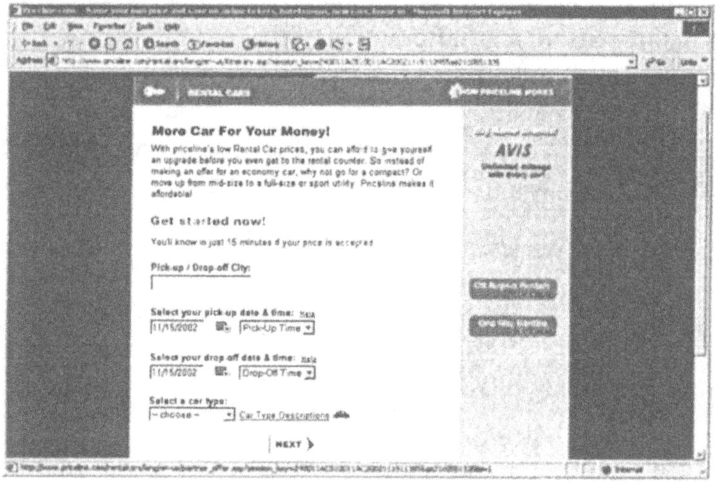

Figure 2 – Online car rental Shopbots

When finally the user manages to find some offers, s/he has to reason and decide for one of them or none. It might be the case that by evaluating the choices available, the user considers that all offers are overpriced compared to the prices of the cars and decide not to buy the service. Therefore, instead of hiring a car, the individual might decides to use local public transport, or a taxi service[8].

From the car-rental company viewpoint, it is a lost sale. If only the car-rental company new how individuals evaluate the different attributes of the service (car rental), they would try to show alternative options from the customer perspective. Maybe showing an offer with a better car would give the correct balance between the price of the car and the rental value.

We are assuming that the user evaluates the car rental options considering the price of the vehicle being hired and alternative ways of transport (such as public transport and taxi). Other decision models for this

[8] Decision processes vary according to the individual and the situation. If the individual is shopping for a company, the decision model might be different.

service would consider car-size within an acceptable price-range. In the car rental business, companies are often bounded to specific carmakers. Therefore, they also have tied prices and conditions down. A web portal offering such services would benefit from our approach by considering the customer's decision-making attributes to evaluate all choices available before showing the best ones to the customer.

4. DISCUSSION

Current personalization schemes are mainly focused on delivering contents, which are either similar to the users' profiles or that are recommended by their peers. Information on e-shoppers (e.g., history, profile) is used to feed the personalization scheme. These schemes do not address ways to improve the company's decision strategies (such as product design), or how it could help the e-shopper's choice decision-making process. Helping the e-shopper in this decisive moment can increase the company's sales. That is often a tactic used in physical stores where the sales person has a decisive impact on the choice.

The implications of not having such personalization approach are the customers' increasing level of dissatisfaction with online shopping causing companies to lose sales and still missing the opportunity of gathering fundamental customer information to run the business.

This paper proposes a theoretical approach of how the problem could be tackled by using customers' choice decision-making process in a personalization scheme. We have based our methodology on the economic theory of customer behaviour and the possibility of getting interactive data on customers stated preference.

The methodology discussed in this paper has the potential to improve substantially the relevance of the results shown to the e-shopper in an e-commerce session. The proposed approach has advantages over current Internet personalization schemes. The first advantage concerns to help the e-shopper in his or her choice decision-making process. It makes a previous evaluation of the alternatives using the parameters the e-shopper would. The benefits of this approach reduce the time the e-shopper is online trying to find suitable offers and evaluating them. The reach of search would be enlarged. The evaluation of the alternatives based on the utility function might prompt advantageous options that the e-shopper would not consider otherwise. We believe that this system would have the potential to reduce disappointments with online shopping. Using the approach shown in this paper it is possible to know the trade-offs individuals make between the characteristics of a product or service and use that to find options that best suits the e-shopper. Options would be shown on the bases of the likelihood of being chosen.

The second advantage will benefit the companies' decision process without any additional cost. Online companies will be able to know how much the e-shopper would pay for a slightly different product, such as a roomier seat in an airplane or a car with side airbags or all wheel drive. Usually, the information on customers gathered over the Internet is mostly used to feed the Internet personalization system. With our approach it can be used to give substantial insight on the way customers evaluate different characteristics of products and services. Companies can use this information to design their product so as to better target their customer's needs.

However, using the approach presented in this paper may present some challenges. The approach main input is data from an interactive SP game that demands customers' time. Customers have to be convinced that providing answers to the interactive game will give them a better service. An easy-of-use interface can help to overcome this problem by reducing the cognitive effort needed for the task. As an example, an adaptation of the Shneiderman's work on Dynamic Queries (Williamson & Shneiderman, 1992) could be employed.

Additionally, there are situations where the cost-benefit of employing the methodology may vary. First, the customer may not be interested in wasting time to take part in a SP game that evaluates low priced products or services. Second, the company could employ this methodology only to the most profitable or high priced 20% products, which often represents approximately 80% of the company's profit.

Furthermore, to implement the personalization based on discrete choice modeling, companies first need to investigate the usual ways their customers decide buying their products or services, what sort of decision process they perform, which attributes and variables they consider, etc. Consequently, additional research on product specific choice decision-models can be developed in order to fully implement the methodology.

5. CONCLUSION

This paper discusses a methodology that uses economic theory on discrete choices to link e-shoppers' decision-making process to available options. The main input to the proposed methodology is the discrete choice data, which will be collected from an interactive Stated Preference "game" that the e-shopper agrees to participate. The data is then used to calibrate a Logit model that will reveal the trade-offs the e-shopper employed in his or her choice decision-making process. Afterwards, these results are employed to search for the available options and calculate their values as the customer would. Options with high value utility are then shown to the e-shopper.

The benefits of using the methodology are twofold: it has the potential to increase customer satisfaction and therefore the likelihood of sales and revisits; the information on customer's trade-offs gives the company insights of how to improve the business (such as product design, promotion, forecasting).

REFERENCES

Aha, D. W. The Omnipresence of Case-Based Reasoning in Science and Application. Knowledge-Based Systems, 1998, 11(5-6), 261-273.

Ardissono, L. and A. Godoy: Tailoring the Interaction with Users in Web Stores. User Modeling and User-Adapted Interaction, 2000; 10: 251-303.

Ben-Akiva, M. & Lerman, R.:Discrete choice analysis. Massachusetts Institute of Technology. USA, 1985.

De Carvalho, M.: A comparison of Neural Networks and Econometric Discrete Choice Models in Transport. Unpublished PhD Thesis. The University of Leeds, UK,1998.

Domingue, J.; Carvalho, M.; Jaicheng Tan; Stutt, A.; Pertusson, H.: Alice: Assisting Online Shoppers through Ontologies and Novel Interface Metaphors. To appear in the proceedings of the 13th International Conference on Knowledge Engineering and Knowledge Management (EKAW02), Spain, 2002.

Dörr, M., Guarino, N., López, M. F., Stefanova, E. S. M., Tate, A.: State of the Art in Content Standards. OntoWeb, 2001. Available at http://www.ontoweb.org/download/deliverables/D3.1.pdf.

Fowkes, T.; Shinghal, N.: The Leeds Adaptive Stated Preference Methodology. Institute for Transport Studies, University of Leeds, Working Paper 558, 2001.

Gruber, T. R.: A Translation Approach to Portable Ontology Specifications. Knowledge Acquisition, 1993; 5(2).

Hurley, G., Wilson, D. C.: DubLet: An Online CBR System for Rental Property Recommendation, LNAI 2080, edited by D. W. Aha and I. Watson. New York, Springer Verlag, 2001; 660-674.

Kanjas, S.:Collaborative Filtering and Recommendation Systems. VTT Information Technology Research Report, 2002.

Louviere, (1988). Analyzing decision making: Metric conjoint analysis. Sage Publications.

McGinty, L. and Smith, B.:Comparison-Based Recommendation. ECCBR, 2002; LNAI 2416, 575-589.

Miyahara and Okamoto. Collaborative information filtering towards distributed cooperative work/learning environment. Meeting the challenge of the new technologies, Proceedings of the Eighth International PEG Conference, Bulgaria, 1997; pp. 260-267.

Pearmain, D., Kroes, E.: Stated preference techniques: a guide to practice. Steer Davies & Gleave UK, 1990.

Williamson, C., Shneiderman, B.: The dynamic HomeFinder: Evaluating dynamic queries in a real-estate information exploration system. Proceedings ACM SIGIR '92.Copenhagen, June 21-24, 1992; 338- 346.

[i]www-ksl.stanford.edu/kst/what-is-an-ontology.html

Chapter 18

XML ALONE IS NOT SUFFICIENT FOR EFFECTIVE WEBEDI

Fábio Ghignatti Beckenkamp and Wolfgang Pree

Abstract: WebEDI relies on the Internet infrastructure for exchanging documents among companies. Typically these documents are orders and invoices and can be exchanged by directly integrating the companies' ERP systems or via Web application upload, download or typing of documents. XML is considered as a state-of-the-art way to reduce the complexity of managing the different data formats. Nevertheless, the XML standard alone does not offer the required semantics. When integrating a few companies, it is possible to hard code the semantics in the applications that process the documents. However, when integrating numerous different companies such a solution does not scale. The semantics has to be captured in a more flexible and scalable way. To cope with this necessity, simple ontologies are required to augment the data description.

Key words: XML, WebEDI, Semantic Web, B2B, Marketplaces

1. INTRODUCTION - THE MERCADOR B2B SOLUTION

Mercador is a B2B company focused on the retail market and has built an infrastructure that allows companies of different sizes to be integrated through the Internet. Mercador implements the concept of a hub. The hub main services are the documents transference, translation and validation. The transference of documents is done from one company to another via the Internet using the Mercador integration servers as the connection hub. When a company sends a document to another, the document is sent to the Mercador hub first, the hub processes the document to validate and translate it to the receiver format. The delivery of the document is also done over the

Internet. Therefore the hub implements a set of services that make easier the integration of many with many companies. The solution has tools that support the integration of companies at different levels of automation.

The very basic infrastructure for companies integration is the Internet. The idea is to bring to the Internet the possibility of exchanging the documents that form the supply process. Those documents, like a protocol, permit the companies to exchange information about their business so as to guarantee the fast execution of the supply process. Mercador has developed a set of Internet functionalities based on the traditional EDI (Electronic Data Interchange) process. Called WebEDI, the system allows companies to exchange business documents using the Internet as the connection infrastructure. So that WebEDI is the transmission of standard electronic documents among companies or business organizations using the Internet.

The transformation and validation of the documents format is an important service that Mercador provides for its customers. One of the most complex tasks of integrating companies is the agreement on the data formats and meaning. Mercador implements, at the hub, services that validate and translate the customer's documents to the Mercador standard for each type of document. From the Mercador standard, the document is then translated to the recipient format. By doing this, Mercador is able to manage the data differences among the different companies. Therefore Mercador implements, at the hub, the necessary knowledge to facilitate the exchange of documents among many with many companies. The hub performs the necessary semantic process of the documents. The transformation and validation of documents is an important service that the hub offers to its customers and which minimizes the difficulties of integrating with many companies.

The Next section explains the XML strengths and limitations when applied to the WebEDI.

1.1 The XML use in WebEDI and its limitations

XML has being increasingly used as a standard for data interchange for applications over the Web. Its widespread adoption has particular significance for enterprise application integration, which is the case of the WebEDI application. It is important, in this context, to understand the strengths and limitations of XML (Yee, A.- eai.ebizq.net).

1.1.1 XML's Strengths

1. ***Powerful meta language***. Provides a mechanism by which other markup languages can be developed for specialized needs or business domains.

Examples are: Chemical Markup Language (CML) and Artificial Intelligence Markup Language (AIML).

2. ***Simplicity***. The document structure and contained data are usually readable and easy to understand.
3. ***Separation of content and presentation format***. The document separates its content and presentation information clearly.
4. ***Common open standard***. It has no adherence to proprietary technologies such as browser, editor or interpreter. In contrast, it has been adopted across multiple industries.

XML has limitations with regard to application integration that are relevant to enumerate.

1.1.2 XML's Limitations

1. ***Inefficiencies of text-based documents***. The document representation in text files can be very large due to repetitions of data structures. This may be inefficient when transferring the document over the network.
2. ***No data transformation facilities***. XML standards have been developed for business-specific exchange formats applied in the B2B space, such as RosettaNet (www.rosettanet.org). However, the adoption of such standards is still not a reality and the transformation of data among disparate systems is necessary. XML is simply one of many data formats that exist; the transformation of data from one format to another is still a primary challenge.
3. ***Absence of content-based routing***. The automation of the integrated business process requires content-based routing and rules. The XML file does not contain the required content and business rules to automatically deliver the document to the proper destination. This work has to be done by an extra system.
4. ***Limited semantic interpretation***. XML provides the ability to create specialized tags that describe a particular entity or behavior; however, the semantic interpretation of what the data represents is outside the scope of the XML document. The meaning of tags and data within the document has to be agreed among the integrated parties.

The use of the XML can be tricky even when applying its strengths. For instance, the use of an important characteristic, such as flexibility, can lead to problems when uncoordinated efforts to design schemas using XML produce incompatible results. It is easy to create XML schemas that are only usable in a single application. This is typical when competing industry groups develop different schemas for all members of that industry.

Integration brokers are specialized in implementing a coordinated integration effort and to complement and compensate the given XML limitations listed above. The cited limitation 1 (Inefficiencies of text-based documents) can be overcome by the use of binary files. The integration broker implements a parser based on the XML Document Type Definition (DTD) that is able to read the XML document, parses and compiles it into binary format that is transferred. The processing (parse and transformation) of those large documents may also be very CPU intensive.

Limitation 2 (No data transformation facilities) is not an exclusivity of the XML standard. The real world is full of different standards provided by different industry sectors, integration brokers or simple system developers. Using XML helps minimize these differences but does not eliminate them. The integration broker implements, at the hub, a transformation process that always includes its XML standard for the given transferred documents. The hub has services that implement the transformation of each given format to the broker and from the broker to the given format. This allows the hub to apply necessary data transformations that guarantee that the document, once in the broker XML format, has reached a certain level of standardization of its data structures.

In the case of limitation 3 above (Absence of content-based routing), the integration brokers are required to implement the document delivery by reading the document content and applying the appropriate routing business rules. The simple association among the involved exchange parties can be represented by the description of knowledge like: supplier X supplies the retail A and B. This information is not contained in the XML document; it is implicit to the document existence. But in the case of a many to many exchange, not all participants are willing to exchange data with all the participants. The deliver of a document has to be confirmed by the existence of a valid relationship among the companies contained in the document. Therefore, the broker shall represent this information somehow and somewhere.

The limitation 4 (*Limited semantic interpretation*) has direct impact on WebEDI documents exchange that requires knowledge about the partner's business rules such as:

1. The correct interpretation of data attributes.
2. The application of partners business rules when processing documents

Case 1 suggests that the information contained in a document has to be interpreted in the same way by both sides, the sender and the receiver. This requires the representation of the abstractions that complement the information about the document-contained data in order to assure its proper process. This is called the semantics of attributes. Examples are:

1. Tax – For both companies it is a numeric value but the sender considers Tax a value from 0 to 100 and represents a percentage. However, the

receiver considers Tax a value that represents a monetary quantity (the already calculated tax value).

2. Package type – Both companies have the same meaning for it, but the domain each one accepts is different. There should be an agreement among the companies to reduce the domain to a unique set.
3. Discount – Again both companies have the same meaning for the attribute but the range for each one is different. There should be an agreement on the range or a formula to appropriately convert the values.

There may be many examples like the ones above regarding the attributes meaning. Mercador implements a database where the schema for the attributes of each document from each company is represented. Before starting the companies operation, there is a system that confronts the schemas of both companies finding the unconformities like the ones above. After an agreement, the rules are stored to be evaluated during transformation.

Case 2 requires knowledge representation of business rules about the partner's relation to the specific document type. One example is that different buyers require different invoice information from the supplier. Another example is the negotiation conditions among partners that are given by groups of discounts to different buyers. Below are some practical examples of rules that have to be considered in such cases:

1. If attribute A has a valid value, then attribute B does not have to be informed.
2. If attribute A has a value less than X, then attribute B must have a value above Y.
3. If the buying quantity is above X, apply discount Z to all order items.
4. If buyer J belongs to the group L, then apply discount Z.

The integration broker typically solves the limitations of semantics by implementing the business knowledge. In this case, the broker becomes the container of the knowledge. Next session explores the alternatives to include semantics in the XML documents.

1.2 Including semantic in the WebEDI

The limitations stated above can be solved in many ways. Depending on the tools and software architecture, the B2B integration company (broker) may solve the necessary semantics by simple custom made solutions. However, the goal is having a solution that scales to an environment of many to many companies exchanging documents and where each company provides the necessary document semantics. The B2B partners should be able to understand the documents doing the integration automatically.

Following there will be an explanation of the pragmatic solution adopted by Mercador to solve the cited limitations. In addition, the new technologies, which are popping up to build a Web containing the semantics, will be presented. These technologies can be extremely useful to the B2B integration.

1.1.3 Mercador actual solution

1.2.1.1 Document routing

Mercador easily solves the document routing once it is a hub. The companies that make part of the Mercador hub always send documents to their partners through the Mercador hub. So the connectivity among the companies is solved by the knowledge the hub holds about the companies connectivity characteristics. Mercador implements in database the necessary information to route the messages from and to the companies. There are no direct connections among the companies to be controlled.

The companies that route documents using the Mercador hub are required to send the documents using one of the three different connectivity tools:

1. WebEDI application – Accessing the WebEDI application the user can do: Document typing – The sender types the document in a special form on the Mercador WebEDI application at the web site; Web site download/upload – The sender uses the Mercador web site to download/upload files over the Internet to the hub;
2. Mercador Client – The Client is an application that is installed at the company computing environment and that is used to send/receive files to/form the Mercador Hub using the Internet;
3. FTP service – The customers may also send or receive files from/to the Mercador hub using an FTP service.

In the first alternative, the sender is identified by the WebEDI user authentication at the Mercador site. In the second case, the user is identified by an authentication service provided especially for the Client and in the last case, the Mercador system does the FTP connection to the company FTP server.

The document receiver, in the case of document typing, is selected from a list of companies, and in the last cases it is collected from the sent content of the sent document. In any case, the document can only be routed to the addressee if there is a valid commercial relation among the companies. Thus, this is the first business logic that Mercador implements at the hub, the so-called *commercial relation* is maintained in database and identifies which companies are prepared to exchange documents. Figure 1 shows the ER model for the commercial relation. Note that for each type of document there must be a valid relationship. A company is only able to send documents to another one that has an explicit commercial relationship for the given type of document.

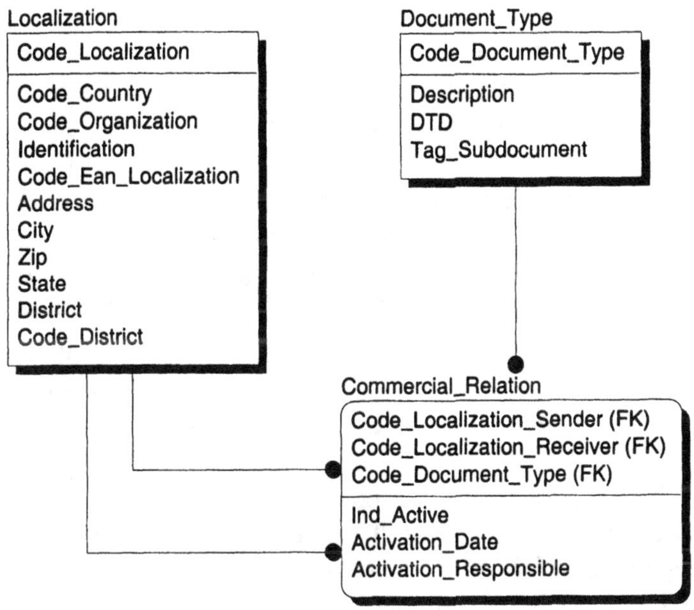

Figure 1 - Commercial relation

The connection using the Mercador Client requires no extra information at the hub. However, the Client has to be configured to connect to the hub. The Client is responsible for requiring the connection to the hub and sending new documents or requiring the available documents at the hub. First, the Client has to be configured with the appropriate IP address (or DNS) of the Mercador server hub and has to be configured to authenticate at the hub. Second, the Client runs at the company computer environment behind firewalls and it must be appropriately configured to cope with the necessary security restrictions of the company network. Usually the Client connects to the Mercador hub using the HTTP protocol on port 80 or the HTTPS protocol on port 443 not being a security problem for the companies.

The opposite occurs for the FTP service. The Mercador hub is responsible for having the login information on the company FTP server. This information is typically FTP server IP address, user and password. Besides this, the hub shall implement a schedule to connect to the FTP server according to the business necessities of the company. For instance, it must connect each 30 minutes from 6:00 to 15:00. Such kind of information belongs to the semantics of the customer business and has to be maintained from the Mercador hub. Once again it is maintained at the Mercador WebEDI database.

1.1.3.1 Semantic interpretation

The Mercador hub implements the integration among the companies by transforming the document from the sender format to the addressee format and by providing the connectivity among them. To accomplish the document transformation, Mercador specifies a standard XML format for each type of document. Whenever a document is sent using the hub, it is first transformed for the Mercador XML standard format. By doing this, Mercador reduces the complexity of mapping the divergences among many companies. The premise is that if a document can be transformed for the Mercador standard, it is possible to transform it to the format of the addressees that are already connected to the hub. To complete the job, the document already transformed for the Mercador standard is then transformed to the addressee format.

The WebEDI application server is responsible for coordinating the document transformation process. The integration server is responsible for performing the transformations and for the connection services for the Mercador Client and the FTP services.

Below are the steps during the B2B transformation of the document:

1. Sender schema validation – Used to validate the document structure. For each company and document type, it is implemented at the integration server a document schema that is used to validate the incoming document.
2. Sender transformation service – Transforms the document for the Mercador XML standard. Any rules regarding correct transformation of the sender document and the Mercador Standard is evaluated at this moment. The rules regarding the business relation between the sender and the receiver are also applied at this moment (for instance, the commercial relation is evaluated). Therefore, the document is transformed for the Mercador XML standard only if it applies all the necessary validation.
3. Receiver transformation validation – Transforms the document from the Mercador XML standard for the receiver format. Once in the Mercador XML format, the document is ready to be transformed for the receiver format. There should exist no restrictions or rules to be applied any longer.

A previous work on the details about the document attributes and semantics has to be done in order to assure that this two-step transformation will work. It is necessary to guarantee that the information transformed from one side to the other keeps the same meaning. An integration expert does this work in a careful mapping process supported by a visual tool for documents layout description.

When a new company joins the hub, it is supposed to be connected to hundreds of companies for exchanging documents. A complete mapping of its documents has to be done and their information has to be in agreement with all the companies it will integrate. It is not practical to do the complete mapping of each type of document it supports to each of the related companies by hand. The goal should be to reduce as much as possible the dependence of human interference on the mapping of divergences among the integrated companies.

This description of the document structure is done in a tool developed by Mercador called DSD – Document Structure Descriptor - where the layout of each company is described in a visual manner. This mapping starts by describing each document attribute in order to model the document structure, the attributes types and possible values associated to the document. Figure 2 shows the ER diagram for the DSD.

The Mercador integration expert constructs this description. Once the description is finished, the DSD tool generates a divergence report from the new description and the Mercador standard description. It also generates a divergence report from the new layout and the layout of the companies that the new one will integrate. These reports guarantee that the attributes in both companies have the same type, valid values and the same meaning. The fact that the attribute is related to an attribute in the Mercador standard, guarantees the meaning.

Any necessary modifications in the layout of the new company have to be done in agreement with the company to be integrated. Usually, the number of divergences decreases as the company integrates with more companies. Once the divergences are solved, the DSD tool is used for automatically generating a XML schema, which is used to validate the attributes of the new file layout on the integration server at runtime.

The DSD tool solves the correct interpretation of the data attributes among companies (limitation 4, case 1) but is not able to solve more complex business rules (limitation 4, case 2). Actually Mercador solves business rules by hard coding them in the WebEDI business components and transformation services. But this solution is not acceptable because it has poor scalability and hard maintenance. The next session explains some alternatives to implementing the remaining semantics.

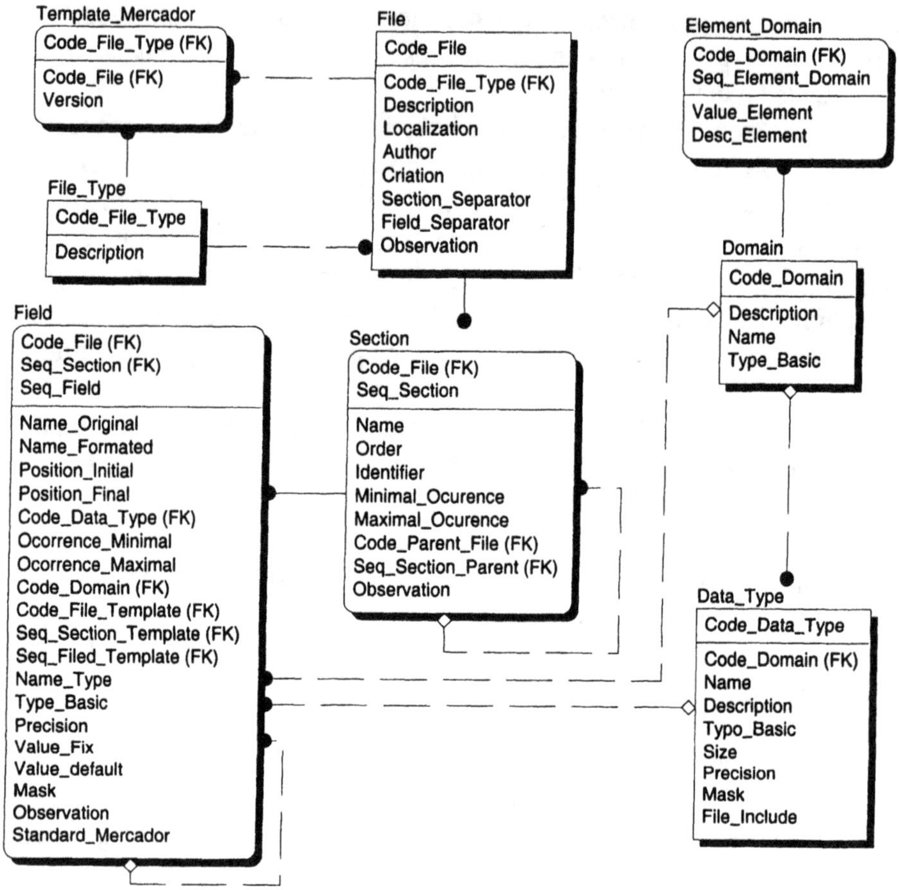

Figure 2 – Document Structure Descriptor ER diagram

1.3 Conclusion and Possible future solution

It is possible to improve the DSD tool to include the description of more sophisticated semantic rules. Those rules could be exported to the WebEDI application in form of integration services, business components, or even be stored in the database to be evaluated whenever necessary.

A promising way of implementing such semantics is the use of the concepts developed for the Semantic Web (Berners-Lee T et al. 2001, Henry, K. 2002). The Semantic Web is the concept of having documents containing semantics on the web and which can be interpreted by Web enabled mechanisms such as agents or web services. The idea is that the information on the web is given explicit meaning. Thus, it is possible to build automatic information processing and integration.

As already explained, XML is useful for defining customized tagging schemes. Complementarily, the RDF's (Resource Description Framework – RDF, 1999 – www.w3c.org) is a flexible approach to representing data giving it semantic meaning. The RDF Schema is a lightweight language to provide basic structures such as classes and properties. RDF is very simple, being very similar to a basic directed graph. It defines how to describe the relationships among resources in terms of named properties and values. The RDF properties represent traditional attribute-value pairs and may also represent relationships among resources. Besides this, RDF is made available on the web by using a serialization syntax based on XML. RDF represents the abstract model and XML provides the concrete textual representation of the model. There are several ways to represent the same RDF data model in XML.

Nevertheless, RDF should include some extra capacities such as data types, consistent expression for enumerations, etc. For the sake of having a more expressive language, a Web ontology language has been defined given the XML and RDF capabilities. The Web Ontology Working Group (WebOnt – www.w3c.org; Clark, K. G. 2002) has been given the task of developing an ontology vocabulary for use in the Semantic Web. This vocabulary should allow the explicit representation of terms and the relationships between entities in this vocabulary. WebOnt intends to create the standardized markup language within which users can formally define specific ontologies for use on the Web. For this purpose, WebOnt is standardizing the DAML+OIL (DAML+OIL, 2001 – www.w3c.org) ontology language.

DAML is the DARPA Agent Markup Language, which is a simple language for expressing more sophisticated RDF class definitions than permitted by RDFS. The DAML group has also added characteristics from the OIL (Ontology Inference Layer), which is a language that uses constructions from the frame-based AI to provide a more sophisticated classification mechanism. (Ouellet R., Ogbuji O., 2002)

Given such knowledge representation language based on XML, the necessary semantics for the WebEDI solution can be completed in a straight-forward way. The SDS tool may still be used to facilitate the building of the knowledge. The whole information about the documents could be stored in a XML file containing the document XML schema to represent the document data structure and the DAML+OIL language definitions to represent the necessary semantics in the same file. In case of storing the complete knowledge into files, the SDS tool would have to be able to do reverse engineering on the persisted files. Storing the knowledge on database would then be optional.

REFERENCES

1. Berners-Lee T, Hendler J. and Lassila O. The Semantic Web. Scientific American, May 2001.
2. Berners-Lee, T. Business Model for the Semantic Web. (www.w3c.com)
3. Clark, K. G. If Ontology, Then Knowledge: Catching Up With WebOnt. May 01, 2002. (www.xml.com)
4. DAML+OIL (March 2001) Reference Description W3C Note. (www.w3c.org)
5. Henry, K. Predicting how Ontologies for the Semantic Web will evolve. Communications of the ACM. February 2002 Volume 45, Number 2.
6. Ouellet R., Ogbuji O. *DAML Reference.* May 01, 2002. (www.xml.com)
7. Ouellet R., Ogbuji O. *Introduction to DAML: Part I.* January 30, 2002. (www.xml.com)
8. Requirements for a Web Ontology Language W3C Working Draft - 07 March 2002. (www.w3c.org)
9. Yee, A. Using XML in Application Integration (eai.ebizq.net)
10. Resource Description Framework (RDF) - Model and Syntax Specification. W3C Recommendation 22 February 1999.

Chapter 19

INSTITUTIONAL WEBSITES PERSONALIZATION USING MACRO AND MICRO USER PROFILES

Paulo Sergio Rodrigues Lima and Marcelo Soares Pimenta

Abstract: Institutional sites usually have a great volume of documents and a heavy navigation structure, requiring complex users browsing behavior. In order to allow users to find quickly and easily some information relevant to their usage it is imperative rethinking the structure of the site and the design of each individual page with relevant links to information concerning each user role, i.e., to be able to customize which information is more adequate for each user. This paper presents a personalization approach specifically designed for institutional sites, supporting personalization in two complementary levels, called respectively macro and micro user profiles. Basically, the macro user profile aims to model features related to organizational roles that are common to groups of users within the same organizational unit (for example, department, division, etc) whereas the micro user profile aims to concern preferences of specific individuals. The paper also presents the main aspects of our personalization approach, step by step explanation of an associated personalization engine and some preliminary results of an actual case study using the prototype built for its validation.

Key words: Institutional Web Sites, Personalization, Macro and Micro User Profiles.

1. INTRODUCTION

In the last years, the number of institutional Websites is growing exponentially: more and more enterprises and institutions have made available a lot of Web-based information and applications (WBI&A) for their suppliers, partners, clients, managers and employees. In fact, WBI&A need only a browser to user access, independently of user location. In addition, WBI&A provide an unified and at same time dynamic way to connect enterprise information systems, use web mail tools, obtain enterprise documents, etc. Particularly, institutional sites usually have a great volume of documents and a heavy navigation structure, requiring a complex users browsing behavior. In order to allow users to find quickly and easily some information relevant to their usage it is imperative rethinking the structure of the site and the design of each individual page with relevant links to information concerning each user role, i.e., to be able to customize which information is more adequate for clients, which is more adequate for suppliers, which is more adequate for managers, which is more adequate for employees, and so on. This is a great challenge: how to make available all enterprise WBI&A according to each kind of user in a simple, fast but adequate way [3].

Institutional sites are a very promising field for applying personalization techniques because they are composed of a great volume of high-diversity information pages, and significant number of well-defined organizational roles that can be used to characterize user profiles. The use of personalization techniques aims to improve the Web site usability since a personalized site automatically customize the interface considering the user profile [7] and each user may think that the site was designed specifically for him/her [4]. However, despite their importance, institutional sites are not so often studied.

This paper presents a personalization approach specifically designed for institutional sites, supporting personalization in two complementary levels, called respectively macro and micro level. It is structured as follows. Section 2 discusses some aspects of personalization based on user profiles and introduces the macro user profile level and the micro user profile level. Section 3 resumes the personalization process. Section 4 describes the personalization approach, illustrated by means some actual examples extracted from a real case study where our approach was applied. Some preliminary results of its evaluation in actual usage conditions with actual users are discussed in the section 5. Finally the section 6 presents the conclusion.

2. PERSONALIZATION AND USER PROFILES

The word profile comes from Latin "*pro filare*" that means to map in general way, and can be considered the outline of a set of object or people characteristics [1].

For general purposes, typical characteristics of user profiles include age, scholarship, profession, etc. However, if we focus on website-related characteristics, more specific and sometimes more technological information are needed like for example browser type and version, operating system, IP address, etc. E-commerce sites usually register behavior pattern (last bought products, preferred music styles or book subjects, etc) of a visitor in order to generate dynamic pages focused on its interests and preferences or to offer discount or special promotions in a future visit.

In addition, during user profiling for institutional websites we may be concerned with some user role or job information related aspects like organizational units hierarchy, organizational function, job related activities, information access policy, etc.

We assume that an important goal for institutional websites is to provide quick and easy access to applications related to user's organizational unit and to information concerned job activities. There is an evident progress in recent years with respect to aesthetics of sites and easiness of building websites with the availability of simple (commercial or not) WYSIWYG tools [6]. Despite this progress, even experienced users have many difficulties in finding information over web due bad design. Personalization leaves users from time-consuming searches for daily-used information. The goal of customized content is to focus on information and links directly related to a specific role or a specific activity. Obviously, any other kind of information related to other activities or roles is accessible by browsing. In summary, direct access is provided to content directly related to daily work activities and complementary non-frequently-used content may be obtained by browsing.

3. PERSONALIZATION PROCESS: OVERVIEW

Before applying personalization techniques at institutional websites, it is very important to define clearly: a) the personalization goals; b) the user groups and institutional roles; c) the appropriate personalization approach to be applied to the institution and d) the personalization level desired. Next these aspects are discussed in more details.

Defining the personalization goals. In order to a better choice of personalization strategies and techniques, the personalization goals should be defined. Our assumption is: in institutional sites the main goal is to provide quick and easy access to information concerned user job activities and applications related to users' organizational unit. For achieving such goal, the site structure and content should be rearranged.

Knowing the institution and modeling users. It is important to know the user as well the institution organizational structure before personalization. The basic process that allows personalizing the attendance to the user is to know more about him [9]. Models can be used to predict the user behavior, to diagnose his errors and to assist him, and they can help the designer in the personalization process [5].

In institutional sites the designer can begin studying the institution organization in order to group the users with the same interests according to the job inside the enterprise organization chart.

A good user comprehension is essential for a good user community classification, and it can contribute in the personalization process, providing:

- A clear view of institution culture;
- Experience and institution functional knowledge;
- Knowledge of process, resources and system's institution.

Selecting personalization levels. After to know the institution structure and her users, the designer need to select the personalization levels he wants to apply to the site. These levels can begin in a generic way, from the user groups creation with common interests up to a detail level that provide an individual preferences of each user.

Defining content and navigation structure according scope and user profiles. For each particular user profiles defined for the institution, the designer should define a particular content and a particular navigation structure, i.e., the designer should select which pages (or which frames) and which links are enabled and visible for such particular user. The idea is to create sub-sites based on each user profile identified and to make standalone tests with the users by profile groups. After the validation of each sub site, these contents and links can be stored in the database for further usage according to the user profile.

Elaboration of navigation map by profile. After defined the content that will be compose the profiles, the designer can elaborate the site map concerning each user identified and validated profile, thus, this map can be used by users and always will be linked to user profile.

4. THE PERSONALIZATION APPROACH

Since we are specially interested in providing customized information for each organization role, we adopt a variation of previously and well known established profile-based personalization mode. Typically, in traditional approaches the elements displayed are chosen according to the data concerned to the **one** user profile. Our personalization approach makes use of **two** complementary user profile levels: the first, called **macro profile**, is based on user group and the second, called **micro profile**, is based on individual preferences of each user.

4.1 Macro Profiles

We assume that typically users are very attached to WBI&A of their organizational units, and consequently we can identify a profile with common features applied to each unit member. This group profile is called **macro profile.** The macro user profile aims to model features, characteristics and job-related relevant information related to organizational roles that are common to groups of users within the same organizational unit (for example, department, division, etc) or working in the same project. In addition, the macro level is mandatory and it allows to adapt website defining both content and navigation according each users group or role.

It is not necessary form filling by users for the macro profile definition. In fact, forms usage usually repulses a user from the site and it only will be achieved after an established reliance with the site [12].In our approach, the user identification is obtained by means the authentication procedure (logon) of the site (or institution network). After the validation of user Id, the content (and links) concerned to the macro user profile will be assembled and displayed.

To exemplify the macro user profile usage, lets consider a organization chart (see figure 1). Users from National Administrative Department and the users from National Informatics Department and National Financial Department will see different website content considering they are members of different departments.

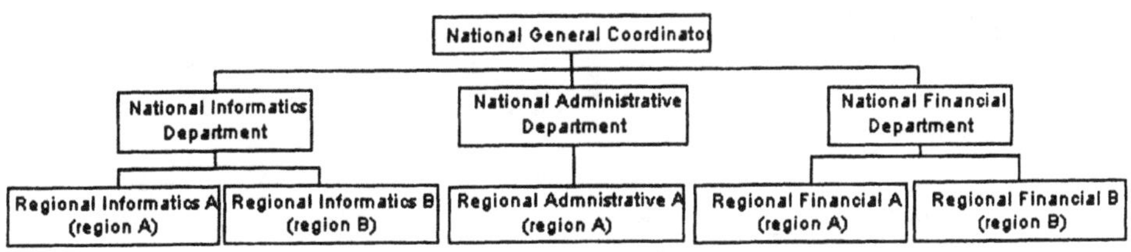

Figure 1. An Enterprise organization chart

The approach was developed in order to make available page layouts and contents totally dedicated and oriented to specific activities of each department. The user interface and content definitions are stored in a database allowing that the site management be decentralized: each department can has an own site management providing his own content.

The layout of personalized Web pages are pre-defined as composed by five main sections: header – including search mechanism (top), messages (bottom), personalized vertical menu (left), department topics (center), general news (right).

The figure 2 shows one interface generated by the personalization engine using this predefined layout. Basically, it shows the website as it is displayed to an user (called Mr. Paulo Lima) associated to user profile "Informatics Department Employee" - an employee of a Regional Informatics unit of Pará Brazilian State (region A in the organizational chart in the figure 1). The content may change in the Area 1 and Area 2, depending on macro user profile.In this view, the section named **Area 1** displays all the links available to Mr. Lima´s usage. Clearly, the decision of link availability (or not) is totally based on macro profile specification. Additionally, the section named **Area 2** contains selected news considered relevant for members of Informatics Department (ID). In the bottom of **Area 2** we can note two links: the left link is a "more"-like link (see "Ver mais destaques") in order to scrolling the ID news; the right link (see "Destaques da Coordenação de Informática") links to upper-level (see figure 2) relevant news, i.e., news from upper-level organizational unit named "Coordenação de Informática". The Area 3 concerns micro level definition and it will be explained later.

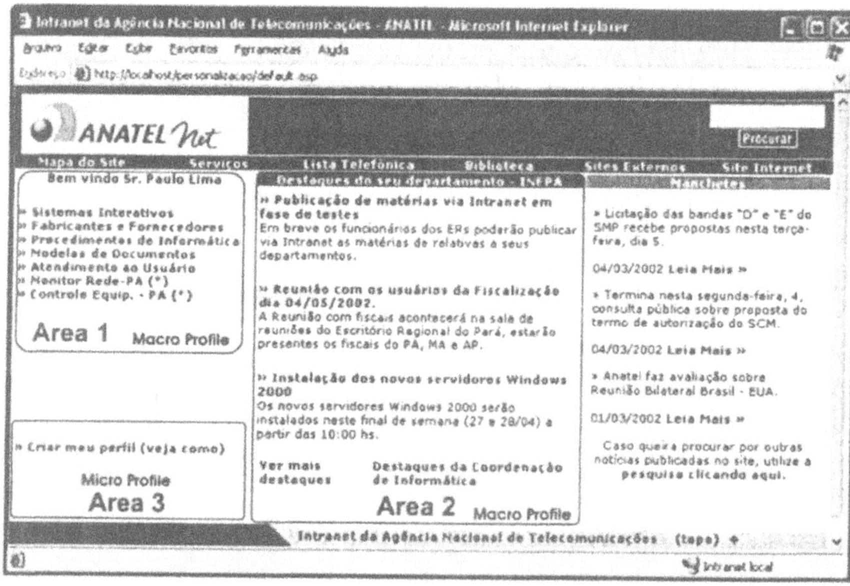

Figure 2. Macro Level Personalized View – Pará Regional Informatics Department Employee

By the other hand, another user – for example associated to another user profile like Regional Informatics Manager - has other macro level personalized view. In particular, some links displayed are distinct, and some of them may be displayed in a new order arrangement, defined in a more appropriate way to this different user profile.

Considering that for most of institutions we can use an organization chart to represent the organizational unit's hierarchy, our personalization approach makes use of this feature, exploring the inheritance concept: each organizational unit can to inherit the upper-level unit's content. For better understanding the figure 3a shows an example of hierarchy, specifying four-levels hierarchy and relationships between the organizational units.

The simple inheritance hierarchy concept is implemented when an lower-level organizational unit inherits content from upper-level units. The figure 3b illustrates an inheritance example using the hierarchy suggested in the figure 3a. In this example the user linked to organizational unit F will inherit the available links in the units A (link A1, link A2, link A3), B (A1,A2,A3 plus B1,B2), and D (A1,A2,A3,B1,B2 plus D1, D2), respectively. In other words, he/she can browse all pages containing inherited information.

The content related to each macro profile is defined by the site administrator. Each department defines a set of "relevant" content, that is stored in the database. The personalization engine "recognizes" the user by means the authentication procedure (logon) and retrieves the appropriate content.

Obviously, the content associated to each macro profile cannot be modified directly by the user, because the content is shared by many users in order to achieve some general department goals. Content changes related the macro profiles are made always by the website administrator.

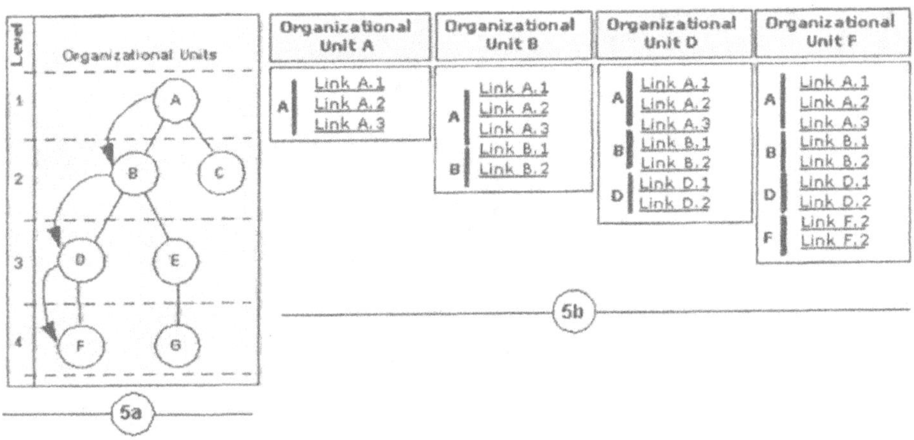

Figures 3a and 3b. Organizational Units hierarchy and OrganizationalUnits Links inheritance with the macro user profile

4.2 Micro Profiles

In addition to macro level group-oriented personalization schema, the user may create his/her individual profile, called **micro profile**. The user can configure this micro profile any time, and this configuration is optional: the user himself/herself decides about his/her personal preferences related to bookmarking, schedule and virtual disk facilities.

To enabling the micro profile the user has to select the link named **making my profile** that appears usually in Area 3. Three options may be chosen individually by the user (see figure 4). These options are:

My favorites: allowing to include her favorite links (with a short description) in a bookmarking list.

My notebook: allowing to manage personal, group and corporate events in order to control individual calendaring and scheduling.

My documents: allowing the storage and further retrieval of files (standards documents, manuals, memos, reports, etc.), independently of user location.

The user enables the options associated to the micro profile selecting the link "Create my profile" (see "Criar meu perfil (veja como)" in the area 3 at figure 2). These options will be displayed in his/her home page by means the pop-up window "Making my profile", chosen in the option "Making my profile". The figure 4 illustrates a simple situation where the pop-window is displayed and the options "My favorites" and "My documents" are enabled. The user can modify his/her options any time by selecting the option "Change my Profile".

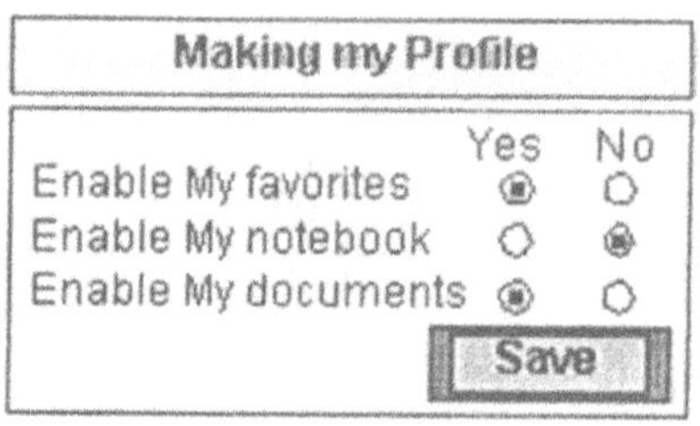

Figure 4. Enabling the micro profile

4.3 Personalization Engine: Architecture and step by step behavior

The approach described above is supported by a personalization engine, a software component which is responsible for providing all functionality related to personalization and user profile treatment. By didactical purposes the personalization engine behavior is presented in sequence of steps. Figure 5 illustrates all steps, considering an successful personalization situation (i.e. existent exceptions handling is not discussed here). The sequence is: 1)The user is identified either by net login or via site login; 2) The web server gets the user id (for example, the net login); 3) The personalization engine recovers the user id by web server scripting execution; 4)The personalization engine sends the user id to database server by web server scripting execution; 5) The database server does the query in the macro profile table using the user id; 6)The database server recovers the macro profile content; 7) The database server does the query in the micro profile table using the user id from the user identification; 8) The database server recovers the micro profile content; 9)The database server sends to the personalization engine the personalized content (micro and macro profile); 10) The personalization engine provides the personalized page by web sever scripting execution; 11) The personalized page is showed to the user.

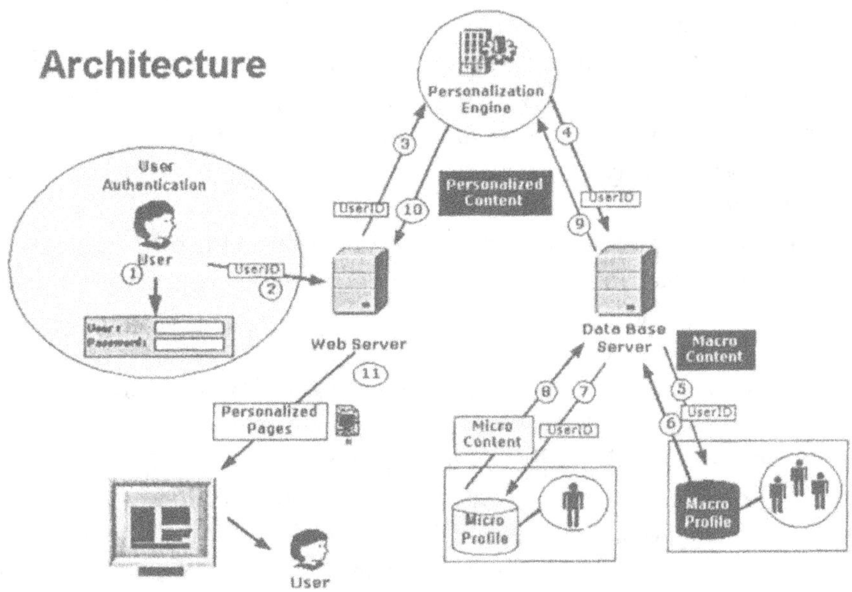

Figure 5. Personalization engine architecture

5. CASE STUDY: EVALUATION AND DISCUSSION

Our personalization approach has been applied in actual conditions at Brazilian Telecommunications National Agency – Anatel, even the definitions are made aiming to turn this method so generic as possible.

A prototype of personalization engine was built and the personalized Anatel website was made available for usage by actual users. Taking into accounting some well-known websites subjective evaluation methods from HCI and Ergonomics literature [2,7,10,11,13], we apply user-satisfaction-oriented interviews in two different moments: i) before the beginning of personalization activities, concerning the non-personalized site; and ii) after finished the final personalized site, according our profile-based approach.

In order to capture rich comments and critics (and also to avoid non-recorded opinions), users was invited to answer all questions *in loco*, immediately after site usage. The interview contains few questions (4), takes only a few minutes to answer, its data analysis is quick and easy, and the whole process costs very little.

The questions were conceived to identify:

a) degree of correspondence between principal homepage menu items and principal user job activities; **Question A**: "Are the menu items available representing the most frequently activities you execute during your job?"
b) convenience of information ordering and structuring for user job activities, clearly to verify if the most frequently used information are easily visualized; **Question B**: "Is the content structured in a appropriate way concerning information you need for your activities?"
c) adequacy of site navigation structure, asking the user if information is *easy-to-find*. In fact, navigation complexity increases very quickly on larger web sites. broken links, ghost pages, long paths and complex navigation are frequently reported on usability testing as symptoms of poor navigation design. **Question C:** "Are the links available for website navigation well organized?"
d) if users would wish a job-oriented personalized view of site (only in the first "before-personalization" interview); **Question D:** "Would you wish a job oriented personalized view of site?"
e) if users are satisfied with the personalized approach implemented (only in the second "after-personalization" interview). **Question E:** "Are you satisfied with the new personalized site?"

The table 1 summarizes the quantitative percentages of interview results. Questions A, B, C, D and E are the questions presented above. The columns "Before Personalization" and "After Personalization" refer to two different moments of interview.

Although a simple examination of the satisfaction interview results seems very successful (80% of our users are satisfied !! – see Question E), the fact is that 20% of your customers are indifferent, i.e. *not completely satisfied.* Perhaps their indifference means they aren't angry enough to be truly "dissatisfied" but they didn't feel good enough to consider themselves satisfied. For this, in addition to the answers, the interview gives users the opportunity to provide comments and we have to examine the comments seriously.

Table 1. Percentages of user-satisfaction-oriented interviews

	Before Personalization			After Personalization		
	Yes	No	Indifferent	Yes	No	Indifferent
Question A	21%	79%	0%	80%	0%	20%
Question B	29%	71%	0%	60%	0%	40%
Question C	21%	58%	21%	80%	0%	20%
Question D	79%	0%	21%	-	-	-
Question E	-	-	-	80%	0%	20%

People often comment about more than one thing, so it was a good practice to break each comment into positive and the negative aspects. We would commonly see many of the same things identified as issues in the quantitative portion of the interview like effective improvements in website navigation schema and information structure organization. Often, the comments provide you with more specific information about how to fix problems yet identified. However, it is not uncommon to uncover completely new issues by reviewing comments. The main defeat identified by the users in the personalization process was the **lack of user involvement** in the personalization policy and practice definition. This specific comment is very helpful in understanding the users needs and considering them in future modifications of personalization process. It will be also taking in account in future development practices and future interviews will specifically ask about such issue as well.

6. CONCLUSION

As institutional websites become more sophisticated, website administrators need to deal with more efficient technologies and approaches in order to face their complexity and diversity. One example is the need to deal with the personalization, which is an increasingly important topic.

Our approach for personalization based on user profiles is an example of a very promising technique because it enables the management of a great volume of institutional job-related information pages. Consequently, the relevant information is automatically customized for each user depending on his/her organizational role (macro profile) and individual preferences for configuration (micro profile). However, we acknowledge our personalization approach may be used to complement other personalization useful and well-known approaches and not to replace them.

Our personalization approach has been applied in actual case study and the results obtained from evaluation methods until now express the relative success of our work but we intend to continue our research towards better improvements. It is better to acknowledge we can involve employees affected by the website in the improvement process itself.

In a further work, for example, we intend to develop an authoring tool that allows each user group administrator to define the content of his group with no direct interference of site designer or administrator. More (qualitative and quantitative) assessment experiments are also needed to evaluate the scalability of our approach.

REFERENCES

1. Davisor. Whitepaper: "Profiling Basics". 04/03/2001. Available at: http://www.davisor.com/main/whitepapers/profbasics.html
2. Dong, J.; Martin, S. Interactive Usage of Customer Satisfaction Surveys to Assess an Evolving Web Site. Proceedings of the 6th Conference on Human Factors and the Web, Austin, TX, USA, June 19, 2000.
3. Greening, D. R. Tracking Users. Web Techniques, July1999.
4. IBM - HIGH VOLUME WEB SITE TEAM. Web Site Personalization. 2000. Available at: <http://www7b.boulder.ibm.com/wsdd/library/techarticles/hvws/personalize.html>.
5. Kass, Robert, Tim Finin., A General User Modeling Facility, Proceedings of CHI'88 - Conference of Human Factors in Computing Systems, Washington D.D., May, 1988.
6. Lynch, P. J; Hortson, S. Web Style Guide: Basic Design Principles for Creating. Web Sites. Yale Univ Press. 1999. 164 p.
7. Nielsen, Jakob. Designing Web Usability: The Practice of Simplicity. New Riders Publishing; 2000.
8. Perkowitz, M.; Etzioni, O. Adaptative Web Sites: An AI Challenge. In: International Joint Conference on Intelligence, 15., 1997. Proceedings[S.l.:s.n.], 1997.
9. Perkowitz, M.; Etzioni, O. Towards Adaptative Web Sites: Conceptual Framework and Case Study. In: International World Wide Web Conference, 8., 1999. Proceedings... [S.l.:s.n.], 1999. p. 167-180.
10. Rubin, J. Handbook of Usability Testing: How to Plan, Design and Conduct Effective Tests. John Wiley & Sons. New York. 1994. 330 p.
11. Scholtz, J.; Laskowski, S.; Downey, L. Developing Usability Tools and Techniques For Designing and Testing Web Sites. 4th Conference on Human Factors & the Web, June 5, 1998,Basking Ridge, USA (1998).
12. Udi Manber, Ash Patel, and John Robison, Experience with Personalization on Yahoo!, Communications of the ACM, special issue on Personalization, 43 (8), August, 2000.
13. Winckler, M.A.; Pimenta,M.S; Palanque, P.; Farenc, C.; Usability Evaluation Methods: What is still missing for the WWW? In: Proc. of 9th International Conference on Human-Computer Interaction, HCII2001, New Orleans USA, August 5-10, 2001.

SECTION 7
Advanced Platforms and Grid Computing

Chapter 20

THE GRID: AN ENABLING INFRASTRUCTURE FOR FUTURE E-BUSINESS, E-COMMERCE AND E-GOVERNMENT APPLICATIONS

Fabrício Silva, Hermes Senger

Abstract: In this paper we discuss the utilization of grid computing platforms as an enabling infrastructure for e-commerce, e-government and e-business applications. First fundamental concepts related with grid computing are presented, as well as a description of the evolution of grid computing platforms from pioneer projects to third generation systems. Then we identify a set of services that can be provided by grid computing platforms which will be fundamental for future e-commerce, e-business and e-government applications, such as database access and integration and knowledge discovery services. We also discuss why the grid may be the platform of choice for providing such services in a geographically distributed area.

Key words: Grid computing, e-business, e-commerce, e-government

1. FUNDAMENTAL CONCEPTS

A Grid is a system that coordinates heterogeneous, geographically distributed resources that are not subject to centralized control, using standard, open, general-purpose protocols and interfaces to deliver non-trivial qualities of service [17]. Grid platforms enable the sharing, selection and aggregation of a variety of resources including supercomputers, servers, workstations, storage systems, data sources and specialized devices that are geographically distributed and owned by different organizations [18]. Grids became possible in the last few years due to the great improvement in performance and the reduction in cost of both computer networks and microprocessors

The grid is characterized by four main aspects [2]:

- *Multiple administrative domains and autonomy.* Grid resources are geographically *distributed* across multiple administrative domains and owned by different organizations.
- *Heterogeneity. A grid involves a multiplicity of resources that are heterogeneous in nature and will encompass a vast range of technologies.*
- *Scalability. A grid may grow from few integrated resources to millions. This raises the problem of potential performance degradation as the size of the grid increases.*
- *Dynamicity or adaptability. In a Grid, resource failure is the rule rather that the exception. Resource managers and applications should adapt themselves in order to use the available resources efficiently.*

There has been considerable effort over the past few years in the development and implementation of platforms for grid computing, such as Condor [30], Legion[26] and Globus[25], to name a few. Condor supports the execution of applications making use of a single machine, pools of machines or machines already in a grid. Condor leverages the exploitation of idle cycles to provide a number of relevant services, such as task checkpoint and task migration. Legion implements an object model for grid computing. Within Legion, everything is represented as objects, from users, programs, storage, processors to special devices. Globus implements a set of infrastructure services to support the execution of applications on computational grids. The collection of services includes discovery and allocation of resources, security, information services, file transfer, and others. The Globus project has strongly influenced the evolution of research on grid computing, by proposing a set of standard mechanisms for creating, managing and exchanging information among service entities.

Originally the Grid has focused on the integration of computing resources, but as more of the possibilities for grid solutions have been realized this resource view has become complex. The same computing resource needs to be accessed and viewed very differently depending upon the application. It becomes much easier viewing the Grid as a collection of services and abstracting away from the underlying resources [31]. Beyond that, one can view the grid as a platform capable of integrating resources in different levels, from computing resources to metadata and knowledge resources.

One of the main reason for that change was the recent proposition of the Open Grid Services Architecture (OGSA) [22], and the associated Open Grid Service Infrastructure specification [38]. The Open Grid Services Architecture (OGSA) has been proposed as an enabling infrastructure for systems and applications that require the integration and management of services within distributed, heterogeneous, dynamic "virtual organizations"[18]—whether within industry, e-science, or e-business. The Open grid service architecture define conventions and WSDL interfaces related to grid services, a potentially transient stateful service instance supporting reliable and secure invocation, lifetime management, notification, policy management, credential management and virtualization.

In the following section we describe in more detail the evolution of grid computing, from the first pioneer projects to the definition of the OGSA specification. Section 3 discusses the application of grid computing in e-business, e-commerce and e-government applications by describing services that can be made available through a grid platform. We believe that these services will play an important role in the infra-structure necessary for delivering the quality of service needed for e-commerce, e-business and e-government applications in the 21st century.

2. THE EVOLUTION OF THE GRID

In this section we provide a short overview of the evolution of grid computing, from its beginning around the late 1980s, until nowadays, focusing on the current perspective. Essentially, grid computing emerged from the need to share resources among individuals, and institutions, in a secure, flexible and coordinated way [2]. Although these main ideas are somehow present in the most of grid projects, there are some differences between what grid computing is meant to be in the past, and what are the main ideas which are currently driving its evolution. De Roure et. al. [12] depicts an evolution scenario in which grid projects can be classified into three generations, which we describe below.

2.1 The First Generation

From the beginning until the middle of 1990s, grid computing initiatives were driven to support a specific group of high performance applications with computational power. This is the first generation, and its major representatives include the FAFNER [15] and the I-WAY [19] projects. FAFNER was created as though a consortium led by Bellcore Labs., Syracuse University and Co-Operating Systems, and created a hierarchical network of web servers to coordinate a distributed effort for factoring big numbers. The main objective was testing the limits of security of the RSA encryption algorithm. The I-WAY project was started in 1995, to implement a high performance wide area backbone to connect seventeen U.S. high performance computing centers. In this project, attempts were made to provide a standardized solution for issues concerning security, scalability and heterogeneity, supporting the execution of large and complex applications. Both of these projects were led to provide computational resources to a range of high performance applications and influenced many other initiatives.

2.2 The Second Generation

The following years have witnessed a plethora of projects and initiatives launched to confront the challenges which feature grid computing systems, including those mentioned in section 1. Because grid applications can make use of a number of multi-institutional resources potentially ranging from dozens to thousands or millions, heterogeneity may become a big challenge. To cope with this, the middleware has emerged as a solution by implementing programming models which provide transparency from architecture details, such as network, operating system, computer hardware, programming languages and physical location. To enable scaling up to thousands or millions of geographically dispersed resources without degrading performance, applications were required to deal with higher latencies, and explore locality of resources they use. Also, because scalability involves traversing multiple organizational boundaries, much effort was issued to support authentication and trust issues. Furthermore, to cope with frequent resource failures, applications had to tailor their behavior to function without losing their functionalities and with maximum performance. Globus [21] and Legion [26] are representatives of second generation middleware for grid computing.

Globus [21] has been developed under an U.S. multi institutional project, to provide a software infrastructure that enables applications to handle distributed computing resources as a single virtual machine. Globus supports a set of basic infrastructure services with well defined APIs. Security, file transfer, resource location and resource management are examples of services implemented in Globus. It implements a multi-layered architecture, in which applications may be composed by aggregating lower level services provided by the Globus infrastructure, by specific tools, or by other applications. Most of these services have been realized in Globus Toolkit 2 (GT2), which is one of the most deployed software for building computational grids. Currently, Globus Toolkit 3 (GT3) provides a reference implementation for OGSA (see section 1) architecture and services, a standard created by the Global Grid Forum, which defines a set of Grid Services as specializations of Web Services to implement the grid.

Legion [26] is an entirely object-oriented system developed at the University of Virginia, which provides an infrastructure for building computational grids. Issues concerning scalability, heterogeneity, security, and resource management are addressed to provide a software infrastructure that enables the execution of applications that handle a great number of resources geographically distributed. Legion maintains the local administrative autonomy over the resource usage by letting local administrators to specify resource sharing policies. Legion supports a unique and persistent namespace for publishing resources and information to the applications. In 1998, the Applied Metacomputing Corp. was created to exploit Legion commercially and in 2001 it was relaunched as Avaki [1].

Furthermore, many other initiatives have been delivered to address specific issues or particular approaches related to grid computing. Some of them focus on handling from dozens to millions of multi-institutional resources. An example is Condor [13], a project started in 1985 at Univ. of Wisconsin at Madison as though a scheduler for idle resources for locally distributed computing system, which evolved to a grid solution to execute high throughput applications. Condor major features include detection and exploitation of computational resources, job management, and reliable executions through checkpointing, migration and restarting of failed jobs. Many other systems provide capabilities for resource management, and support the specification of resource scheduling policies, resource sharing policies, job requirements, execution monitoring, and fault tolerance. Examples include the Load Sharing Facility (LSF) [46] by Platform Computing Corp. [36], the Sun Grid Engine (SGE) [41], and the Portable Batch System (PBS) [42]. In addition, storage can be viewed as a resource. The Storage Resource Broker (SRB) [37] supports homogeneous access to a wide range of storage devices.

Peer-to-peer computing [8] provides a very plausible treat for scalability and fault tolerance issues through decentralization. With peer-to-peer, computers can share resources such as spare computing cycles, storage capacity, and databases, through the network without incurring in bottlenecks as it happens with the traditional client/server model. Some examples of P2P based distributed storage systems in the grid context include the FARSITE, the OceanStore, the Self-Certifying File System (SFS), and the PAST [12]. Also, web technologies have been used to build grid portals, which enable scientists and researchers of specific interest communities to access grid resources. A grid portal may provide user authentication, resource scheduling facilities, and access to remote information. Examples of grid portals include the NPACI HotPage (https://hotpage.npaci.edu/), the SDSC Grid Port Toolkit (http://gridport.npaci.edu/), and the Grid Portal Development Kit (http://doesciencegrid.org /projects/GPDK/).

2.3 The Third Generation

Although problems which are inherent to geographically distributed systems (e.g. scalability, heterogeneity, security and adaptability) are still motivating many research activities, a new vision of grid computing is emerging, which goes beyond viewing the grid in terms of its enabling technologies. Instead, a more holistic view of the grid is emerging as other evolution aspects became apparent. This new envisioning of the grid is much more concerned with requirements, i.e., the features needed to support known applications such as e-Science, and what features can enable new applications which were not initially targeted to run on the grid [18], such as e-Business and e-Government.

As more applications have been implemented and deployed, the need for reuse application and information components became evident. Also, service orientation and use of metadata are increasingly present in grid systems. By combining these two features, grid resources can be assembled in a flexible manner to compose grid applications. Metadata are needed to cope with the automated discovery of functionality and availability of a great number of heterogeneous and geographically distributed resources. It is important as grid scales beyond the human capacity to manage and assemble resources manually. In addition, service orientation allows independence from the programming model, because the only public information about services are their interfaces and public metadata. Together, these features leverage the opportunities for system integration.

The service oriented vision of the grid approximates to other related research communities, such as the Web Services and Service Oriented Architectures. The World Wide Web Consortium (W3C) are leading the specification and deployment of standards (e.g. SOAP, WSDL, UDDI and RDF) to support a service oriented approach. Other important developments provide support for the process level. The IBM's Web Services Flow Language (WSFL) [43] allow workflows to be viewed as services, and Microsoft's XLANG [44] supports transactions that involve multiple web services. There are also initiatives to cope with the design of web services systems, such as the Web Services Modeling Framework [45]. As one can notice, there is a common way for the evolution of web services and grid systems. The emergence of the OGSA architecture is an evidence of the converging interests of grid computing and web services communities. The OGSA specification is an initiative launched by IBM and the Globus Project, which are currently being led in the context of Global Grid Forum (GGF) [24], a community that aggregates over four hundred organizations and five thousands individuals in order to promote and support the development of grid technologies and applications. At the mainstream of this evolution is the concept of Grid Services, which are grid enabled specializations of Web Services. Grid Services are entities independently created and deployed by Virtual Organizations [18]. Thus, the specification and deployment of standards is a keystone which can leverage both interoperability and competition among applications, toolkits and base services, implemented by different organizations. Interoperability allows building grid systems by mixing, matching, and aggregating different and competing implementations. Such degree of integration can be though as a major characteristic of third generation, since it was not achieved during the second generation of grid systems.

Also, automation and dependability are important features of third generation grids. The multitude and heterogeneity of resources in a grid will prevent humans to efficiently manage resource aggregation and usage. Third generation grids must be able to discover, bind, and use a great number of geographically distributed resources. Thus, automated configuration capabilities based on widely available metadata must be realized. Also, the multitude and distribution of grid resources make failures to happen so frequently that it can be though as the rule, not the exception. To cope with this, applications must be able to recover automatically from malfunction. Also, changes in the status of grid resources are so prevalent that automated reconfiguration capabilities are needed. All of these capabilities can only be implemented whether pervasive and robust information services are supported, which can provide detailed information about grid resources and their current status. Some of these aspects are closely related to the IBM's envision of 'autonomic computing' [28]. This term alludes to the autonomic nervous system, which is capable to control heart beats and body temperature, freeing the conscious brain dealing directly with these and other important low level functions. Thus, third generation grid systems need self-organizing, self-healing, and self-optimization capabilities to free applications to concentrate in the business logic.

3. GRID COMPUTING IN E-BUSINESS, E-COMMERCE AND E-GOVERNMENT

In this section we describe some services that can be made available thorough the grid, and that will be fundamental for the deployment of the next generation of e-business, e-commerce and e-government applications. These services provide integration of different types of resources in a level of transparency beyond the capabilities of current technology. Examples of such services are: Database access and integration, Knowledge discovery services (Data mining), distributed transaction services, and several other services which can take advantage of features of third generation grid computing platforms, such as automated reconfiguration capabilities. In this section we will focus mainly in two services: database access and integration and knowledge discovery.

In the following we discuss the current status of the research related to the implementation of these services using the grid, and the advantages of the grid as the platform of choice for the deployment of such services.

3.1 Database Access and Integration on the Grid

One service of major importance for many e-business, e-commerce and e-government applications is a virtual, uniform access to a set of heterogeneous, geographically dispersed databases. The fundamental value proposition for a Grid in general, and a database access service in particular, is virtualization [14], or transparent access to dispersed data sources. For data intensive applications, the following set of transparencies are relevant [10]:

Heterogeneity Transparency - An application accessing a data source should be independent of the actual implementation of that source, so that both can evolve independent of each other.

Name Transparency - with grid sources, applications should not even *specify* data objects explicitly. Instead, data access should be via *logical domains*, qualified by *constraints on attributes* of the desired object.

Ownership & Costing Transparency - As far as possible, applications should be spared from separately negotiating for access to individual sources located on different domains, whether in terms of access authorization, or in terms of resource usage costs.

Parallelism Transparency- An application processing data on a grid should automatically get the benefits of parallel execution over grid nodes.

Distribution Transparency - Applications should be able to maintain distributed data in a unified fashion.

The level of transparency described above is far beyond the capabilities of the current data management technologies. Recently several sets of services for integrating databases into the grid have been proposed recently [39,10,9]. The objective is to provide at least a subset of the level of transparency just described.

The OGSA platform is a natural candidate for the deployment of such services. For instance, the nature of the Grid offers interesting new possibilities for areas such as distributed query processing [39]. Once a query has been compiled, distributed grid resources can be acquired on demand for running the distributed query execution middleware. The choice of which resource to allocate can be made based on parameters such as performance and price. Once the acquired resources are no longer necessary, soft state control can be used to automatically deallocate resources, making them available for other queries. In fact, there is currently a great deal of effort to propose and implement a distributed database access and integration service based on the OGSA and OGSI specifications.

In the international area, efforts towards this objective are concentrated in the Global Grid Forum [24] and on the OGSA-DAI project [32], the former financed by the British government.

3.2 Data Mining on the Grid

Knowledge discovery in databases is the non-trivial process of identifying valid, novel, potentially useful, and ultimately understandable patterns in data [23]. Data Mining (DM) is a step in this process that centers on the automated discovery of new facts and relationships in data. DM consists of three basic steps: data preparation, information discovery and analysis of the mining algorithm output [4]. All these steps exploit huge amounts of data and are computationally expensive. In this sense, several techniques have been proposed to improve the performance of DM applications, such as parallel processing [27] and implementations based on cluster of workstations [3].

Some recent work suggest that grids are natural platforms for developing high performance data mining services [7,27,33]. More specifically, Orlando et al. [33] describe the application of two data mining algorithms in the Knowledge Grid architecture [7]: the DCP Algorithm and the K-means Algorithm. However, other Data Mining techniques can take advantage of a grid infrastructure, as described, for instance, in [27].

3.3 Example of Utilization: an e-government Scenario

In this subsection we present a possible utilization scenario of the services just described, adapted from [10]: A government agency periodically mines patient records from various hospitals in a geographic defined area to detect biohazards. The patient records will be located in several, geographically dispersed, and heterogeneous databases. A grid-enabled database access service can deliver a union of the related records from those hospitals, qualified by symptom, for instance. This set of records will be used as input to a knowledge-discovery service on the grid that will mine the records, looking for patterns that would indicate the occurrence of a biohazard, such as the contamination of a river.

3.4 Autonomic Computing and the Grid

As mentioned in section 2.3, autonomic computing capabilities comprise important requirements which are driving the evolution of third generation grids. In this evolution scenario, grid services may be implemented as autonomic components capable to interact to each other [35]. Each service may be specified by a set of output services it can provide, a set of input services and resources it requires, and a set of policies to control its behavior and interaction to other services. In this scenario, self-configuration capabilities can be supported to simplify and reduce human interaction in configuration, version control, and maintenance tasks for complex e-business and e-commerce complex systems. Self-optimization capabilities allow services to be provided with better quality, by optimizing usage of resources and other services required. Also, service unavailability may represent significant loss for companies. By means of self-healing capabilities, a grid service could find a substitute for a required service or resource which has become unavailable, and reconfigure itself to go on working. In addition, self-protection capabilities can allow large and complex systems to self-defend automatically against malicious attacks and cascade failures. By taking self-healing measures and using early warning, systems can prevent from system wide failures.

Autonomic computing did not start with grid computing. However, many commonalities have been identified by their related communities. And finally, these new capabilities and functionalities which can be achieved by the cooperation between these two communities may significantly benefit e-commerce, e-business and e-government applications.

4. CONCLUSIONS

In this paper we presented fundamental concepts related with grid computing and the evolution of the grid as a platform aimed at the integration of resources at different levels. We discussed some services that can be made available through a grid platform and the importance of such services in an infra-structure capable of delivering the quality of service needed for e-business, e-commerce and e-government applications in the 21st century. We believe that while technology evolves it will become clear that other kinds of services will have the grid as the platform of choice for deployment, with the development of capabilities related to third generation grid computing platforms.

BIBLIOGRAPHY

1. Avaki, http://www.avaki.com
2. Baker M., Buyya R., Laforenza D. Grids and Grid Technologies for wide area distributed computing. Software – Practice and Experience 2002;32:1437-1466.
3. Baraglia R., Laforenza D., Orlando S., Palmerini P., Perego R., Implementation Issues in the Design of I/O Intensive Data Mining Applications on Clusters of Workstations. Proceedings of the 3rd Workshop on High Performance Data Mining, IEEE IPDPS'2000; Cancun, Mexico. Los Alamitos: Computer Society Press, 2000.
4. Berman F. From Teragrid to knowledge grid. Communications of the ACM 2001;44(11):27-8.
5. Bigus, Joseph P. *Data Mining with Neural Networks*. USA, McGraw-Hill, 1996.
6. Berendsen H. J. C., van der Spoel D., van Drunen R. GROMACS: A message-passing parallel molecular dynamics implementation. Comp. Phys. Comm.1995;91: 43–56.
7. Canataro M., Talia D. The Knowledge Grid. Communications of the ACM 2003;46(1):89-93.
8. Clark D. Face-to-Face with Peer-to-Peer Networking. Computer 2001;34(1):18-21.
9. Database Access and Integration Services Working Group: http://www.cs.man.ac.uk/grid-db/
10. DAIS Working Group. Services for Data Access and Data Processing on Grids. Available at http://www.gridforum.org/Meetings/ggf7/drafts/DAIS_DataServices.pdf
11. De Roure D., Jennings N.R., Shadbolt N.R. "The Semantic Grid: A Future e-Science Infrastructure". In *Grid Computing: Making the Global Infrastructure a Reality*, Berman F., Fox G., Hey T. Eds. Wiley Series in Comm. Networking and Distrib. Syst. West Sussex: John Wiley& Sons, 2003.
12. De Roure D., Baker M.A., Jennings N. R., Shadbolt N.R. "The evolution of the Grid". In *Grid Computing: Making the Global Infrastructure a Reality*, Berman F., Fox G., Hey T. Eds. Wiley Series in Communications Networking and Distributed Systems. West Sussex, England: John Wiley& Sons, 2003.
13. Douglas Thain, Todd Tannenbaum, and Miron Livny, "Condor and the Grid", in Fran Berman, Anthony J.G. Hey, Geoffrey Fox, editors, Grid Computing: Making The Global Infrastructure a Reality, John Wiley, 2003.
14. Ellis C., Gibbs S., Rein, G. Groupware. Some issues and experiences. Communications of the ACM 1991;34(1):38-58.
15. The FAFNER Web Page http://www.npac.syr.edu/factoring.html
16. Fayyad U.M., Shapiro G.P., Smyth P. "From Data Mining to Knowledge Discovery: An Overview". In: Advances in Knowledge Discovery and Data Mining, Fayyad U.M., Piatetsky-Shapiro G., Smyth P., Uthurusamy R., Eds.. MIT Press, 1996.
17. I.Foster. What is the Grid? A Three Point Checklist. *GRIDToday*, July 20, 2002. Retrieved October 2003 from http://www.gridtoday.com/02/0722/100136.html
18. Foster I., Kesselman C., Tuecke S. The Anatomy of the Grid: Enabling Scalable Virtual Organizations. International Journal of Supercomp. Applications 2001;15(3), 2001.
19. Foster I., Geisler J., Nickless W., Smith W., Tuecke S. Software infrastructure for the I-WAY high performance distributed computing experiment. Proceedings of 5th IEEE Symposium on High Performance Distributed Computing, 1997. Los Alamitos: IEEE Computer Society Press, 1997.
20. Foster I., Kesselman C. (editors). *The Grid: Blueprint for a New Computing Infrastructure*. Morgan Kaufmann Publishers, 1999.
21. Foster I., Kesselman C. Globus: A Metacomputing Infrastructure Toolkit. International Journal of Supercomputer Applications 1997;11(2):115-128.

22. Foster I., Kesselman C., Nick J., Tuecke S. The Physiology of the Grid: An Open Grid Services Architecture for Distributed Systems Integration. Presented at Global Grid Forum, June 2002. (Online Version http://www.gridforum.org/ogsi-wg/drafts/ogsa_draft2.9_2002-06-22.pdf)
23. Freitas A.A., Lavington S.H., *Mining Very Large Databases with Parallel Processing*. Kluwer Academic Publishers, 1998.
24. Global Grid Forum Web Page: http://www.gridforum.org/
25. The Globus Alliance Web Page: http://www.globus.org/
26. Grimshaw, A. and Wulf, W.. Legion: The next logical step toward the world-wide virtual computer. Communications of the ACM 1997;40(1):39–45.
27. Hinke H., Novotny J. Data Mining on NASA's Information Power Grid. Proc. HPDC 2000; Pittsburgh, USA. Los Alamitos; IEEE Computer Society Press, 2000.
28. Kephart J.O., Chess D.M. The vision of autonomic computing. Computer 2003;36(1):41-50.
29. Litiu R., Prakash A. Developing adaptive groupware systems using a mobile component framework. Proceedings of the ACM 2000 Conf. on Comp. Supported Cooperative Work (CSCW'00); December 2-6 2000; Philadelphia USA. New York: ACM Press, 2000.
30. Litzkow M., Livny M., Mutka M. Condor–A Hunter of Idle Workstations. Proceedings of the 8th International Conference of Distributed Computing Systems (ICDCS'88). Los Alamitos: IEEE Computer Society Press,1988.
31. Magowan J. A view on relational data on the grid. Proceedings of IEEE IPDPS'03. 2003 Apr 22-26; Nice. Los Alamitos: IEEE Computer Society Press, 2003.
32. Open Grid Services Architecture Data Access and Integration. http://www.ogsa-dai.org.uk/
33. Orlando S., Palmerini P., Perego R., Silvestri F. Scheduling high performance data mining tasks on a data grid environment, Proceedings of Intl. Conf. Euro-Par 2002. 2002 August 27-30; Paderborn, Germany. LNCS 2400. Springer-Verlag, 2002.
34. Palen L., Salzman M., Youngs E. Going wireless: behavior & practice of new mobile phone users. Proceedings of the ACM 2000 Conf. on Comp. Supported Cooperative Work (CSCW'00); December 2-6 2000; Philadelphia USA. New York: ACM Press, 2000.
35. Pattnaik P., Ekanadham K., Jann J. "Autonomic computing and the Grid". In *Grid Computing: Making the Global Infrastructure a Reality*, Berman F., Fox G., Hey T. Eds. Wiley Series in Communications Networking and Distributed Systems. West Sussex, England: John Wiley& Sons, 2003.
36. Platform Computing Corp. at http://www.platform.com
37. Rajasekar A.K., Moore R.W. Data and metadata collections for scientific applications. Proceedings of the European High Performance Computing Conference 2001. Amsterdam, Holland, 2001.
38. Tuecke S., Czajkowski K. , Foster I., Frey J., Graham S., Kesselman C., Snelling D., Vanderbilt P. Open Grid Services Infrastructure (OGSI), 2003. (Online Version http://www.gridforum.org/Meetings/ggf7/drafts/draft-ggf-ogsi-gridservice-23 2003-02-17.pdf)
39. Watson, Paul "Databases and the Grid". In *Grid Computing: Making the Global Infrastructure a Reality*, Berman F., Fox G., Hey T. Eds. Wiley Series in Comm. Networking and Distrib. Syst . West Sussex, England: John Wiley& Sons, 2003.
40. World Wide Web Consortium: http://www.w3.org/
41. http://www.sun.com/software/gridware/
42. http://www.openpbs.org

43. Web Services Flow Language (WSFL) Version 1.0. IBM. Available at: http://www-4.ibm.com/software/solutions/Webservices/pdf/WSFL.pdf
44. Web Services for Business Process Design, http://www.gotdotnet.com/team/xml_wsspecs/xlang-c/default.htm
45. Web Services Modelling Framework, http://www.cs.vu.nl/~dieter/wese/
46. ZHOU S. LSF: load sharing in large-scale heterogeneous distributed systems. Proceedings of the IEEE Workshop on Cluster Computing 1992. Los Alamitos, IEEE Computer Society Press, 1992.

Chapter 21

INTER-ORGANIZATIONAL E-SERVICES ACCOUNTING MANAGEMENT ON COMPUTATIONAL GRIDS

Franco Arcieri[1], Fabio Fioravanti[2], Enrico Nardelli[1,3] and Maurizio Talamo[1]

(1) NESTOR - Laboratorio Sperimentale per la Sicurezza e la Certificazione di Servizi Telematici Multimediali - Università di Roma "Tor Vergata"

(2) Dipartimento di Informatica - Università dell'Aquila

(3) Istituto di Analisi dei Sistemi ed Informatica "Antonio Ruberti" - Roma, CNR

Abstract: Accounting management is of strategic importance for a successful uptake of computational Grid technology within the user community. Computational Grid is one the most important paradigms for distributed computing and high-performance e-service provision. In this paper we present an architecture for accounting management of e-services on computational Grids which fully meets both the reliability and security requirements for accounting management architectures defined by the Internet Engineering Task Force (IETF). The presented solution, based on previous work successfully deployed in many italian public administrations, nicely fits with the overall Grid architectures and features a clear separation between management of the service and its control.

Key words: Certification of E-Services, Accounting Management, Grid Accounting.

1. INTRODUCTION

Grid-aware accounting management is concerned with the generation, communication and processing of data related to the consumption of resources used during a computation requested by a Virtual Organization on a Computational Grid. Resources which can be considered for accounting can be very heterogeneous, ranging from physical devices (like mass-storage devices or CPUs) to network bandwidth and process activity. Furthermore, security and reliability requirements of accounting management protocols greatly vary depending on the intended use of collected resource consumption data. Indeed, accounting data can be used for performing activities which are inherently imperfect, like capacity and trend analysis, or activities requiring a higher degree of precision, like charging users for resources usage (*billing*) or verifying the correctness of a procedure (*auditing*). For example, moderate packet loss can be tolerated when predicting future trends in resource usage, while it becomes unacceptable in usage-sensitive billing where it may cause revenue loss. Moreover, given that Computational Grids naturally span over multiple domains, and accounting data are exchanged between different organizations, reliability and security requirements must be stronger than in the intra-domain setting. Accounting management is of strategic importance for the development of computational Grids ([Foster et al., 2001]), where members of autonomous and independently operated organizations join to form a Virtual Organization (VO). In this scenario, members of a VO share a common set of resources and applications which cooperate in order to provide a complex service resulting from the composition of several component subservices. Of course, these interactions take place in a strictly controlled way , and one of the most important requirements is to be able to certify the actual provision of a component subservice to the requesting client, both for billing purposes and for conformance to service level agreements. Since component subservices are usually already existing and based on legacy IT systems, any solution to this certification problem should feature low invasiveness and strong independence from the application level.

Unfortunately, approaches to accounting management which have been presented in the literature (see, for example, [Aboba et al., 2000] and [Rigney, 1997,Case et al., 2002,Carrel and Grant, 1997]) propose solutions which require agreement on a common set of standards and protocols. Therefore, they are not suitable to be easily adopted in Grid environment because they require that all involved organizations modify their legacy IT systems, thus raising a serious organizational problem, not to mention additional setup and maintenance costs. On the contrary, our solution is based on the analysis of network traffic and the reconstruction of information flows related to e-service provision. In particular, by aggregating and correlating requests and replies related to the same service invocation, our architecture allows one both to monitor the performance of the whole system or of some of its nodes, and to certify service supplying, thus providing invaluable support in case of legal disputes.

We show that our solution satisfies the reliability and security requirements for accounting management defined by the Internet Engineering Task Force (IETF) in [Aboba et al., 2000] while featuring low invasiveness. Thus, it can be rapidly and effectively deployed on very large networks with no changes in the software infrastructure, as it is being demonstrated by its successful use in the realization of a number of inter-organizational e-services in the Italian Public Administration (PA) (see [Arcieri et al., 1999,Arcieri et al., 2002,Arcieri et al., 2001a,Talamo et al., 1999] for details).

The paper is organized as follows. In Section 2 we discuss requirements for accounting management in an inter-organizational framework for e-service provision. Our architectural solution is described in Section 3 and its possible extension to certification of web services is briefly discussed in Section 4.

2. REQUIREMENTS FOR INTER-ORGANIZATIONAL E-SERVICES ACCOUNTING MANAGEMENT FOR GRIDS

In an inter-organizational decentralized model, like that used on computational Grids, autonomous organizations exchange information over the network in order to provide a complex service to the end-user. The overall service thus results from the cooperation among several component subservices, each supplied by a (partially) independent organization which autonomously manages its legacy information systems and its policies about information security and dissemination.

Each subservice provider must then both carry out assigned institutional activities in an effective and efficient way and cooperate with *service providers*, which are in charge of providing the overall service to the end-user. A service provider is an organization which is distinct from the organizations providing the component subservices, and is the unique responsible for provision of a specific service. In terms of the architecture of a computational Grid, subservices are resources and service providers are members of a VO. Additionally, from a different perspective, we may also consider service providers as resources and end-users as members of another VO.

An example from the Italian PA is the support that Ministries of Health and of Labour and the National Institute for Social Welfare have to provide, in a scenario where relevant changes are introduced in the national welfare system, both to the Government (in evaluating various options for changes) and to citizens (to keep them up-to-date with their rights and duties according to their pension scheme).

In this scenario, we can identify additional requirements which have not been considered in the IETF Reference Architecture proposed in [Aboba et al., 2000]. An ideal solution to the inter-organizational certification problem would require the involved organizations to agree on a common set of protocols and standards like those being developed by the IETF, or proposed by the Open Grid Services Architecture (OGSA) research group in [Foster et al., 2002] and to modify their IT systems accordingly.

Unfortunately, in the inter-organizational scenario presented above this approach is unfeasible for the following reasons. Agreement on a common set of standards is typically not scalable: each organization naturally tends to use a different set of protocols for communication with each other cooperating organization, thus leading to a clearly unmanageable proliferation of "standards". Also, the adoption of a unique standard cannot be imposed by a hierarchically superior organization in a scenario in which involved organizations are financially and operationally autonomous.

Indeed, as we have already mentioned, component subservices are provided by existing legacy IT systems which are very difficult or impossible to modify, and are thus regarded as black-boxes. Even if existing IT systems can actually be modified, this practice is not economically profitable because of the additional financial resources required both for implementation and for maintenance.

3. AN ACCOUNTING MANAGEMENT ARCHITECTURE FOR E-SERVICES IN AN INTER-ORGANIZATIONAL SCENARIO

We now illustrate our reference architecture and we show that:

(i) it can be regarded as an instance of the IETF Reference Architecture presented in [Aboba et al., 2000] for accounting management,

(ii) it meets the requirements for auditing applications as they have been defined in [Aboba et al., 2000], and

(iii) it meets the additional requirements for accounting management in an inter-organizational framework.

The architecture we propose for certification of e-services provision is based on the following main entities: *network probes*, the *probes controller* and the *certification server* (see Figure 1).

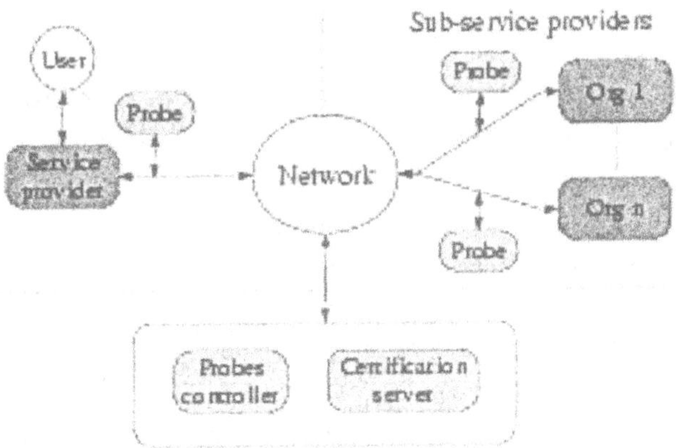

Figure 1: The architecture for certification of e-services in an inter-organizational scenario.

Network probes (or *probes*, for short) are devices installed at service and subservice providers which monitor and analyze network traffic related to a specific service for which they have been configured. Probes reconstruct information flows from acknowledged TCP segments related to the same service request and send reconstructed application-level data (e.g. XML) to the probes controller for further processing. Probes are equipped with an UPS (Uninterruptable Power Supply) unit, self-diagnosis tools and largely redundant disk space, thus providing support for archival accounting and increasing robustness of the accounting management process against data loss and hardware failures. They are fully scalable because only service specific traffic is monitored. Probes have no terminal devices which can be used to access their internal resources, but they are fully configurable from remote by using SNMP and a proprietary protocol based on UDP.

With respect to the context of the IETF Reference Architecture ([Aboba et al., 2000]), network probes play the role of network devices: they are located at organizations sites and send accounting data to a central server. From a functional point of view, however, they are also similar to accounting servers because they aggregate accounting information and generate session records, although also the architecture presented in [Aboba et al., 2000] provides for this eventuality.

Network traffic generated by network probes is not very high because only synthetic information, possibly compressed, is transmitted over the network, thus reducing resource consumption. This approach also feature high scalability in terms of computing power because accounting data processing is spread among several probes and the load on the probes controller is reduced. Notice that no change in existing IT systems is needed for probes to function properly and thus our architectural solution features low invasiveness.

In order to better understand how network probes reconstruct application level messages exchanged by two hosts on the Internet we will now shortly review the behaviour of the TCP/IP protocol. When a process running at the application level (e.g. an HTTP client or HTTP server) wants to send a message (e.g. an HTTP request or response) to a process running on a different host, it passes the application message to the underlying transport protocol along with the destination host address and port number. In the TCP/IP protocol suite there are two choices for the transport protocol: *(i)* TCP which is a reliable and connection-oriented protocol, and *(ii)* UDP which is unreliable and connectionless. In the following we will focus on TCP because it is the transport protocol adopted by most application level protocols (e.g. HTTP, SMTP, FTP, etc.). The case for UDP is even simpler. The TCP layer, on receiving an application message, splits it in smaller parts whose size depends on a TCP parameter called MSS (Maximum Segment Size). Each of this smaller parts, together with an header containing checksums, acknowledgments and information about congestion control, constitutes a TCP *segment*. TCP segments are delivered to the underlying IP layer which is in charge of delivering them to destination, possibly over multiple IP *packets*.

Network probes intercepts IP packets travelling on the network, recognize acknowledged TCP segments and reconstruct application messages. Recall that TCP is a reliable protocol, that is, parties involved in a TCP communication notify each other about the correct receival of data. In summary, the activity of network probes is similar to that of the TCP and IP protocol implementations running at the host they are installed at. For example, a network probe installed at host A and configured for monitoring the traffic between A and a different host B, processes network traffic from B to A as if it was the intended recipient of the application message.

The *probes controller* aggregates and correlates data about sub-services provision related to the same service request in different organizations. Correlation of exchanged information flows also guarantees protection against replay attacks: for example, the probes will ignore a provider claiming to have provided a service which was not requested by any user. Confidentiality can also be ensured by encrypting local storage, if desired. Correlated data are then forwarded to the certification server. The probes controller is thus, in the IETF architectural framework, playing the role of the accounting server both from the architectural and the functional points of view.

Communication between network probes and the probes controller is based on TCP and features data object integrity. Authentication and confidentiality are also guaranteed by use of public-key cryptography. The data collection model is event-driven with support for batching and scheduling thus featuring high reliability and scalability, and is fully programmable.

The *certification server* receives correlated data from the probes controller and stores them in a relational DBMS. Correlated data can then be used for monitoring the performance of the system, locating bottlenecks, generating detailed reports and statistics, and, most importantly, for generating certificates of service provision. Therefore the certification server plays, both functionally and architecturally, the role of the billing server in the IETF reference architecture.

Since the communication model between network probes and the probes controller can be programmed, it is also possible to improve resilience against faults:

(i) by making network probes contact failover probes controllers in case that the attempt to contact the primary controller fails, and

(ii) by implementing scheduling algorithms which are more suitable for use in some applications, like those with strict constraints on processing delay.

Of course, the same arguments also apply to communication between the probes controller and the certification server.

An additional advantage of our solution is that its neutrality both from a technical and from an organizational point of view allows a trusted third party on the same network to carry out the monitoring and certification tasks, thus providing a clear separation between *management* of the service and its *control*, and thus avoiding risks and conflict associated with cases where a same organizations plays a multiplicity of roles. Notice that the organizational issues presented above have not been addressed in [Aboba et al., 2000], which just requires stronger confidentiality of information exchanged by inter-domain accounting applications.

The architecture here presented has been deployed for certifying e-service provision in the following systems in the Italian Public Administration e-Government initiative ([Arcieri et al., 2001b,Mecella and Batini, 2001]):

1. SICC (Sistema di Interscambio Catasto Comuni) ([Arcieri et al., 1999,Talamo et al., 1999]) is a system for exchanging cadastral data between the following entities: Ministry of Finance, Municipalities, Notaries and Certified Land Surveyors. The system is accessible nation-wide through a Web-based interface since September 1998. The effectiveness of its use is demonstrated by the number of administrative transactions successfully completed through the system in year 2001: they are 800.000 per month, corresponding to 60% of the overall transactions related to cadastral services

accomplished every year in Italy. Note that certificates related to ownership, location, geometry and value of real estates, are mandatory in estate's selling transactions and their issue is subject to a fee, paid for by the buyer. In the month of September 2001 the total value of Ministry of Finance's income deriving from certificates issued through this system corresponds, on a yearly basis, to the 20% of the overall sum cashed by the Ministry for cadastral services, which is estimated for the year 2001 to amount to 78 million euros (roughly 70 million US dollars[1]).

2. SIM (Sistema Informativo della Montagna) ([Arcieri et al., 2001a]) is a system for providing services in various fields (cadaster, labour, pensions, public registry of personal data) to Italian citizens living in mountain areas. Its design started in 1998 and the overall financial effort has been, until now, of about 52 million euros (roughly 47 million US dollars). Nowadays it is currently being used as a fully operational system in about one thousand operating centers, all over Italy, serving more than 10 million inhabitants, more than 4000 of the about 8000 municipalities and covering more than 54% of the Italian territory. Its extension to the whole country is currently under implementation.

4. MONITORING AND CERTIFICATION OF WEB SERVICES

The problem of enabling interoperability between e-business applications on the World Wide Web is currently being addressed by using XML-based ([Bray et al., 2000]) standards like the Web Service Definition Language (WSDL) ([Christensen et al., 2001]) and the Simple Object Access Protocol (SOAP) ([Box et al., 2000]). These technologies provide a framework within which it is possible to expose existing network applications in a uniform and elegant manner.

A WSDL document contains the definition of the message exchange pattern between a service requester and a service provider (one-way, request/response, or publish/subscribe), together with the definition of the structure of the messages, the message data types and the bindings to concrete network protocols (HTTP GET/POST, SOAP, MIME) to be used for invoking the service itself. Messages exchanged by the service requester and provider are typically formatted according to the SOAP protocol. SOAP messages are XML documents consisting of three parts: (*i*) an envelope

[1]at the exchange rate of about 0.9 US dollars per 1 euro

describing the message content and the rules for processing it, (*ii*) an optional header for extending a message with new features like authentication and transaction management, and (*iii*) a body containing data related to service requests or responses. Although HTTP is used as the main network protocol, SOAP can potentially be used in combination with a variety of other protocols, like FTP, SMTP or RPC.

In a Web Service architecture ([Boot et al., 2002]), if a service requester wants to use a web service, it must first obtain the WSDL document containing the service description, either from the service provider itself or from a discovery agency, that is a network-accessible catalog where service providers publish their service descriptions, like the Universal Description, Discovery and Integration of Web Services (UDDI) ([Bellwood et al., 2002]). In order to successfully complete the service invocation, interaction between requester and provider must obey to the specifications contained in the WSDL document which describes the service. As long as the parties involved in service provision adhere to the same service description, the software systems actually providing the Web services can be implemented by using any technical solution, ranging from Java Servlets to Application Server Pages.

In order to certify web services provision by using the architectural solution presented in Section 3, network probes must be installed both at the service requester and at the service provider sites, and configured for analyzing and reconstructing exchanged messages as specified in the WSDL document describing the service. Furthermore, since a service can be invoked several times between the same requester and provider, even during a short time interval, exchanged data should also contain information allowing one to exactly identify the service invocation (like timestamps or sequence numbers), thus allowing the probes controller to reconstruct the message exchange pattern related to the same service provision. For example, if SOAP is used as the underlying network protocol, session information can be included in the header of all SOAP messages exchanged by the service requester and provider in a service provision.

Since our solution is based on reconstruction of application-level data, different technical solutions must be used for allowing it to operate on encrypted communication channels. However, our solution can be deployed even in case that exchanged information is partially encrypted. Indeed, in most cases by analyzing only the high-level structure of messages one is able to provide a certificate of service provision. Selective encryption of messages is also preferable from a computational point of view because encryption and decryption require additional computational power, and current specifications for Web services security ([Atkinson et al., 2002]) are following this direction, by requiring that an encrypted SOAP message is still a valid SOAP message.

5. CONCLUSIONS AND FUTURE WORK

In this paper we have described a specific system architecture (and its implementation) which:

(i) fully meets the reliability and security requirements for accounting management architectures defined in [Aboba et al., 2000] by the Internet Engineering Task Force (IETF),

(ii) satisfies additional requirements related to specific organizational constraints deriving from an inter-organizational distributed framework for e-service provision, and

(iii) can be used for monitoring resource usage on computational Grids both for billing and certification purposes.

Using the terminology of [Aboba et al., 2000], certification of e-service provision is an auditing application, since certifying e-service provision consists in the verification of the correctness of a procedure.

All IETF requirements are met, since network probes and the probes controller support archival accounting, data integrity is ensured both at packet and object level, authentication and confidentiality are guaranteed by use of public-key based technology, protection against replay attacks is provided by the correlation of requests and service provisions performed by the probes controller. Our architecture features both physical and functional independence from the application level and is thus a non-invasive solution for monitoring and certifying e-services supplying in an inter-organizational setting.

The neutrality of the proposed solution both from a technical and from an organizational point of view allows a trusted third party on the same network to carry out the monitoring and certification tasks, providing a clear separation between *management* of the service and its *control*, and thus avoiding the risky multiplicity of roles played by the organizations.

As a development of the solution presented in this paper, we plan to address in more detail the following issues: (*i*) how to extend the proposed solution for certifying web services provision, and (*ii*) how to expose the certification process itself as a web service.

REFERENCES

[Aboba et al., 2000] Aboba, B., Arkko, J., and Harrington, D. (2000). Introduction to accounting management. RFC 2975.

[Arcieri et al., 1999] Arcieri, F., Cammino, C., Nardelli, E., Talamo, M., and Venza, A. (1999). The italian cadastral information system: a real-life spatio-temporal DBMS. In Bölhen, M., Jensen, C., and Scholl, M., editors, *Workshop on Spatio-Temporal Database*

Management (STDBM'99), pages 79–99, Edinburgh, Scotland, U.K. Lecture Notes in Computer Science vol.1678, Springer-Verlag.

[Arcieri et al., 2002] Arcieri, F., Cappadozzi, E., Naggar, P., Nardelli, E., and Talamo, M. (2002). Coherence maintainance in cooperative information systems: the Access Key Warehouse approach. *International Journal of Cooperative Information Systems*, 11(1–2): 175–200. A preliminary version was published in the Proceedings of the 4th International Conference on Cooperative Information Systems (CoopIS'99).

[Arcieri et al., 2001a] Arcieri, F., Cappadozzi, E., Nardelli, E., and Talamo, M. (2001a). SIM: a working example of an e-government service infrastructure for mountain communities. In *Workshop Electronic Government (DEXA-eGov'01), associated to the 2001 Conference on Databases and Expert System Applications (DEXA'01)*, pages 407–411, Munich, Germany. IEEE Computer Society Press.

[Arcieri et al., 2001b] Arcieri, F., Giaccio, R., Nardelli, E., and Talamo, M. (2001b). A framework for inter-organizational public administration network services. In *International Conference on Advances in Infrastructure for Electronic Business, Science, and Education on the Internet (SSGRR'01)*, L'Aquila, Italy. IEEE Computer Society Press.

[Atkinson et al., 2002] Atkinson, B., Della Libera, G., Hada, S., Hondo, M., Hallam-Baker, P., Klein, J., LaMacchia, B., Leach, P., Manferdelli, J., Maruyama, H., Nadalin, A., Nagaratnam, N., Prafullchandra, H., Shewchuk, J., and Simon, D. (2002). Web services security (WS-Security).

[Bellwood et al., 2002] Bellwood, T., Clement, L., Ehnebuske, D., Hately, A., Hondo, M., Husband, Y., Januszewski, K., Lee, S., McKee, B., Munter, J., and von Riegen, C. (2002). Universal description, discovery and integration of web services (UDDI) version 3. http: //uddi.org/pubs/uddi_v3.htm.

[Boot et al., 2002] Boot, D., Champion, M., Ferris, C., McCabe, F., Newcomer, E., and Orchard, D. (2002). Web services architecture. http: //www.w3.org/TR/ws-arch.

[Box et al., 2000] Box, D., Ehnebuske, D., Kakivaya, G., Layman, A., Mendelsohn, N., Frystyk Nielsen, H., Thatte, S., and Winer, D. (2000). Simple object access protocol (soap) 1.1. http: //www.w3.org/TR/SOAP.

[Bray et al., 2000] Bray, T., Paoli, J., Sperberg-McQueen, C. M., and Maler, E. (2000). eXtensible Markup Language (XML) 1.0 (Second Edition). http: //www.w3.org/TR/REC-xml.

[Carrel and Grant, 1997] Carrel, D. and Grant, L. (1997). The TACACS+ protocol. Internet Draft draft-grant-tacacs-02.txt.

[Case et al., 2002] Case, J., Mundy, R., Partain, D., and Stewart, B. (2002). Introduction and applicability statements for internet standard management framework. RFC 3410.

[Christensen et al., 2001] Christensen, E., Curbera, F., Meredith, G., and Weerawarana, S. (2001). Web Services Description Language (WSDL) 1.1. http: //www.w3.org/TR/wsdl.

[Foster et al., 2002] Foster, I., Kesselman, C., Nick, J. M., and Tuecke, S. (2002). The physiology of the grid: An open grid services architecture for distributed systems integration. http: //www.globus.org/research/papers/ogsa.pdf.

[Foster et al., 2001] Foster, I., Kesselman, C., and Tuecke, S. (2001). The anatomy of the grid: Enabling scalable virtual organization. *International Journal of Supercomputer Applications*, 15(3): 200–222.

[Mecella and Batini, 2001] Mecella, M. and Batini, C. (2001). A review of the first cooperative projects in the italian e-government initiative. In *1st IFIP Conference on E-Commerce, E-Business, E-Government (I3E-01)*, pages 831–844, Zürich, Switzerland. Kluwer.

[Rigney, 1997] Rigney, C. (1997). RADIUS accounting. RFC 2139.

[Talamo et al., 1999] Talamo, M., Arcieri, F., Conia, G., and Nardelli, E. (1999). SICC: An exchange system for cadastral information. In Güting, R. H., Papadias, D., and Lochovsky, F., editors, *6th International Symposium on Large Spatial Databases (SSD'99)*, pages 360–364, Hong Kong, China. Lecture Notes in Computer Science vol.1651, Springer-Verlag.

Chapter 22

A WEB SERVICES PROVIDER

Jean-Paul BAHSOUN * [1], Bilal CHEBARO [2], Samar TAWBI ♦ [2]

Abstract: In this paper, we define a generic tool 'GenericServ' that offers a 'service providers' platform which facilitates the programming tasks of web application development. The system's architecture is generalized to propose three patterns for business applications' development. The paper is divided into two major parts. In the first one, we expose the motivation for the definition of the service provider, where we emphasize the architecture of the system and the arguments to choose such architecture. In the second part, we define the patterns based on this generic service provider.

Key words: Web development, Web services, Genericity, Patterns.

1. THE WEB EVOLUTION

The Internet became an essential media, especially the World Wide Web that is recognized and used by enterprises, government agencies and the wide public. The information flow that uses the World Wide Web represents today more than two thirds of the overall Internet traffic [2]. In the last five years, the Online Computer Library Center research determined that the public web has more than doubled in size, increasing from 1,450,000 sites in 1998 to over three million in 2002 [15]. More than 55% of web site creators are not programming specialists. The web has become now a requirement for everyone in all domains. User types are multiplying, the requirements are increasing and the web sites are more and more complex. For this reason, a web services provider, facilitating the web applications development,

♦ Scholar from the CNRSL.
* Names are in alphabetical order.

becomes critical, especially if it is a generic one that can be deployed in many areas on the web. This paper presents a solution to simplify the development of web applications.

The paper is divided into two major parts: in the first one (section 2) we will outline the motivation for this generic service provider's development. Then, we will propose a general architecture for our system, taking into consideration the different existing architectural styles that can be used in this application. In the following part (section 3), we define a set of patterns, based on this tool, and generalized to the development of business applications.

2. GENERICSERV: A GENERIC SERVICE PROVIDER FOR WEB DEVELOPMENT

2.1 Presentation

Since the needs for sophisticated web sites is more and more relevant, especially for the functionality that is related to business domains like commerce, education and banking, a generic service provider 'GenericServ' was proposed in [23]. It is a generic tool or server that offers a platform providing the web site designers the ability to store, access and process the information. Its generic nature makes it independent of business domains. It defines a protocol for service creation and a general architecture that can be extended to deploy concrete servers in Web domains as article publishing, e-commerce, e-mailing and education, by defining the set of functionalities needed in these areas.

Many tools were proposed to facilitate the web site development, from the HTML editors to content managers. But the first category of tools appears to be too basic for sophisticated needs. This is especially true for the functionalities that are related to business domains. On another hand, even though the content managers are very helpful for easy web development, the users of such tools should have the product on their working environment. So their update and upgrade are not automatically accessible by the users. Meanwhile, approaches that have been taken in the distributed data communications and the Internet development areas like [5][8][9][10] face two main obstacles for web application developers: They still require deep programming skills and they impose a very specialized development environment. In this context, we propose a generic service provider as a platform used to create servers offering complex functionalities that Web application creators could call from their pages (HTML or XML). The services offered are remotely called from the web pages, and are executed in real time, when the page is accessed. Such server has a simple interface to use as it is dedicated to non-specialists.

Thereby, the generic service provider constitutes an abstract architecture that proposes the management of client sessions, services and data storage as well as a template for the definition and the interface to use the services. In fact, this nucleus serves as a base for concrete servers extension. Instantiating a server will consist of implementing the functionalities needed in a certain domain, as services according to a template imposed by the generic service provider. These services always follow the same interface so that the provider could take them into consideration automatically. In addition, this allows easy evolution and maintenance of the system and complete transparency to end-users. This tool defines also the cooperation means between the developers and the graphical designers. It is responsible of merging the two efforts through a common interface.

We should note that although it could; in general, the concrete service provider does not provide hosting services for web sites. In fact, it constitutes an intermediate layer between the end-user (i.e. the navigator) and the hosting server of the web site (see figure 1).

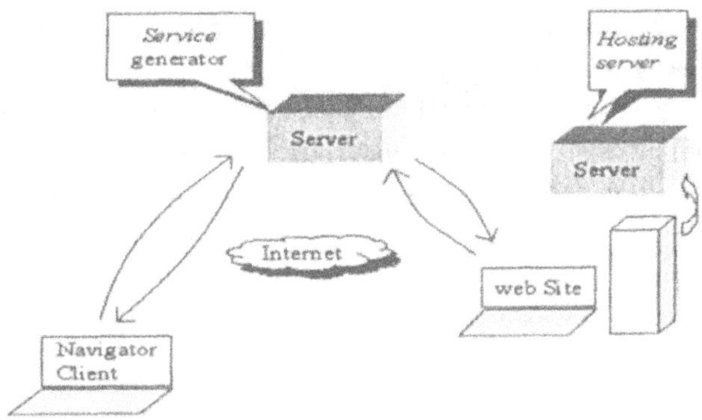

Figure 1. The concrete provider operation constitutes an intermediate layer between a navigator and a hosting server

2.2 The System's General Architecture

The system explained above aims at providing a simpler and reusable way to achieve the goal of implementing Web sites covering several services and domains. Therefore, the architecture design should support the required software system qualities such as robustness, adaptability, reusability and maintainability [3][4]. So in the case of this services provider, the architecture has to guarantee its flexibility, easy reuse and evolution.

Patterns have become now a must that software designers and developers should use or at least try to apply systematically in all phases of their work. There have been patterns applied to the analysis phase [11]. Patterns have also been applied at the architectural analysis phase of the software development process. In this level, patterns propose an abstract representation of the system's architecture [1] [4]. They are also called architectural styles [21] [20]. A number of architectural styles [12] was defined serving certain categories of software systems like repository models [6], layered architectures [19], client server architectures [17] and others.

The major concern in 'GenericServ' was to choose the architectural style that is the most suitable for this system. Since each one of the known styles [12] [18] has some drawbacks to apply to our system, we have opted for a heterogeneous architecture using a combination of the three architectural styles listed above.

2.2.1 A heterogeneous architecture

The provider will be deployed on a three-tier client/server architecture that has the advantage to split the deployment to three levels, which are appropriate to the deployment of the system.

In this architecture, every server based on GenericServ (the generic service provider) will consist of the three tiers: client, processing and database (Fig.2). The other styles are used to define the internal architecture of the different tiers.

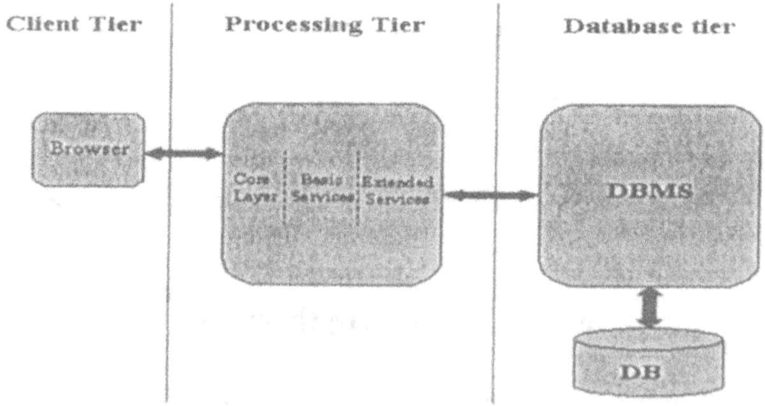

Figure 2. The 3-tiers architecture of the server

In the following we provide more details on each one of the tiers forming the architecture.

- The ***Client tier*** represents the interface that the web applications' creator uses for accessing the services offered by the provider. This is a thin client with almost no processing responsibilities to free the web application developers from programming tasks. So, it does not restrict any special software platform in order to free the users from any programming effort. Thus, the web site creators will just need the list of the services available with their parameters, if any.
- The ***Database tier*** centralizes the data related to the applications in a repository database managed by a DBMS. It is accessed through the services offered to users, but stays completely transparent in order to allow independent updates and changes. It is accessed through an interface implementing the DAO pattern [13]. This makes the deployment of the Database server changeable and upgradeable independently of the other parts of the system.
- The ***Processing tier*** constitutes the services provider itself. It is deployed as the middle-tier containing all processing, client management, testing and security. Its internal architecture consists of three opaque layers designed in a strongly modular object-oriented architecture in order to take the advantages of the independence that the layered style offers.

'GenericServ' standardizes the management of the services and defines a communication protocol between the different actors of the application. Thus, around this core, a set of services could be defined for a specific domain to form a useful server, like the one created for online article publishing needs 'PubliWeb' [23]. In the next paragraph, we will focus on the structure of this tier as it is the central one and it contains all the processing tasks of the system.

2.3 The processing tier architecture – three layers –

In the processing tier of every concrete server, we distinguish three layers (Fig. 3):

1. The **core layer** It constitutes the kernel of the system that includes general processing. This core layer defines and implements the management of the users and services in a generic way, in order to be independent from the specified servers business domain. It proposes a skeleton for all services that will be defined on this server. It contains four modules:
   - **Connection manager:** It constitutes the entry point of the server (its interface of use). It manages the connection requests, the verification,

if the requests include the necessary parameters, and the interaction between the client and the server.

- **Analyzer:** It analyses users' requests. It must be transparent in order not to put too many constraints on the users. Its role is to capture the service calls, to verify them syntactically and to extract their attributes. Then it sends these calls to the "Service Manager" that will delegate them to the corresponding services.
- **Service manager:** It dispatches the calls, arriving from the analyzer, to the appropriate services after verifying their existence and their dependencies with other services. This is achieved by keeping a temporary trace of the running services that will be consulted for any dependency tests.
- **Session Manager:** It handles the client session's management. It also safeguards the state of the client's work during the session.

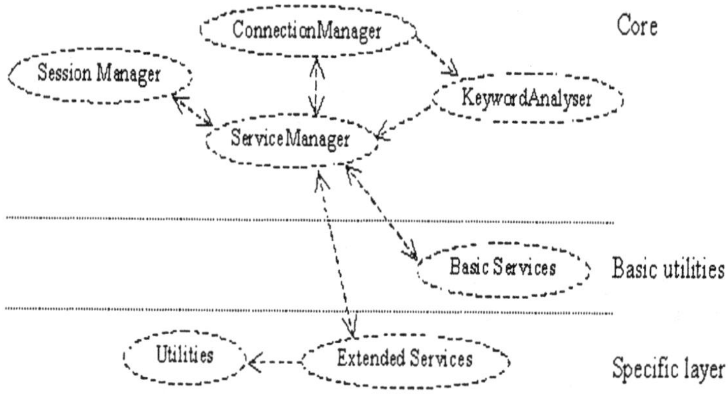

Figure 3. GenericServ based services providers' architecture.

2. The **Basic Utilities**: this layer contains basic services that can be present in all service providers independently of their domains. It contains, for example, services to communicate with the database tier. Other common services like client authentification, security or any other domainless functionnality can be created in this layer in order to encourage reuse and avoid redunduncy.
3. The **Specific Layer** represents the set of services proposed in a precise domain like article publishing. Some examples are online article submission, article viewing and article acceptance or rejection. A module 'Utilities' is defined in this layer covering redundant parameters and functions that are necessary for the services' processing.

The processing tier uses also the Reflection pattern defined in [6] as a way to divide the system into two levels: a Base level and a Meta level. The base level is responsible for the computation that stays stable through the concrete server's lifetime. It contains the core and the basic utilities layers. The Meta level contains the varying part of the system. It is the services layer. The separation of the server into these three layers allows an instantiation of the generic provider into concrete servers with a minimum of effort, since this task consists of implementing services according to the prototype imposed by GenericServ.

3. A CATALOGUE OF PATTERNS

In this section, we define a set of patterns inspired from GenericServ and generalized to business applications. The first pattern proposes a *service provider* to help the non-experimented programmers to perform sophisticated development in a certain domain. In the second pattern '*Generic Server*', we propose genericity in order to respond to the permanent evolution of users' needs and to make the server functionality independent from the business domain of the services it provides. And finally, the last pattern defines an opaque '*modular layered architecture*' for this type of applications. These patterns are discussed in the following sections.

3.1 The Services provider pattern

Context

Applications in a same domain may have lots of common functionalities with each other, but different designs and user interfaces. Redoing the same development several times is a waste of time and effort. In addition, an application needs different type of actors (analyzers, graphic designers, developers...). And sometimes the cooperation between these actors may cause certain problems.

One would like to create an application in a certain domain. Therefore, the creator of the application may not be very well experienced in programming, although he wants to conceive a robust and performant application.

Problem

You would like to minimize the redundancy of the functionalities in order to reduce the development efforts. So, you are trying to reuse functions that have been developed in other applications. On another hand, you have to guarantee a good cooperation between the different actors of the application without affecting the robustness and the efficiency of the applications. And essentially, you want to offer some functions created by experienced developers to non-specialists or simply less experienced people.

Solution

Create a service provider offering the common functionalities needed by the applications in a certain domain, in the form of services that could be used by the applications' creators (Fig.4). This services provider could be used by different applications in the same domain. It has to have a simple interface for the users. This interface must be transparent to the user services in order to allow the developer to call the services without the need to know the structure of the provider. The system could be deployed as a server in a client/server structure, where the applications using the services will act as clients of this server; or, for example, as an application where the users will choose their services in some kind of graphical interface.

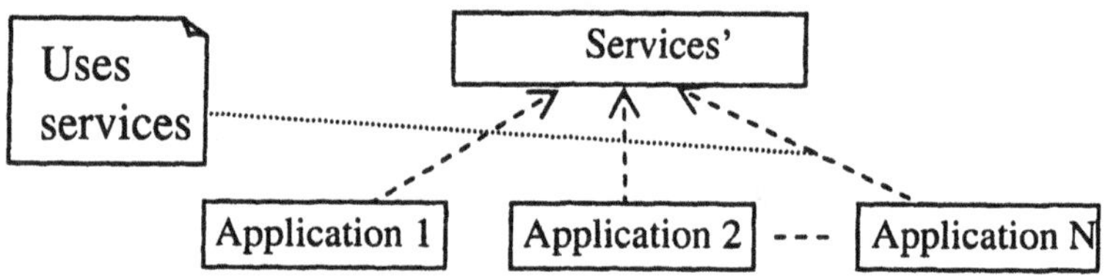

Figure 4. *The Service provider structure*

Known uses

The html page editors, the IDLs (Interface Definition Languages), the development environments for programming languages like Borland C++, Symantec Visual Studio, etc.

Consequences

- The Services Provider pattern will free the application creator from most of the development tasks, so it makes it possible for a large group of non well-experienced programmers to implement relatively sophisticated applications.
- There is a complete separation between the different actors of an application while keeping an efficient communication between them.

Related patterns

Generic Server (see next section) is the generic version of this pattern. Our pattern is inspired from Technical Infrastructure [14] that proposes a solution to system complexity by encapsulating the computing tasks from the application developer.

3.2 The Generic Server pattern

Context

You are designing a services provider application, but you need different types of services in different domains. In addition, not all of the services are foreseen at the time the application is created. You want your system to be extensible during its lifetime, and to easily support updates. New services will continuously be added responding to new requirements. Realizing a service provider will soon be insufficient because of the continuously increasing needs. On another hand, the extensions and updates of the services built on the system must not affect the system's users. These users do not want to redo their work each time the provider's developers make changes to the system.

Problem

How do you design such an application in a way to ensure the flexibility and extensibility of the system?

What could guarantee the harmony of the work between all actors of the application? They must stay independent from each other without affecting the performance of the system.

Solution

Conceive a generic application in such a way that it standardizes the handling of its services and the communication with users. This could be achieved by defining from one side, the protocol of communication with the users, and from the other side, a prototype of services definition. As for the generic service provider 'GenericServ' (Fig.5)

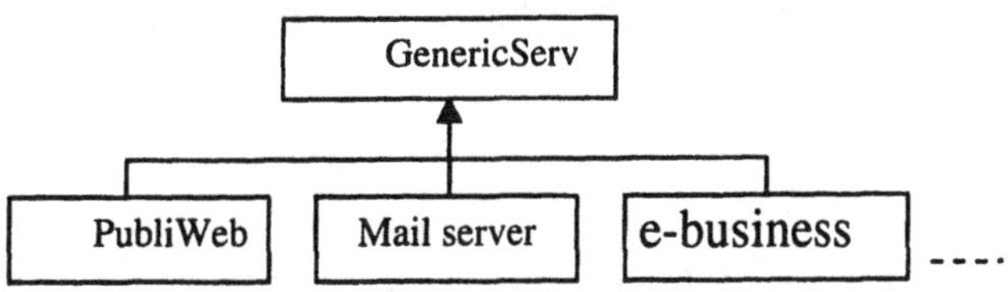

Figure 5 . GenericServ, a generic web server

Known uses

In the domain of distributed applications development, the distributed object norms have produced a communication layer as in CORBA [8] and DCOM [5], in order to free the developers of such applications from the implementation of communication functionality. Another known use of this pattern is libraries, like CASTOR [7], that illustrates a generic source generator offering services to map between Java objects and XML schemas.

Consequences

- Applying the Generic Server pattern makes the application more extensible and more flexible. The standardization of the services allows the developers work independently from the control and management of the services.
- The generic nature of this pattern makes it domain independent, so it will be applicable to more categories of applications.
- Future and unexpected needs could be added automatically thanks to the standard manipulation of the services.

Related patterns

The Generic Server could be structured using several architectural patterns like the Three Layered Architecture [18] and Pedestal [16]. But we suggest that it will be structured according to the 'Encapsulated modular layers' pattern explained below.

3.3 The Encapsulated Modular Layers pattern

Context

You want to conceive a generic application. But now you need to build a good architecture. This architecture should reinforce the genericity. You should keep the actors of the application independent from each other in order to maintain a generic domain independent system.

Problem

You must find the best way to realize the genericity of your system in a flexible architecture.

Solution

Create a four-layer architecture where you separate the processing core from the services offered by the application. Therefore, conceive each layer in a modular object-oriented structure. The layers must be opaque, so that they can evolve independently from each other. Simplify the interfaces between the layers in order to simplify modules updates.

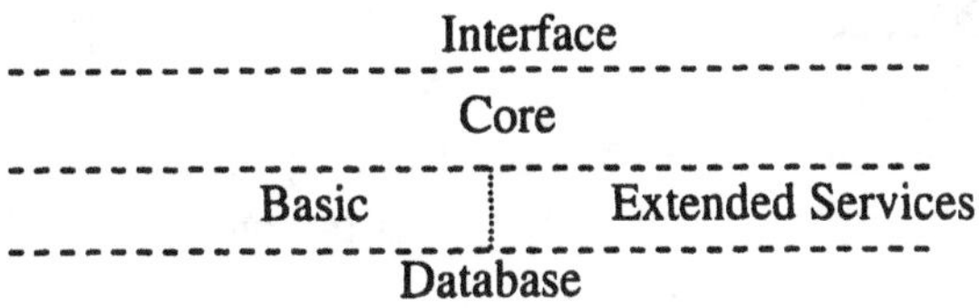

Figure 6. Encapsulated modular layers structure

The first layer contains the interface that is visible to users, and encapsulates all layers behind. The second layer is the core of the server. It includes general processing, like clients management and services control. In this layer, the generic aspect of the application will be defined, because it will contain service management and manipulation, but it is defined independently from their business domains. It defines a standard way to deal with services. The communication with the users and other layers is done via interfaces encapsulating its internal functionnality. So, it will be completely opaque to the user and to the other layers. A third layer that contains the services offered by the application is defined. It is divided into two modules: 'basic services' that are common to all domains (can be called domainless services), and 'extended' services with their utilities that are domain dependant. The two central layers are the common layers to all types of servers since they are domain independent. And finally, the last layer contains the database management (Fig.6).

Known uses

The Amoeba operating system [22] consists of a kernel providing basic services for processes and network communications, memory management and I/O services. But the difference is that in the 'Encapsulated opaque Layers' pattern, we restrict the services to have a predefined skeleton in order to generalize their management.

Consequences

- The opaque Layers with the encapsulating interfaces emphasize the reusability of the system.
- The evolutions of the four layers are independent from each other due to their opacity.

Related patterns

This pattern gives the development structure of an application of the type 'Generic Server'. The services offered by an application specifying the 'Encapsulated Opaque Layers' can follow the structure of the Service Prototype pattern [24]. It is a specialization of 'Layered Architecture' pattern [19].

4. CONCLUSION

We have presented in this paper an approach to solve the issues encountered in web applications development that are related to the complexity of the new sophisticated functionalities needed in this space. We have focused on the web development domains since it involves a large variety of persons with very different programming expertise. So, we have proposed a generic tool for developing web applications. The architectural style used for this tool is a heterogeneous one that combines advantages gained from different basic styles. Then, based on this tool we have defined three patterns for business applications' development. The first proposes a service provider to avoid redundant functionalities and to facilitate the development. The second gives a generic aspect to the provider when used for several or evolving domains. And the third defines architecture for this type of tools emphasizing modularity, reuse and easy update.

REFERENCES

1. Abowd, G.; Bass, L.; Kazman, R.; & Webb, M. SAAM: A Method for Analyzing the Properties of Software Architectures, 81-90. Proceedings of the 16th International Conference on Software Engineering. Sorrento, Italy. CA: IEEE Computer Society Press, 1994.
2. Abramatic, J.F., Dévelopement technique de l'Internet, W3C, http://www.w3c.org, 1999.
3. Aksit, M., Bergmans, L., Berg van den K., Broek, van den P., Rensink, A., Noutash, A., & Tekinerdogan, B., Towards Quality-Oriented Software Engineering, to be published in Software Architectures and Component Technology: The State of the Art in Research and Practice, M. Aksit (Ed.), Kluwer Academic Publishers, January 2000.
4. Bass, L., Clements, P., & Kazman, R. Software Architecture in Practice, Addison-Wesley 1998.
5. Brown, N. & Kindel, C., Distributed Computing Object Model Protocol – DCOM/1.0, www.grimes.demon.co.uk/DCOM/DCOMspec.htm, 2002
6. Castro, J. & Mylopoulos, J., Information Systems Analysis and Design, 2001.
7. castor.exolab.org, 2002.
8. Daniel, J., Au cœur de CORBA (avec Java), Vuibert 2000.
9. Enterprise Java Bean 2.1 Specification, java.sun.com, 2002.
10. Farly, J., Java Distributed Computing, O'REILLY, 1998.
11. Fowler, M.., Analysis Patterns : Reusable Object Models, Addison-Wesley, 1996.
12. Garlan, D., "An Introduction to Software Architecture," Advances in Software Engineering and Knowledge Engineering, Volume I, edited by V.Ambriola and G.Tortora, World Scientific Publishing Company, New Jersey, 1993.
13. http://java.sun.com/blueprints/corej2eepatterns/Patterns/DataAccessObject.html, 2002.
14. Meszaros, G., Archi Patterns: A process pattern language for defining architectures, Pattern Language Of Program Design conference, 1997.
15. The Online Computer Library Center, http://www.oclc.org, 2002.

16. Rubel, B., "Patterns for Generating a Layered Architecture", Pattern Language of Program Design, Vol.1, Addison Wesley, 1995.
17. Sadoski, D. & Comella-Dorda, S., Three Tier Software Architecture, URL : http://www.sei.cmu.edu/str/descriptions/, 2000.
18. Shaw, M., Making Choices: A Comparison of Styles for Software Architecture. IEEE Software 12, 6 27-41, November, 1995.
19. Shaw, M., 'Some Patterns for Software Architectures', Pattern Language Of Program Design, Addison Wesley, 1996.
20. Shaw, M. & Clements, P. A Field Guide to Boxology: Preliminary Classification of Architectural Styles for Software Systems, Proc. COMPSAC97, 1st Int'l Computer Software and Applications Conference, August, 1997.
21. Shaw, M. & Garlan, D. Software Architectures: Perspectives on an Emerging Discipline, Englewood Cliffs, NJ: Prentice-Hall, 1996.
22. Tanenbaum, A.S., Modern Operating Systems, Prentice Hall, 1992.
23. Tawbi, S. & Chebaro, B., GenericServ: A generic server for web application development, web requirements & e-services workshop of the 1st EURASIA conference for Advances in information & communication technology, workshop proceedings, Austrian computer society, 2002.
24. Tawbi, S. & Chebaro, B., Service Providers patterns for client/server applications. Poster in the proceedings of the ICEIS 2003, 5th International Conference On Enterprise Information Systems, IEEE computer society press, 2003.

Chapter 23

USING METAMODELS TO PROMOTE DATA INTEGRATION IN AN E-GOVERNMENT APPLICATION SCENARIO

Adriana Maria C. M. Figueiredo, Aqueo Kamada, Luciano L. Damasceno, Manuel de Jesus Mendes and Marcos Antonio Rodrigues

Abstract: The increasing popularization of Internet has created unprecedented expectations and promises when dealing with a more intelligent use of the available data. This use is based on metadata and on a new dimension of metadata modeling techniques. An initiative supported by OMG culminated in the current MOF – Meta-Object Facility specification, which is an open standard with facilities to definition and manipulation of metadata and metamodel. This paper describes a metadata management system that is based on the MOF specification and presents one usage of this system in an e-government scenario.

Key words: Metadata, Meta-Object Facility, Metadata Management System

1. INTRODUCTION

E-Governments initiatives around of the world are trying to make available to the citizens many of the public services accessible via Web. The most common principles in these initiatives are: i) cooperation among different public administrative units in the sense that they need to rationalize their resources and integrate their software assets; ii) device and user interface independence, meaning that users don't have any limitation on how they can access public services, whether by PC, hand-held, WebTV, mobile phone or any other device; iii) transparent access to the public services, in the sense that the citizen doesn't need to know the different administrative units involved in the execution of a selected service.

Adherent to these principles the main goals of these initiatives are: i) to allow independent public administrative units to cooperate amongst themselves to deliver services; ii) to provide public administrative units with a shared and common semantic description of concepts on which they can cooperate; iii) to provide services that can be deliverable through the coordinated use of legacy systems spread over the public administrative units.

In this context, to aggregate disparate existing services and legacy systems, moving the focus of interoperability from monolithic system to message sequences that cross public administrative units and geographic boundaries, an innovating and effective infrastructure to share and manage metadata is fundamental to service collaboration and cooperation. The incompatibility and proprietary solutions for the metadata administration have been the primary difficulty for a transparent access to data resident in different platforms. The elimination of these problems will accelerate the integration of applications, which require exchange of metadata, such as e-government, data warehousing, e-commerce and information portal applications.

A new dimension of metadata modeling and managing techniques was proposed by OMG with the introduction of the open standards Meta-Object Facility – MOF and XML Metadata Interchange – XMI [OMG 2003] that provide a comprehensive way to express and represent metadata, distributed access and standard data exchange mechanism.

This work presents the development of a set of tools in order to provide an effective and consistent framework to manage metadata based on the standards MOF and XMI.

In section 1, we address the motivation of this work. In section 2, the main foundations of MOF and XMI are presented. The architecture and tools that comprise the metadata management system are detailed in section 3. Section 4 describes the use of the implementation in an *e*-government scenario. Finally, in section 5, final considerations and future work are presented.

2. META-OBJECT FACILITY

The Meta-Object Facility (MOF) [OMG 2003] is a model driven distributed object framework for defining, managing and integrating metadata in software systems. To achieve its objectives, MOF defines an abstract language to describe metamodels and a generic framework for managing, in terms of repository, the metadata described by the metamodel.

The core of the MOF approach to metadata management is openness. The objective is to provide a framework that supports many existing metadata specification propositions and standards, and that allows new ones to be added as needed. To achieve this goal, MOF uses a four-layer metamodeling architecture.

Figure 1 illustrates the four-layer architecture adopted by MOF. A major difference between this architecture and other metamodeling architectures [MDC 1999, EIA 1994] is the fact that the meta-metamodel is defined using its own metamodeling constructs.

The top layer, M3, is the meta-metamodel level and corresponds to the MOF Model, the only model residing in this layer. The MOF Model forms the base for the metamodeling architecture and its main responsibility is to define the language to specify metamodels. Examples of meta-objects in the meta-metamodeling layer are: MOF::Class, MOF::Association and MOF::Attribute.

The next layer, M2, is populated with metamodels instantiated from the MOF Model. UML and CWM [OMG 2003] are examples of standard metamodels defined by OMG to the software development domain. However, M2 layer is not restricted to standard metamodels. New metamodels can be defined and existing ones can be extended in order to accomplish specific contexts' requirements. Examples of meta-objects in the metamodeling layer are: UML::Class, UML::Attribute, CWM::Table and CWM::Record.

Layer M1 corresponds to metadata layer and it consists of models representing a domain of information. Examples in the model layer are the class "Person" and the attribute "Profession".

Finally, in layer M0 reside instances or data described by M1 models. Examples: "Ester", "teacher".

The MOF is intended to support a wide range of usage patterns and applications. Two distinct viewpoints are provided for possible usage patterns:

Modeling Viewpoint: it is the designer's viewpoint that uses the MOF to define information models for a particular domain. The model is then used to drive subsequent software design and implementation steps.

Data Viewpoint: it is the programmer's viewpoint that uses a meta-level to get understanding of an element in the lower level.

In addition to the open architecture that allows the definition of metamodels from different domains, the MOF specifies standard mappings to expose instances of MOF compliant metamodels with Corba IDL and Java interfaces. The prime purpose of these mappings is to define interfaces for information models described in terms of the MOF Model. These interfaces ensure structural and logical consistency in manipulating the metadata described by a metamodel.

The XML-based Metadata Interchange format (XMI) [OMG 2003] is another OMG's specification proposed to enable easy interchange of MOF based metadata in distributed heterogeneous environments. The XMI specification basically defines the mapping of a MOF metamodel to XML Document Type Definition (DTD) or XML Schema. The generated XML DTD or XML Schema is used to encode the metadata into XML documents and to decode XML documents and reconstructs the metadata. Any metadata repository or tool that can encode and decode XMI streams can exchange metadata with other repositories or tools with the same capability. Applications can be made more intelligent and automatic by adding metadata information to the runtime environment through the use of XMI documents, transferred over the Internet and validated and interpreted at a remote location.

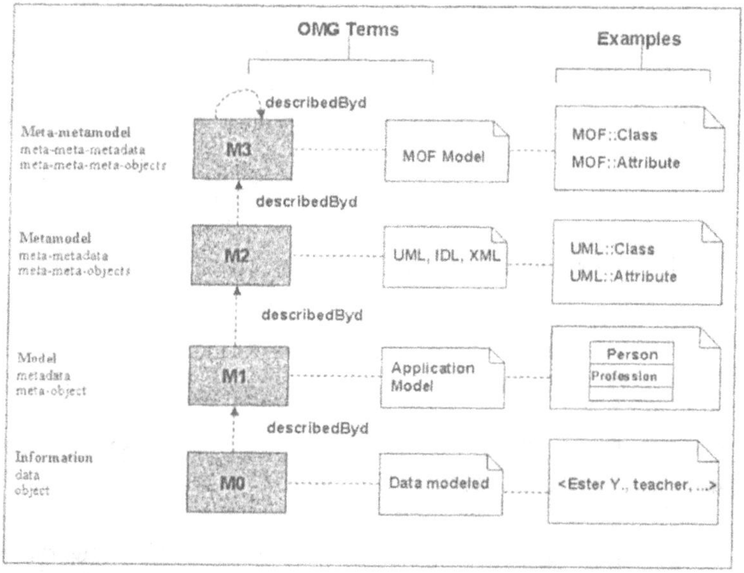

Figure 1 – *MOF Metamodeling Architecture*

3. GRM – A METADATA MANAGEMENT SYSTEM

The GRM system is part of a pilot project being developed at Research Center *Renato Archer* (CenPRA) with the main objective of creating a platform to support the development of Internet collaborative services in the government domain.

A central component of this platform is a metadata repository where complex artifacts like business rules and collaboration models can be stored. GRM is the system being developed to fulfill this key requirement. It is based on the MOF architecture and is expected to be used in both viewpoints described in section 2. From *the modeling* viewpoint, GRM can stores metamodels and corresponding models. From the data viewpoint, metamodels can be accessed to get the understanding of heterogeneous metadata.

3.1 GRM ARCHITECTURE

GRM's architecture is illustrated in Figure 2. The components of this architecture are based on the repository system architecture presented in [Bernstein, 1998] and they are described next.

Metamodeling Tools: these are client softwares that access the repository manager in order to allow the definition and manipulation of the artifacts (i.e. metamodels) stored in the repository. Examples of tools are graphical editors, XMI importer/exporter and compilers. Language generators are a very important kind of tool for GRM. In this case, a generator maps, automatically, a metamodel to MOF standard mappings such as XMI, Corba or Java, generating XML DTD, XML Schema, Corba and Java interfaces and their respective implementation;

Repository Manager: this component and the information model form the repository engine. Services are provided to access and manage the repository itself and the metamodels it stores. Features like version control, check-in/check-out, and access control and configuration management are required for a shared use of the repository;

Information Model: the information model specifies the structure and semantics of the metamodels stored in the system;

Persistence: the artifacts manipulated by the repository manager are stored in databases (RDBMS, OODBMS) and/or files system.

In the bottom of Figure 2 are presented boxes with some of the most relevant tools of GRM repository system. On the left side are the metamodeling tools and on the right side tools for modeling and metamodeling. In the next sub-section we describe with more details the implementation of the MODL compiler and the XML DTD generator.

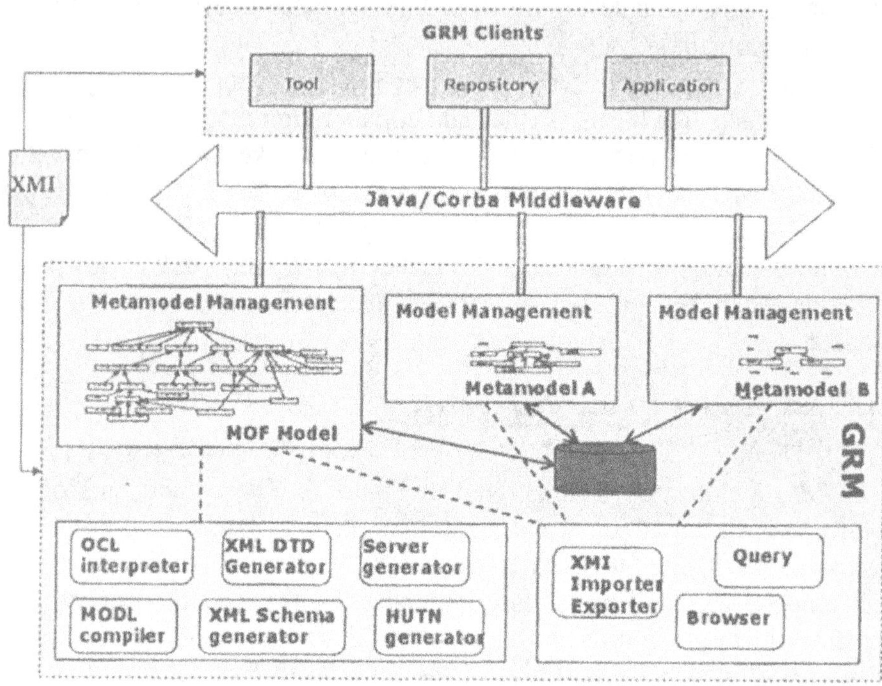

Figure 2– GRM Architecture

3.2 IMPLEMENTATION DETAILS

GRM has been realized using the Open Source Complex Information Manager (CIM) [Unisys 2002] developed by Unisys Corporation. CIM is an implementation of the MOF 1.4 and JMI specifications [JCP 2002]. Its goal is to provide a platform independent metadata infrastructure for developing model driven tools and applications suites. CIM standard edition is available with the following features: a GUI-based administrative tool for configuring and managing the CIM, Java interfaces generator and metamodel server generator, XMI importer/exporter, access control and persistence using XMI format files.

Our objective is to improve this open source, adding the following features:

- an MODL compiler;
- a generic metadata browser to metamodels and their respective models;
- a generic way to describe models using the Human Understandable Text Notation – HUTN. This notation is also generated from a metamodel.
- persistence in a relational database;

- XML DTD and XML Schema generators;
- Corba IDL generator;
- an Object Constraint Language (OCL) interpreter to evaluate the constraints used to enhance the semantics of a metamodel;
- query mechanisms, allowing to search the repository for specific elements.

At the moment, we have implemented the MODL compiler and the XML DTD generator. We have decided for the implementation of an MODL compiler because CIM comes with only one-way to populate the repository: import of metamodels expressed in the XMI format. XMI was specified to be a machine exchange format and it is neither succinct, nor easily readable or writable.

Meta-Object Definition Language (MODL) [DSTC 2001] is a textual language for specifying MOF metamodels and it provides an alternative way to populate the repository. The MODL compiler was implemented in the Java programming language. The tools JavaCC [JavaCC 2002] and JJTree [JJTree 2002] were used in the lexical analysis and in the generation of the syntax tree, respectively.

Figure 3 illustrates the different approaches used by the XMI importer and the MODL compiler, regarding meta-objects generation and external references. The importer tool reads an XMI file (1), generates a DOM tree (2) that is used by a generator (3) to create meta-objects (4 and 5), that is, MOF instances.

The MODL compiler approach is similar to the XMI importer. The two differences are in step 2 and they are presented next.

1. the DOM tree is generated (2c) from the syntax tree (2a) created during the compilation of the metamodel; and
2. The MOF model is accessed (2b) in order to check external references used in the metamodel expressed in MODL.

Figure 3 illustrates an example, where metamodel C is inserted in the GRM repository using the XMI importer and the MODL compiler. The metamodel C imports constructs defined in metamodels A and B and when it is mapped to XMI format, metamodels A and B are mapped as well. If metamodels A and B do not exist in the repository at the moment that C is inserted, A and B are also inserted. In the case of the MODL compiler, external references, expressed in the *import* command, are checked as the DOM tree is generated. The MOF Model interfaces are accessed in order to get the information about the contents of the repository. In the case, metamodels A and B do not exist, when C is compiled, an exception is raised.

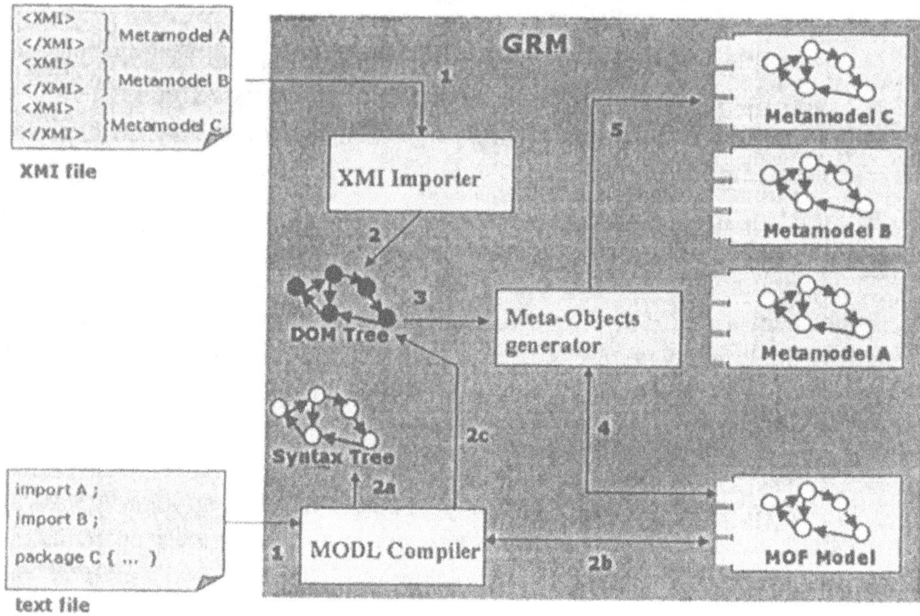

Figure 3 – Meta-objects generation in GRM system

Although the CIM comes with functionalities to import/export XMI files, the software does not generate XML DTD for a specific metamodel and, therefore, it is not possible to validate the syntax and the semantics of models instantiated from the metamodel. An XML DTD generator was implemented making heavy use of MOF Reflective Module. This module is defined in the MOF specification and its objective is to provide an introspection mechanism common to all metamodels. The XML DTD generator tool navigates through the metamodel, visiting all its constructs and for each one, reflective APIs are used in order to find out the meta-object that describes that construct. According to the rules defined in the XMI specification the construct can be mapped to an XML element or to an XML attribute.

4. AN *E*-GOVERNMENT APPLICATION SCENARIO

Governments around the world are looking for alternatives to deliver services in the Internet. Traditionally, government resources are spread over the public administrative units and a critical problem in this scenario is the challenge of integrating these resources, respecting the autonomy of each single public unit.

At CenPRA, it is being developed a prototype with the main objective of validating an architecture that facilitates the integration of distributed and heterogeneous legacy systems and data. The major components of this architecture are illustrated in figure 4 and described next.

The service platform provides a uniform service access and management environment, across different administrative and technological services. The integration framework, focus of this paper, is comprised of modules based on MOF technology and with the main objective of enabling data and metadata integration. Legacy systems, also referred to as legacy layer, are composed of databases, applications, and systems with enormous value to the organization.

In this scenario, the CWM and EDOC Modules are components generated automatically by the GRM system and based on the CWM and EDOC metamodels, respectively. The code generated implements the interfaces that allow the access and manipulation of models instantiated from CWM or EDOC metamodels.

The CWM specification [OMG 2003] provides metamodels to represent common warehouse metadata such us object-oriented, relational, record, multidimensional, XML data resources and others. Additionally, CWM provides a transformation metamodel with constructs for specifying transformation rules.

The EDOC specification [OMG2003] provides metamodels to specify enterprise-distributed systems using a component-oriented approach. The Entity metamodel is used in our scenario to model entity objects that are representations of concepts of the government domain. For example, a citizen.

CWM Module is in the integration framework and in the legacy layer. In the legacy layer it is responsible for getting the data and transforming it to XMI format. Metadata and data are interchanged between the integration framework and the legacy layer as stream. In the integration framework, transformation models map metadata from legacy layer to an EDOC entity model and resulting entity objects are made available to the service platform.

For the realization of this scenario, it was adopted an open service platform provided by Fokus Institute [Fokus], which is one of CenPRA's partners in the development of the e-government platform. The main objective of Fokus' platform is to enable integration and composition of existing and new application services based on different programming languages, access and services technologies.

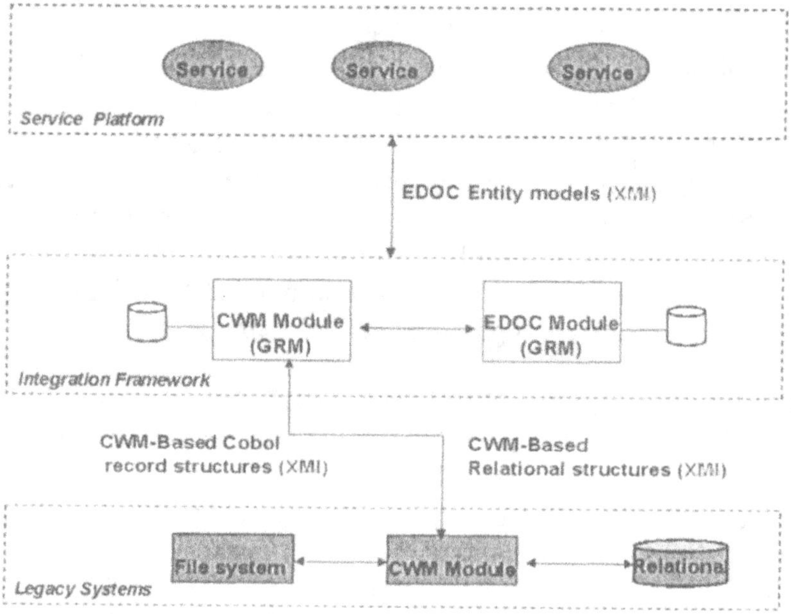

Figure 4– *GRM in an eGovernment Scenario*

5. CONCLUSIONS AND FUTURE WORK

In this paper a metadata management system to support metamodeling and distributed metadata access has been described. CWM and EDOC metamodels have been used to solve the problems posed by integrating legacy data in an e-government scenario. The components of GRM, created around Metadata Repository, such as MODL Compiler, Metamodel XML DTD Generator, Interfaces Generator, Metamodel Server Generator, Metamodel/Model Visual Tool, XMI Import/Export APIs and Query, improve the capabilities to generate metadata that needs to be exchanged among heterogeneous systems in an Internet based collaborative system. One of such improvements can be seen, for example, on MODL Compiler, which provides a flexible alternative way to populate the metadata repository. Another improvement is provided by a XML DTD generator, which navigates among a metamodel and maps its constructs to XML elements and attributes. New components are being implemented to increase the functionalities of the open software. The strength of GRM is in the fact that its inception, and all its components, were developed having in mind three specific requirements: (1) platform and operating system independence; (2) adherence to open standards proposed in the distributed system context; and, (3) use of the orthogonality concept when features are added, applying them to any metamodel.

The use of MOF concepts and CWM and EDOC metamodels create a robust foundation to an extensible and flexible architecture that promotes data integration and, not less relevant; ensure that legacy data quality is maintained.

Further research is ongoing on the basis of the present work. Application Process integration is the next issue to be addressed in the prototype. Research to define a framework that could facilitate the development and composition of Web services is being developed. The first scenario for this framework includes the creation of government ontologies, semantic representation of data and process to support consistent description of information for e-Gov services composition.

REFERENCES

Bernstein P. A (1998) "Repositories and Object oriented Databases", ACM/SIGMOD vol. 17, pages 88-96, 1998 http://www.acm.org/sigs/sigmod/disc/disc99/disc/record/issues/9803/bernstein.pdf accessed in 05/19/2003

DSTC (2001) "dMOF – User Guide (Release 1.1)", Distributed Systems Technology Centre (DSTC), University of Queensland, Brisbane, Australia, June/2001. http://www.dstc.com/Downloads/CORBA/MOF/dMOF1_1.UserGuide.pdf

EIA (1994) "CDIF - CASE Data Interchange Format – Overview", Electronic Industries Association – EAI, January, 1994

Fokus – Fraunhofer Institute for Open Communication Systems http://www.fokus.fraunhofer.de/

JavaCC (2002) "Java Compiler Compiler (JavaCC) - The Java Parser Generator". http://www.webgain.com/products/java_cc/documentation.html

JCP (2002) "Java Metadata Interface (JMI) Specification", version 1.0, JSR40 – Java Community Process, June/2002. http://jcp.org/aboutJava/communityprocess/final/jsr040/index.html

JJTree (2002) "JJTree Reference Documentation". http://www.webgain.com/products/java_cc/jjtree.html

MDC (1999) "Open Information Model", Object Management Group, Needham-MA, USA, April/2002. http://www.omg.org/cgi-bin/doc?ad/99-04-06

OMG (2003) "Catalog of OMG Modeling and Metadata Specifications", v. 1.4, Object Management Group, Needham-MA, USA, October/2003. http://www.omg.org/technology/documents/modeling_spec_catalog.htm

SECTION 8
Cooperation and Integration of e-Services

Chapter 24

A SERVICE ORIENTED APPROACH TO INTERORGANISATIONAL COOPERATION

Christian Zirpins, Winfried Lamersdorf, Giacomo Piccinelli

Abstract: Many E-business applications are based on increased cooperation between various organisational units and partners. System support for such applications can be provided using concepts from the area of *service oriented computing* – thus lifting inter-organisational integration to a higher level of effectiveness and efficiency. *E-services* provide means for modularisation of arbitrary organizational assets into components that can be dynamically offered, discovered, negotiated, accessed, and composed in an open application environment. Technically, E-services are software systems that are implemented on top of conventional information and communication technology. As an important step into that direction, *Web Services* have laid the foundation for interoperable communication between arbitrary systems. This paper introduces an approach to plan, build, and run such application-level services efficiently. Therefore, a fundamental notion of service, originating from distributed systems, is being extended by a specific concept of *cooperative interaction processes*. Accordingly, an application-level service model and corresponding service engineering mechanisms are proposed and realised as system software middleware based on OGSA Web Services and BPEL4WS processes.

Key words: E-Business, Inter-Organisational Integration, Cooperative Interaction Processes, Electronic Services, Workflow, Web-/GRID Service Architecture

1. INTRODUCTION: SERVICE ORIENTED DISTRIBUTED APPLICATIONS

Various application domains like electronic business, -government and -education face *recurrent cooperation scenarios* where a constant change of participants is a predominant characteristic. Typical examples are business-to-business integration problems (Medjahed et al., 2003) that focus on the dynamic relationship between one company and a set of frequently changing partners. Also, situations like flexible outsourcing of business functions or dynamic supply chain management face similar recurring types of cooperation with interchangeable partners. For example, a company might contract out freight logistics to various carriers or forwarding agencies changing over time. The rationale for this kind of relationship is, on the provider-side, to expose new revenue streams (e.g. providing freight logistics on demand) and, on the customer-side, to seek for new efficiencies (e.g. outsourcing freight logistics if profitable) in a form that allows for constant optimization of partnership settings. Strategic planning of cooperation types, tactical preparation of cooperation settings, as well as operational control of functional cooperation are among the main challenges to be tackled here. In more advanced scenarios, the patterns of functional cooperation are often a subject of variation too, because different partners pose different operational requirements that have to be negotiated between the participants beforehand.

For example, a customer cooperates with various carriers that all move goods but impose different procedures of payment. Moreover, when broadening the scope, a party often faces multiple of such cooperative relationships that are in some cases mutually dependent. In order to preserve these dependencies, they have to be made explicit independently from individual partners. For example, a forwarding agency has to ensure that it can move goods of various individual customers by relying on alternating carriers under contract.

1.1 Extending the service notion

Henceforth, the notion of a *service* is used to refer to such recurring cooperation scenarios between changing autonomous participants. In order to substantiate this notion, one can benefit from former work in distributed information systems: Revisiting ODP concepts (ISO/IEC-JTC1/SC21, 1995), we distinguish the constant class of cooperation (*service type*) from changing cases of cooperation (*service instance*). Service instances can vary in the conditions of cooperation referred to as *service properties* (e.g. QoS) that arise from the characteristics of actual participants. Those participants are typed by *roles*, indicating expected cooperative behaviour within service relationships. Providers offer type and properties of instances they are willing to participate in. Clients observe offers of a specific type, choose a provider with respect to service properties and engage in service instances.

Specific interdependencies between services are often referred to as *service composition*. In this case, a participant relates (*composes*) services in which he acts as a provider (*composite services*), to services in which he acts as a client (*service components*), stating how characteristics of the composite service are put down to characteristics of service components. In terms of characteristics, services on application-level are more complex then those found in classical distributed object systems. Apart from the 'semantic' reason (e.g. move goods), the 'syntactic' cooperation process (e.g. customer orders → carrier confirms and ships goods → customer pays) is among the predominant service characteristics. In particular, the focus here is on the interaction patterns, that is, the communication processes between roles.

The field of problems faced by organisations in terms of service participation can be structured into strategic, tactical and operational challenges. On the strategic level, exposing and expressing semantic and syntactic aspects of service types and their interdependencies requires expressive models and systematic design methodologies (*service modelling*) taking under account the (technological and conceptual) context of participants. For example, a forwarding agency needs models to express a) meaning and procedure of a logistics service it provides b) dependencies of the logistics service on a freight service that it uses and c) mappings of the service interactions to its internal business information systems. On the tactical level, service types have to be constantly maintained to keep track with organisational change (*service type adaptation*). Also on this level, partners have to be located for the types of service a participant is interested in as client (*service discovery*) or provider (*service publication*). On the operational level, partners have to be matched (by providers) and chosen (by clients) for service types (*service aggregation*). In some cases, providers additionally have to choose component service types matching the clients of composite services beforehand (*service composition*). During the actual service interaction procedure, terms and conditions of the service have to be ensured (*service coordination and control*). Additional flexibility can be reached by dynamic changes of service instances (*service instance adaptation*).

Generally for all levels, system software middleware is needed to arrange organisational environments of information- and communication technology (ICT) into a *cooperative information system* (Michelis et al., 1997) which realise services and provide support for the various tasks described above.

We refer to such a middleware as *service management system* and to the joint tasks of planning, building, and running of service oriented distributed applications as *service engineering*.

1.2 Current state of technology

Current techniques of service oriented computing are strongly focused on technology. While application-level (i.e. business) service support is out of their scope, they nevertheless pave the way towards it. The emerging Web Service standard (Tsalgatidou and Pilioura, 2002) provides interoperability between heterogeneous systems by leveraging the expressive power of XML to specify operational interfaces that can be accessed using open internet communication. Thus, organisations can externalise their internal information systems as web enabled components. Those components provide interaction endpoints (subsequently called *ports*) to participate in automated inter-organisational cooperation. Concerning cooperation procedure, the service oriented model adopted by Web Services only defines a very basic type of interaction (i.e. 'broker triangle'). However, web service flow standards like BPEL4WS (Curbera et al., 2002) provide the means for individual definitions of basic interaction processes. This is the crossing point to more general research on cooperative, inter-organisational interaction-processes (e.g.(Baïna et al., 2003, Bussler, 2002, Schuster et al., 2000)) and workflow (e.g.(van der Aalst, 1999, Colombo et al., 2002, Chen and Hsu, 2000)), where several practical approaches for application-level services are located (e.g.(Mecella et al., 2001, Perrin et al., 2003, Casati et al., 2001)).

1.3 The Fresco Project

The FRESCO project is about foundational research on service composition (Piccinelli et al., 2003b). Its goal is to develop a framework of concepts and technologies that support organisations in playing the provider role for composite services. As a basis for composition, the focus is on the components first. Subsequently, a fundamental service model was developed that describes basic application-level services as classes of recurring cooperative interactions. The model was then implemented as a generic service engineering environment built on the Web Service family.

In the remaining parts of this paper, the Fresco approach will be detailed:

After the second part sketches a basic blueprint of our service engineering concepts, the third part introduces the Fresco Toolkit implementation. Finally, a summary and an outlook are given.

2. SERVICE ENGINEERING IN FRESCO

Fresco service engineering is based on a model that defines services as structured sets of cooperative interaction procedures. This model implies a specific architecture of service oriented applications that builds on an open, distributed component environment with service-enabling extensions. Subsequently, a service engineering environment provides a concise framework to plan, build, and run such service oriented distributed applications.

2.1 Service Model

The FRESCO Service model (Piccinelli et al., 2003a) defines a view on services that is provision-oriented and service-centric. Cooperation procedures that constitute atomic, self-contained parts of a service-relationship are exposed by so called capabilities. In particular, capabilities represent purpose, interaction logic, and resulting artefacts of the cooperation between organisational roles. Thereby, capabilities define additional coordinative roles that introduce a level of indirection between participating roles. Unlike meta-level protocols, capabilities take the position of first-class participants (i.e. coordinators) that may be just virtually or effectively enforced. A service is made up by a set of such capabilities.

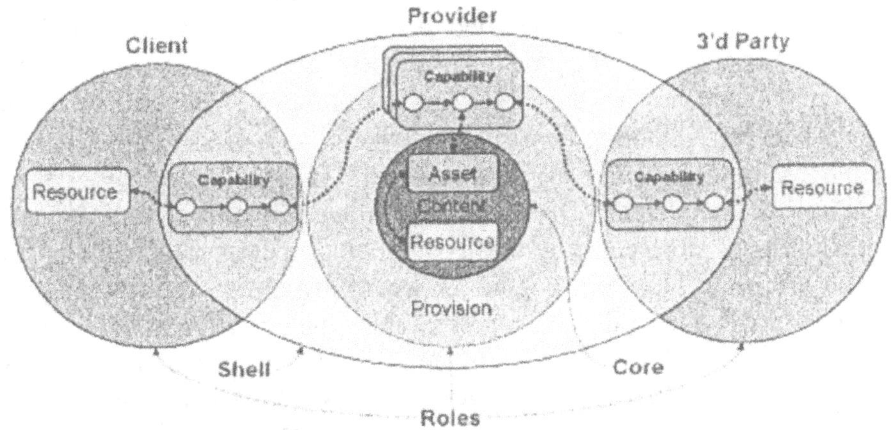

Figure 1. Fresco Service Model

An important feature of the model is a separation of capabilities in terms of service content and -provision. Content reflects the purpose of a service (e.g. moving goods). It is assumed that it arises from specific resources of the provider (e.g. internal processes, knowledge, people, machines, etc.). To represent service content, cooperation procedures, featuring interactions with such resources, are explicitly exposed as meaningful units of content (e.g. transport tracking...) by capabilities referred to as *assets*. Assets are degenerated in the sense that they don't represent cooperative interaction between roles but monologues of the provider (i.e. binding (Bussler, 2002)) that have to be provided to clients indirectly by other capabilities. Assets are grouped into a *service core* representing the complete content.

Provision addresses procedures that drive a service and make content available (e.g. negotiating terms and conditions, incorporating assets, etc.), whereby control is exclusively and proactive. Service provision capabilities (hence called "capabilities") are grouped around core assets in a layer called service shell. Within a shell, capabilities are mutually interrelated and share a common view on roles and provision-relevant information. Interrelations embody the overall behaviour of provision by defining the global interplay of capabilities. A service is fully characterised by defining the basic core and, above all, the enabling shell (Figure 1). Our main focus is on the later. To realise this service model, associated technology has to focus on a) an architecture mapping the service notion to organisational ICT and b) an environment of mechanisms that facilitate service engineering tasks on top of it.

2.2 Service oriented architecture

For technology mapping we define a framework referred to as *service oriented architecture* (SOA). It provides a layer of abstraction that is assumed to wrap around diversified ICT systems in order to provide a homogeneous platform for service management. Service types are defined as *schemas* with respect to the SOA. Service instances can be run in any environment implementing the SOA framework.

In SOA, we assume that all organisational ICT resources of any role (e.g. client's ERP, provider's DBMS...), providing ports for service-related interactions, are represented by means of a homogeneous component model. Shell capabilities appear as glue between ports that reflects purpose, interaction logic and result. We represent this glue using workflow concepts based on the WfMC reference model (WfMC, 2002). Common patterns are prescribed to define capabilities as well as their structuring and interrelations by means of the workflow language XPDL, resulting in a *service schema.*

In particular, a capability maps to a set of workflow schemata describing a self contained unit of interaction. Ontology-associations define the *purpose* of *interaction logic* that emerges from the flow of interaction activities and results in data artefacts. Interaction activities can be defined for a participant (i.e. a role-associated component-port) to express cooperative procedure or for another capability workflow to express capability interrelation. Coherent sets of capability workflows are grouped together into packages with respect to a self contained task (e.g. negotiation capability, payment capability). The shell is given as a top-level package, where each capability is abstracted as a component type itself that realises the enclosed interaction flows and has a specific role assigned to it. Thus, various coordination concepts can be expressed including centralised- (orchestration) and distributed scenarios.

In brief, a schema specifies a partitioned set of highly interrelated components with precise interaction behaviour, where a subset A represents interacting participants and a subset B represents and enforces their interaction patterns. Service engineering is about planning, building and running B based on A.

2.3 Service engineering environment

Our concept of service engineering defines a set of basic engineering mechanisms that allow building customized extensions upon it. Besides *modelling*, the main problems addressed here are *adaptation*, *aggregation*, and *coordination*.

As services are inherently complex, we anticipate that support will be needed for their design, that is, a graphical *service modelling language* and tool, which help developers in creating service schemata. This is supposed to be the initial step of the service lifecycle, performed by the provider role.

Service schema management provides the functionality to process the schema programmatically. Beside storing and retrieving it, adaptation is its vital task. We adopt a rule based approach that provides a precise and systematic way to change schemata automatically. Back in the service lifecycle, the schema is subject of continuous static adaptation until eventually brought to action.

Then, it's the task of *service aggregation management* to create a service instance, based on the schema and a mapping of roles to actual participants. The main problem is to allocate resources of the participants according to the components associated to their roles, thereby optimising resource allocation while guaranteeing a constant and consistent flow of service procedures even when schema or participants change during provision. Initially, at least the provider is known and resources for an initial capability have to be allocated.

Service engines are components that manage the aggregation and coordination of capabilities they realise. The crucial problem is for participants to implement the capabilities of an engine while keeping the service context including associations to other engines and a homogeneous view on roles and data. We propose a generic implementation framework that can be parameterized with executable specifications generated from the schema. When all engines reach a final state the instance expires and the service lifecycle continues with a new round of static schema adaptation.

In addition to the core functions introduced so far, three other mechanisms are considered particularly useful: *service monitoring* to integrate the measurements of distributed sensors deployed throughout the service components into a coherent view of the overall service status, *security management* to allow controlling component access and delegating access privileges, and, finally, *type management* that defines a type system for generic software components and allows discovering compatibility and equivalence between them to support the handling of resources during service design and aggregation. Figure 2 gives an overview of all mechanisms and their respective relations.

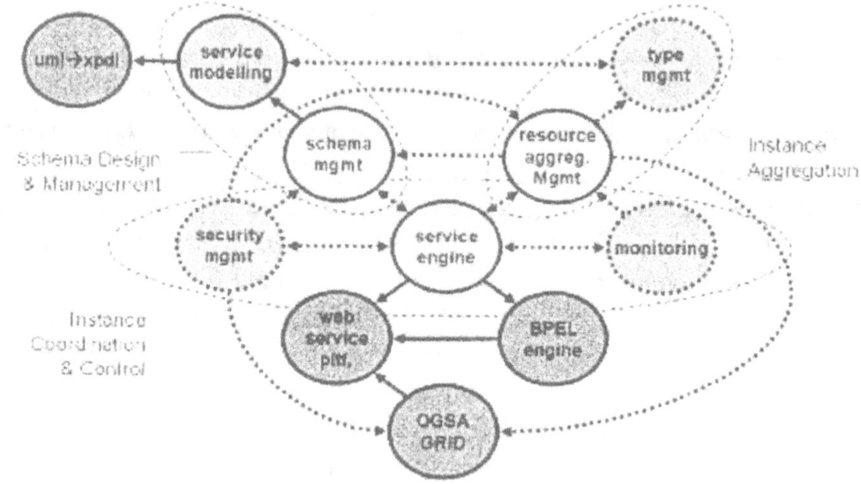

Figure 2. Service Engineering Environment

A vital characteristic of the overall engineering environment lies in the fact that all management mechanisms are first class components themselves. Thus changes can a) be made at provision time and b) arise from capabilities themselves. For example, a capability can lead to dynamic changes of participants (e.g. a new participant is introduced as a result of a brokerage capability) or dynamic schema adaptation (e.g. a payment procedure is changed as the result of a negotiation capability). Note, that this allows extending the service engineering mechanisms by realising them as capabilities.

3. THE FRESCO TOOLKIT

As a proof of concept, we developed the core functions of *service schema design, adaptation, aggregation,* and *coordination* in a prototype environment referred to as the *Fresco Toolkit (FrescoTK) 1*. This implementation is structured into parts related to *service schema* and *instance management*.

3.1 Service schema management

The focus of service schema management in Fresco is on the representation and organisation of interactive procedures that make up a service. A schema defines the complete shell of a service including roles and resources as well as the mapping of procedures to capabilities. More precisely, a service schema is realised as a structured transformable set of abstract workflow definitions. We use XPDL, where generic workflow elements are defined in the context of packages that can again refer to other packages thus allowing the definition of coherent structures. A service schema contains a) a root package representing the service shell and declaratively defining the service capabilities, b) one set of packages for each capability that defines its interactive procedures, and c) one context package that defines a common context of roles, data structures, and resources. The FrescoTK *Schema Manager* component (Figure 3) holds generic specifications of various service schemata and makes them programmatically accessible. Its vital characteristic is the ability to apply a variety of transformations to them that allow for controlled changes of service structures as well as for the logic of interactive procedures. Moreover, it is possible to change the representation of procedures into executable format.

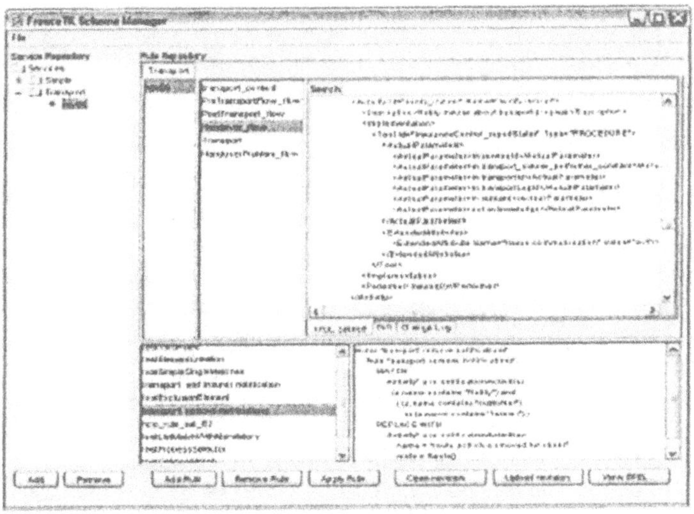

Figure 3. FrescoTK service schema tool

Evolutionary adaptation is supported by means of a language for change rules that are enforced by a rule engine within the schema manager. It allows *matching* arbitrary patterns in XPDL workflow process descriptions and *removing* or *replacing* those matches with newly created process elements into self contained revisions. However, the procedural logic of capability components, given in XPDL, has to be transformed into a format that can be executed by an engine. Those engines are used as active components that enforce provision procedures at runtime (see 3.2). The transformation is based on the fact that most workflow languages share a set of core concepts with common semantics (see (van der Aalst, 2003)). We chose the emerging BPEL4WS standard as our execution format and defined a mapping to it from XPDL (a full coverage can be found in the FrescoTK documentation).

3.2 Service instance management

Service instance management comprises organisation of participants and resources for service instances. The FrescoTK *Aggregator component* evaluates service schemata for involved roles and necessary resources (Figure 4).

During service execution, all roles have to be assigned to participants and each of them has to provide the resources associated with its roles. An individual strategy can be chosen for each service that specifies how to do role assignment and resource creation in terms of schedule and execution model.

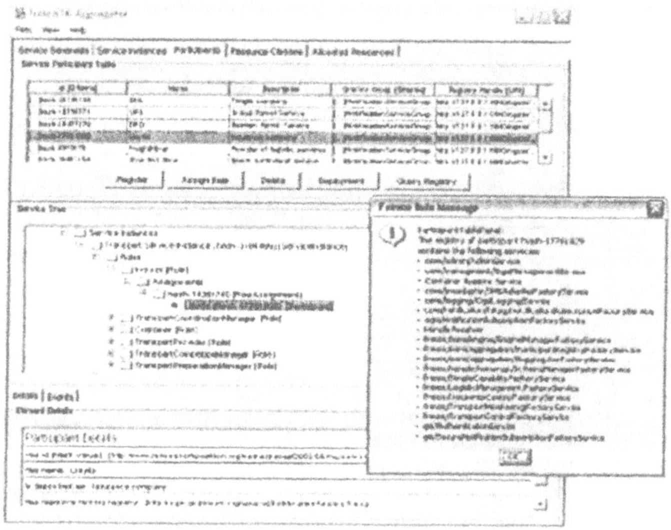

Figure 4. FrescoTK Aggregator

In FrescoTK the SOA is based on OGSA (Foster et al., 2002) as the component model. Thus, service related resources as well as schema management, aggregation, and engine components are built as GRID services. Engine components are realised by a BPEL engine that executes process specifications generated by the schema manager (Figure 5). The engine is wrapped as a grid component by adapters and proxies that are automatically generated for each capability. They bridge the gap between stateless Web Services and long lived Grid Services using the aggregator to resolve references of individual resources.

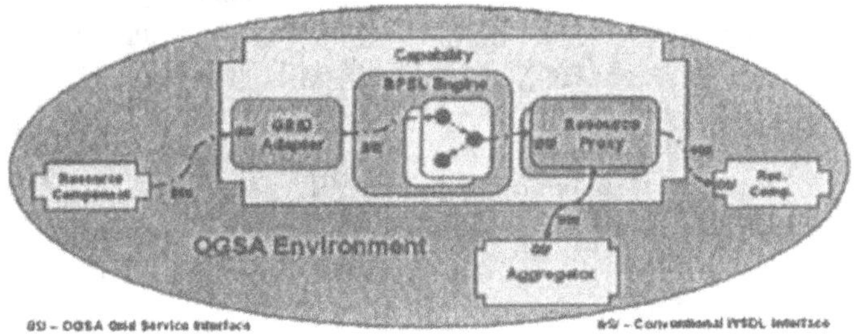

Figure 5. Capability Grid Architecture

4. CONCLUSION

As inter-organisational relationships and cooperation increase in advanced e-business (and other similar) applications both in terms of quantity and of quality, the need for new classes of distributed applications arises that allow their effective and efficient management. In this paper, we focus on recurring cooperation scenarios between changing autonomous participants and system support for them based on service-oriented distributed applications.

While, in such a context, a suitable technological foundation is already in place to interconnect the participants, adequate support for, e.g., planning, building, and running such solutions is still missing. Therefore, we propose a service model based on advanced Web service and Grid Service technology and address a set of problems realising it within a basic service engineering approach. This approach applies process theory and workflow concepts to specify, aggregate, enact, and adapt services as interaction patterns between distributed resources. In particular, we adopt a homogeneous view on resources, coordination-, and engineering mechanisms that allows for a degree of introspection and dynamic self-adaptation.

We claim that this concept is powerful enough to implement complex service scenarios with customized requirements. In future work, we will use the service engineering mechanisms to examine models and mechanisms for service composition that allow relating and connecting the capabilities of composite services to the capabilities of their service components.

REFERENCES

Baïna, K., Tata, S. and Benali, K. (2003) A Model for Process Service Interaction, In Busines Process Management International Conference, BPM 2003, Eindhoven, The Netherlands, June 26-27, 2003. Proceedings(Ed, Weske, M.) Springer, pp. 261 ff.

Bussler, C. (2002) Behavior abstraction in semantic B2B integration, In Conceptual Modeling for New Information Systems Technologies. ER 2001 Workshops. HUMACS, DASWIS, ECOMO, and DAMA. Revised Papers Lecture Notes in Computer Science Vol.2465. 2002(Ed, Hunt, I.) Springer Verlag, Berlin, Germany, pp. 377-89.

Casati, F., Sayal, M. and Ming Chien Shan (2001) *Developing e-services for composing eservices,* In *Advanced Information Systems Engineering. 13th International Conference, CAiSE 2001. Proceedings Lecture Notes in Computer Science Vol.2068. 2001*(Ed, Norrie, M. C.) Springer Verlag, Berlin, Germany, pp. 171-86.

Chen, Q. and Hsu, M. (2000) *Inter-Enterprise Collaborative Business Process Management, HPL-2000-107, Software Technology Laboratory, HP Laboratories Palo Alto*

Colombo, E., Francalanci, C. and Pernici, B. (2002) *Modeling Coordination and Control in Cross-Organizational Workflows,* In *Proc. CoopIS/DOA/ODBASE 2002*(Eds, Meersmann, R. and Tari, Z.) Springer, pp. 91 ff.

Curbera, F., Goland, Y., Klein, J., Leymann, F., Roller, D., Thatte, S. and Weerawarana, S. (2002) *Business Process Execution Language for Web Services, V 1.0, BEA, IBM, Microsoft*

Foster, I., Kesselman, C., Nick, J. and Tuecke, S. (2002) *The Physiology of the Grid: An Open Grid Services Architecture for Distributed Systems Integration, Open Grid Service Infrastructure WG, Global Grid Forum* ISO/IEC-JTC1/SC21 (1995) *Basic Reference Model of Open Distributed Processing -- Part3: Architecture. International Standard, 10746-3, ISO*

Mecella, M., Pernici, B., Rossi, M. and Testi, A. (2001) *A Repository of Workflow Components for Cooperative e-Applications,* In *Proceedings of the 1st IFIP TC8 Working Conference on E-Commerce/E-Business (Salzburg, Austria, 2001)*BICE Press, pp. 73-92.

Medjahed, B., Benatallah, B., Bouguettaya, A., Ngu, A. H. H. and Elmagarmid, A. K. (2003) *Business-to-business interactions: issues and enabling technologies,* The VLDB Journal, (Springer, Online First, April 3, 2003).

Michelis, G. D., Dubois, E., Jarke, M., Matthes, F., Mylopoulos, J., Papazoglou, M. P., Pohl, K., Schmidt, J., Woo, C. and Yu, E. (1997) *Cooperative Information Systems: A Manifesto,* In *Cooperative Information Systems* (Ed, Papazoglou, M., Schlageter, G.) Academic Press.

Perrin, O., Wynen, F., Bitcheva, J. and Godart, C. (2003) *A Model to Support Collaborative Work in Virtual Enterprises,* In *Business Process Management International Conference, BPM 2003, Eindhoven, The Netherlands, June 26-27, 2003. Proceedings*(Eds, Aalst, W. M. P. v. d., Hofstede, A. H. M. t. and Weske, M.) Springer, pp. p. 104 ff.

Piccinelli, G., Zirpins, C. and Gryce, C. (2003a) *A Provision-Centric Model for Electronic Services,* In *IEEE International Workshops on Enabling Technologies: Infrastructure for Collaborative Enterprises (WETICE-2003)*IEEE Computer Society, pp. 113-116.

Piccinelli, G., Zirpins, C. and Lamersdorf, W. (2003b) *The FRESCO Framework: An Overview,* In *2003 Symposium on Applications and the Internet Workshops (SAINT 2003 Workshops)* IEEE Computer Society, pp. 120-123.

Schuster, H., Georgakopoulos, D., Cichocki, A. and Baker, D. (2000) *Modeling and Composing Service-Based and Reference Process-Based Multi-enterprise Processes,* In *Proc CAiSE 2000*(Ed, Bergman, L.) Springer, pp. 247-263.

Tsalgatidou, A. and Pilioura, T. (2002) *An overview of standards and related technology in Web Services,* Distributed and Parallel Databases, 12, 135-62.

van der Aalst, W. M. P. (1999) *Process-oriented architectures for electronic commerce and interorganizational workflow,* Information Systems, 24, 639-71.

van der Aalst, W. M. P. (2003) *Don't go with the flow: Web services composition standards exposed,* IEEE Intelligent Systems, 18.

WfMC, (2002) *Workflow Management Coalition,* http://www.wfmc.org, 1.5.2003

Chapter 25

A DATA AND EVENT ORIENTED WORKFLOW PROCESS DEFINITION METAMODEL COHERENT WITH THE UML PROFILE FOR EDOC SYSTEMS

José A. Soto Mejía

Abstract: This paper presents a plausible mapping between the adopted UML Profile for EDOC systems and one of the submissions to the OMG's Request for Proposal for a Workflow Process Definition Metamodel. Then, given that the proposed by DSTC Workflow Process Definition Metamodel does not consider the specification of events therefore in this paper it is suggested a new workflow process definition metamodel that include these aspects with a clear separation of the data and event oriented control flow dependencies and that is compatible with the UML Profile for EDOC systems. Furthermore, in order to build the proposed in this paper workflow definition metamodel as an UML profile, the new workflow metamodel is derived (stereotyped) from the UML 1.4 Activity Graphs. To introduce a representation that be computationally interpretable by a workflow engine a textual notation is introduced to represent a workflow process definition coherent with the proposed Workflow Process Definition Metamodel. As a proof of concept a simple prototype of a workflow engine able to interpret the textual representation of the workflow process definition was implemented in the Java language.

Key words: Workflow, Process Definition, Metamodels, Enterprise Distributed Object Computing, EDOC, UML profile.

1. INTRODUCTION

The adopted by the Object Management Group (OMG), UML Profile for Enterprise Distributed Object Computing (Heaton, 2002), supports the requirements for driving an object oriented design of an enterprise computing system (a software system that provides support for carrying out an integrated set of business processes across an enterprise) to an implementation in an enterprise distributed computing environment using an enterprise class component model.

It is recognized that a successful implementation of such a system requires the operation of the system to be directly related to the business processes it supports. As it is also recognized that workflow management systems are today essential to corporate organizations that need to automate their business processes since they allow organizations to specify, execute and efficiently monitor their business processes. The specification of those parts of the business process to be automated is done using the model elements of a given workflow process definition metamodel or its corresponding workflow process definition language. In this sense, a few years ago (December 2000) the OMG started the process (Boldt, 2000) to standardize a metamodel and/or a profile which extends the UML for defining workflow processes and more recently the OMG has issued a more general Request for Proposal-RFP- (Cohete, 2003) that solicits submissions that specify a business process definition metamodel, which is platform independent with respect to specific business process definition languages.

To avoid a frequently misunderstanding between business analysts (workflow modelers) and software developers it is important that a clear mapping exists between the workflow process definition metamodel and the enterprise computing system metamodel. With this in mind, in section 2, this paper presents a plausible mapping between the above cited UML Profile for Enterprise Distributed Object Computing (EDOC Profile) and the submitted by DSTC (The Cooperative Research Center for Enterprise Distributed Systems Technology) response (Keaton, 2001) to the OMG's Request for a Proposal for Workflow Process Definition Metamodel (Boldt, 2000).

In Section 3, given that the proposed by DSTC Workflow Process Definition Metamodel does not consider the specification of events then in this paper it is suggested a workflow process definition metamodel that include these aspects (using the sub-profile for events that is part of the EDOC profile), with a clear separation of the data and event oriented control flow dependencies. Furthermore, in order to build the proposed in this paper workflow definition metamodel as an UML profile, the proposed metamodel is derived (stereotyped) from the UML 1.4 Activity Graphs (Heaton, 2001) and the implications of this derivation for the mapped metamodels (EDOC Profile and the submitted by DSTC metamodel) are pointed out. To introduce a notation that is computationally interpretable by a workflow engine, Section 4 introduces sketches of a notation to represent a workflow process definition coherent with the proposed workflow process definition metamodel. The paper ends with a Conclusions and On-going Work Section 5, that sketches the current state of the standardization process going on in the OMG which goes along the approach proposed in this paper

2. MAPPING BETWEEN METAMODELS

Before the mapping between the proposed in this paper workflow definition metamodel and the EDOC metamodel (the UML Profile for Enterprise Distributed Object Computing) is presented, a brief description of the main meta-entities of each of the metamodels is given in Tables 1 and 2. In Table 1, there is a summary of the main concepts related with the business and events profile for EDOC systems and in Table 2 a summary with the main concepts for the DSTC's workflow process definition metamodel. In order to facilitate the comparison and the identification of the source of context, the EDOC terms are presented in **bold** and the DSTC term in *italic.*

Table 1. Main Concepts from the EDOC Profile

ProcessComponent: It represents an active processing unit-"it does something". It defines a set of **Ports** (a **port** defines a point of interaction between **ProcessComponents**) for interaction with other **ProcessComponents** and it has a set of properties that are used to configure the **ProcessComponent** when it is used.
CompoundTask: It defines how to coordinate a set of related **Activities** that, in combination, perform some larger scale activity, ultimately in the context of a **BusinessProcess**. It is also a container of **Activities, DataFlows** between these **Activities**, and the **ProcessRoles** which model bindings to objects required by these **Activities.**
Activity: It represents the execution of a part of a **BusinessProcess** using one of two mechanisms: (i) The creation of a composition of nested **Activities, ProcessRoles** and **DataFlows** described by the **CompoundTask** that the **Activity** references through its uses association. (ii) The execution of some feature of an object bound to a **ProcessRole** instance referred to via the Activity's **performedBy** association. An **Activity**'s PortUsages representing **InputGroups**, which contain **ProcessPortConnectors** representing **ProcessFlowPorts**, are the alternative means by which the **Activity** may supply data to these mechanisms to initiate some action.
DataFlow: It represents a causal relationship in a business process. The source of the **DataFlow** must "happen" before the sink of the **DataFlow**. **DataFlows** also propagate data values between causally related **ProcessPortConnectors.**
ProcessFlowPort: It represents the formal types of inputs to and outputs in the context of a **CompoundTask**. **ProcessPortConnectors** represent the usage of the **ProcessFlowPorts** of a **CompoundTask.**
ProcessPortConnector: It represents the usage of a **ProcessFlowPort** of a **CompoundTask**. A **ProcessPortConnector** may be associated with a source or a sink of a **DataFlow. DataFlows** allow the connection of the **ProcessPortConnectors**, which are owned by a **CompoundTask**.
BusinessProcess: It represents the complete process specification. A **BusinessProcess** defines the **ProcessComponent** view of a process definition that coordinates a set of related **Activities**. It defines a complete business process, which can be invoked, usually using **Ports**, which are connected via **DataFlows** to the **ProcessPortConnectors** of the **Activities**, which it contains. In other words a **BusinessProcess** is an ordinary **ProcessComponent** on the outside, and a **CompoundTask** on the inside.
ProcessMultiPort: It represents a set of related **ProcessFlowPorts** used to describe the inputs and outputs of **CompoundTasks**. They act as a form of correlator for **DataFlows**.
InputGroup: It is a specialization of **ProcessMultiPort**. It is a container for a number of **ProcessFlowPorts**, which are the inputs to a **CompoundTask**, and acts as a form of correlator for

DataFlows.
OutputGroup: It is a specialization of **ProcessMultiPort**. It is a container for a number of **ProcessFlowPorts**, which are the outputs of a **CompoundTask**, and acts as a form of correlator for **DataFlows**.
ExceptionGroup: It represents the outcome of a **CompoundTask** that failed to complete its function.
ProcessRole: It defines a placeholder for concrete **ProcessComponents** that perform an **Activity** or that are used in the performing of an **Activity**. The owner of a **ProcessRole** is a **CompoundTask** and the behavior of the **ProcessRole** becomes part of the behavior of **Activities** to which it is associated.
Publisher: It is a component that exposes a list of **Publications**, and produces **PubSubNotices** accordingly.
Subscriber: It is a role or component that exposes a list of **Subscriptions**, and consumes **PubSubNotices** accordingly.
Publication: It is a declaration of capability and intent to produce a **PubSubNotice**. A Publisher owns it.
Subscription: It is the expression of interest in receiving and capability to receive a **PubSubNotice**. A Subscriber owns it.
EventBasedProcess: It is a **Subscriber** and has **NotificationRules** associated with its **Subscriptions**. It is a **Publisher** and publishes **ProcessEvents**. **ProcessEvents** describe the life cycle of the **EventBasedProcess**.
NotificationRule: It is a rule associated with a **subscription** which determines what should happen within the **EventBasedProcess** holding the **Subscription** when a qualifying **PubSubNotice** is delivered. Optionally an **EventCondition** that requires the delivery of additional events can further guard the **NotificationRule**. A **NotificationRule** is owned by an **EventBasedProcess**
PubSubNotice: It is any data structure that is *announcedBy* a **Publication** and/or *subscribedTo* by a **Subscription**. Instances of **PubSubNotice** are communicated as **DataFlows** from **Publishers** to **Subscribers** based on the **Subscriptions**.
EventNotice: It is any **PubSubNotice** that is triggered by a business event. An **EventNotice** may describe at most one business event.
BusinessEvent: It is any event of business interest that happens within an enterprise. A business event triggers one or more event notices and is described by one or more event notices.
NotificationRule: It is a rule associated with a **Subscription** which determines what should happen within an **EventBasedProcess** holding a **Subscription** when a qualifying **PubSubNotice** is delivered
EventCondition: It identifies a **Subscription** and specifies a **PubSubNotice** instance subset of which one must have been received to satisfy this condition. A **NotificationRule** owns it.

Next Table 2, presents a synthesis with the main concepts that are part of the proposed by the DSTC Workflow Process Definition Metamodel.

Following, the particular characteristics of the mapping between the meta-concepts of the DSTC Workflow Process Definition Metamodel and the EDOC metamodel are explained and Table 3 presents a synthesis of the mentioned mapping.

Table 2. Main Concepts from the DSTC Workflow Definition Metamodel

BusinessProcess: It establishes a context within which sets of business actions, taking place in a prescribed manner, **are** coordinated to achieve some enterprise objective.
Task: It represents a self-contained unit of work. It contains its *InputGroups, OutputGroups* and indirectly *Inputs* and **Outputs**.
CompoundTask: It defines how to coordinate a set of related *Activities* that, in combination, perform some larger scale activity, **ultimately** in the context of a *BusinessProcess.*
Activity: It represents the **execution** of a part of a *BusinessProcess* using one of two mechanisms: (i) The creation of an instance of a *CompoundTask* referred to via the Activity's *definedBy* association or (ii) The execution of some feature of an Object bound to a *BPRole* instance via the Activity's *performedBy* association. The *InputGroups* contained by an *Activity* represent the alternative means by which the *Activity* may supply data to these mechanisms to initiate some action.
DataFlow: It represents a causal relationship in a *business process*. The source of the *DataFlow* must happen before the sink of it. They also propagate data values between causally related *DataElements.*
DataElement (abstract): It **represents** data used in *Task*'s input/output. It has two subtypes *Input* and *Output*. A *DataElement* is contained in a *DataGroup.*
Input: It models a collection of data values of a particular type consumed by a *Task* instance. An *Input* is contained in an *InputGroup.*
Output: It models a collection of data **values** of a particular type produced by a *Task* instance. An *Output* is contained in an *OutputGroup*.
DataGroup (abstract): It represents a set of related *DataElements* used to describe the inputs and outputs of a *Task*. They act as a **form** of correlator for *DataFlows. InputGroup* and *OutputGroup* are concrete subtypes of *DataGroup.*
InputGroup: It models a set of data values required by a *Task* to do some work. When owned by an *Activity* the *InputGroup* models the actual parameters to some behavior of the *Task*. When owned by a *CompoundTask* the *InputGroup* models the formal parameter to some behavior of the *Task.*
OutputGroup: It represents a possible outcome of the execution of a *Task*. It provides data values associated with that outcome. An **OutputGroup** is contained by a *Task.*
ExceptionGroup: It represents the outcome of a *Task* that failed to complete its function.
BRRole: It defines a placeholder for entities **that** perform an Activity or are used in the performing of an Activity. Its 'find' attributes constraints the set of entities that may be bound to the BPRole at run time.

In the profile for EDOC the EDOC's **BusinessProcess** meta entity inherits from the EDOC's **ProcessComponent** meta entity, which defines a set of **Ports** for interaction with other **ProcessComponents** and that has a set of properties that are used to configure the **ProcessComponent** when it is used (see definitions of the mentioned meta-entities in Table 1). Since in the approved by the OMG profile for EDOC systems, EDOC's **CompoundTask** meta-class specializes EDOC's **BusinessProcess** meta-class the former one (EDOC's **CompoundTask** meta-class) inherits the whole semantics associated with the last one (EDOC's **BusinessProcess** meta-class). Given the above mentioned specific semantic inheritance and being conceptually strict a direct mapping (shown in Table 3) among the meta concepts from the two models (The **CompundTask** meta class from the EDOC´s profile and the *CompoundTask* form the DSTC workflow metamodel) should not be plausible.

But, knowing that the definition of a workflow process does not require the details involved with the design of an EDOC system using an enterprise class model we could, just to allow a comparison between the meta entities of the two models, leave temporarily aside the specific EDOC semantic, related with the component design issues, and make the following mapping shown in Table 3.

Table 3. Mapping between the profile for EDOC and the DSTC's Workflow Process Definition Metamodel

Meta-concepts form the Profile for EDOC	DSTC's Workflow Process Definition Metamodel
BusinessProcess	BusinessProcess
CompoundTask	CompoundTask
Activity	Activity
DataFlow	DataFlow
ProcessFlowPort	DataElement
input ProcessPortConnector	Input
output ProcessPortConnector	Output
ProcessMultiPort	DataGroup
InputGroup	InputGroup
OutputGroup	OutputGroup
ExceptionGroup	ExceptionGroup
ProcessRole	BRRole

3. WORKFLOW PROCESS DEFINITION METAMODEL

Given that the proposed by DSTC Workflow Process Definition Metamodel does not consider the specification of events then in this section it is suggested a new workflow process definition metamodel adding these aspects (using the sub-profile for events that is part of the EDOC profile). The new suggested workflow definition metamodel (illustrated in Figure 1) shows a clear separation of the data and event oriented control flow dependencies allowing the specifications of both aspects.

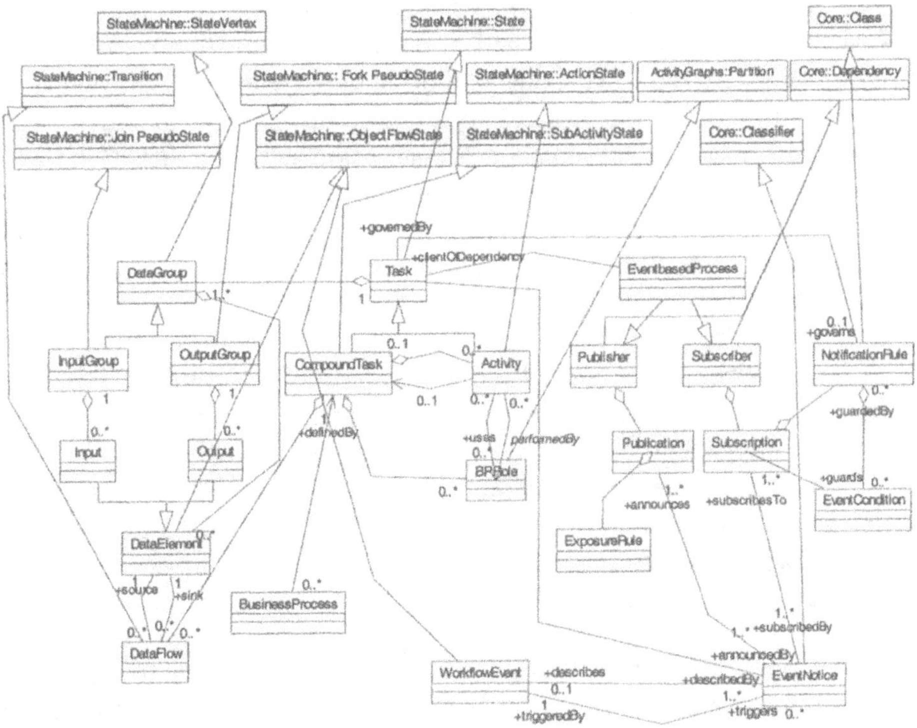

Figure 1. A workflow process definition metamodel with data and event oriented control flow dependencies

In order to derive (stereotype) the proposed in this paper workflow definition metamodel as an UML profile the metamodel is derived from the UML 1.4 Activity Graphs (Heaton, 2001). The use of the UML Activity Graphs semantics to profile the illustrated in Figure 1 metamodel creates difficulties to link an Activity (in the proposed metamodel) with its potential realization. The association of an Activity with the BPRole entity via UML partitions (as in Figure 1) compels us to redefine the DSTC's *BPRole* semantics. The DSTC's *BPRole* has a *type* and an attribute *'find'* to constraint the set of entities to be bound by the *BPRole* at run time. Since, BPRole is a UML's partition and not anymore a class (as it is DSTC's *BPRole*) it is not now possible to associate with it (with BPRole) a *type* and an attribute *'find'* to constraint the set of entities to be bound by the *BPRole* at run time.

If we wanted to be partially coherent with the EDOC metamodel then ProcessRole should subtype an UML meta class. But in this last case (ProcessRole) as a subtype of an UML meta class) it would not be possible to associate in the metamodel depicted in Figure 1 the ProcessRole entity with an Activity, since Activity is being modeled there (see Figure 1) as an UML Action State. If we wanted to be totally compatible with the EDOC profile and used the UML Collaboration as the base to profile the metamodel (as it is the DSTC's approach) then we would have to give up the well accepted way to model business processes using activity graphs.

Since the UML Dependency can associate any two UML Model Elements, then to add to the proposed workflow process definition metamodel all the semantic richness of the EDOC metamodel for Events, a Task (in Figure 1) is associated with an EventBasedProcess that stereotypes UML Dependency.

The BusinessEvent entity from EDOC profile, as defined in Table 1, has been replaced in the metamodel illustrated in Figure 1 by the WorkflowEvent entity with a redefined semantics as follows.

A workflow event is defined in (Hollingsworth, 1999) as the occurrence of a particular situation or condition which has significance to one or more workflows and causes a defined action via the workflow management software. The purpose of this concept is to provide a mechanism, which may be used to co-ordinate or synchronize different processing activities.

A workflow event has two elements (i) a workflow event trigger or cause and (ii) a workflow event action or response. The workflow event trigger is defined as the set of predefined circumstances that causes a particular action to be taken. The workflow event trigger concept, (as the set of predefined conditions under which the workflow event is emitted), is introduced in the metamodel (see Figure 1) through the meta entity ExposureRule.

The workflow event action is defined as the pre-defined response following the trigger condition. These actions may be associated with specific control actions, to be undertaken by the workflow management software. For example, this might result in the workflow management software enabling a particular activity instance to be started.

The set of control actions and associated conditions is introduced in the model through the meta entity NotificationRules (from the EDOC'NotificationRules). To further specify the control actions and associated conditions it is suggested to use behavioral Rules as the ones described in (CIMOSA). These rules have a clear mapping to the six different Workflow Management Coalition routing possibilities among activities (WfMC, 1999). Table 4 below shows in the first column the CIMOSA behavioral rules and in the second the corresponding routing connection with a diagram sample of each one of them.

Table 4. Mapping between CIMOSA behavioral rules and WfMC routing possibilities

CIMOSA Behavioral Rules	WfMC-Routing possibilities
Process triggering rules: they are used to start a process with or without events (without events means that the business process EF1 is just called by a parent process). WHEN (START WITH *event*-1) DO EF1 WHEN (START WITH *event-1* AND *event*-2) DO EF1 WHEN (START) DO EF1	Start of the process.
Sequential rules: they are used to represent branching conditions in the flow of control. ES(EFi) is a pre-defined function returning the ending status (ES) after completion of EFi, ANY is a reserved word representing all the possible ending status of EFi, and end-status-n represents a specific ending status of EFi. WHEN (ES(EF1) = ANY) DO EF2 (also called forced sequential rule) WHEN (ES(EF1) = *end-status*-1) DO EF2 (also called conditional sequential rule)	**single thread:** an activity A2 can be executed after the completion of another A1. 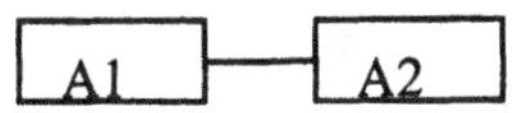 **or-join:** a point within the workflow where two or more alternative activity workflow branches re-converge to a single common activity as the next step within the workflow. 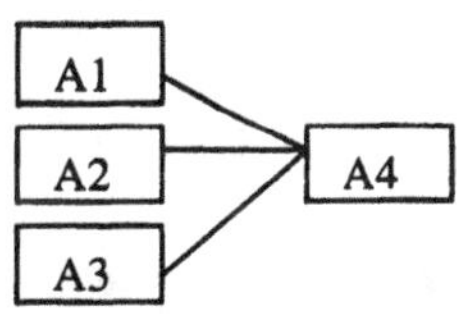 **or-split:** a point within the workflow where a single thread of control makes a decision upon which branch to take when encountered with multiple workflow branches. The decision is specified by transition conditions (e.g. C1, C2 and C3). 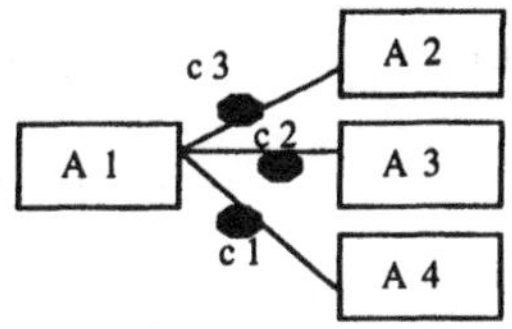

***Spawning rules*:** they are used do represent the parallel execution of EFi; '&' is the parallel operator, and SYNC is a reserved word meaning the simultaneous start of EF2, EF3 and EF4: WHEN (ES(EF1) = *end-status*-1) DO EF2 & EF3 & EF4 (also called asynchronous spawning) WHEN (ES(EF1) = *end-status*-1) DO SYNC(EF2 & EF3 & EF4) (also called synchronous spawning)	**and-split:** a single thread of control splits into two or more threads, which are executed in parallel within the workflow, allowing multiple activities to be executed simultaneously. 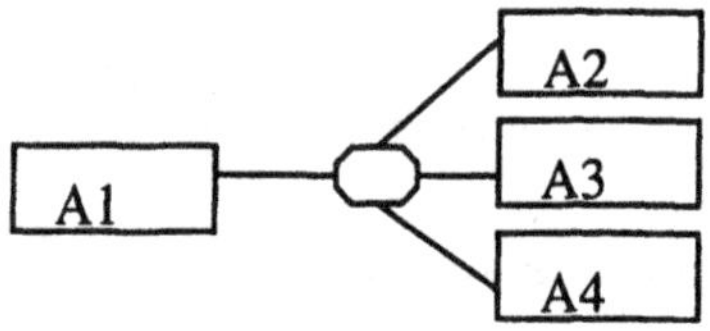
***Rendezvous rules*:** they are used to synchronize the end of spawning rules: WHEN (ES(EF1) = *end-status-1* AND ES(EF2) = *end- status-2* AND ES(EF3) = *end-status*-3) DO EF4	**and-join:** a point in the workflow where two or more parallel executing activities converge into a single common thread of control. It is a synchronization point in the workflow. 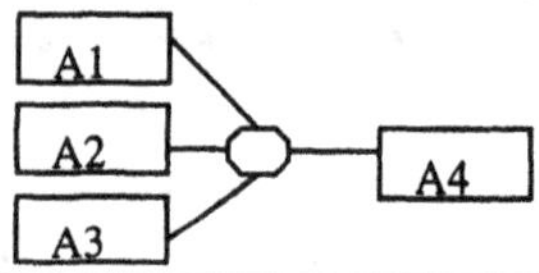
***Loop rules*:** they are used to execute the same EFi iteratively: WHEN (ES(EF1) = *loop-value*) DO EF1 WHEN (ES(EF1) = *exit-value*) DO Efx	**Iteration:** a workflow activity cycle involving the repetitive execution of one (or more) workflow activity(s) until a condition is met. C1 and C2 are transition conditions. 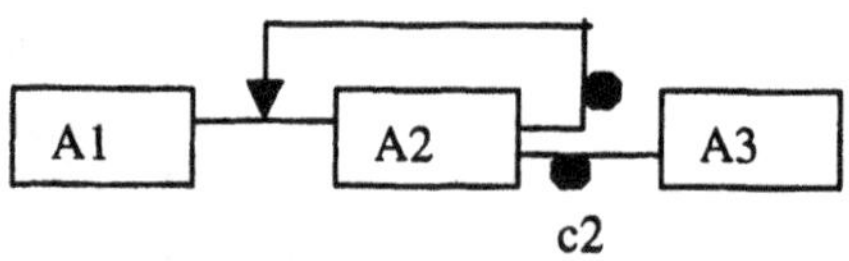
***Process end rules*:** they are used to indicate the end of the process: WHEN (ES(EF1) = *end-status*-1) DO FINISH	End of the process

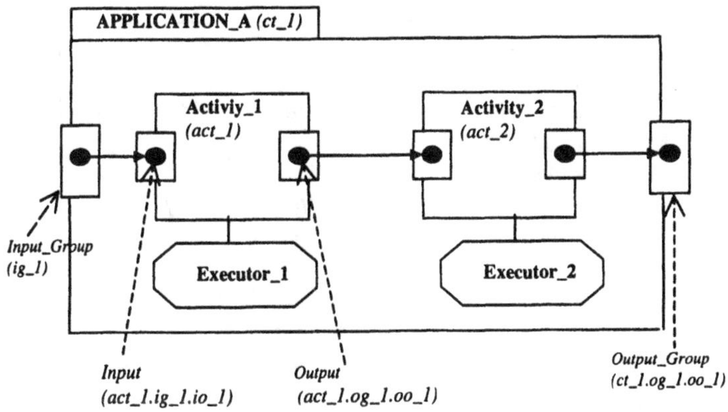

Figure 2. A process definition using an EDOC compatible notation

4. THE TEXTUAL NOTATION

To introduce a notation that be computationally interpretable by a workflow engine, this section introduces sketches of a textual notation to represent a workflow process definition coherent with the proposed workflow process definition metamodel illustrated in Figure 1. As a matter of example, lets assume we have the application of Figure 2, depicted using a compatible EDOC notation. The application is made up of two simple activities linked in strict sequence order and executed by Executor_1 and Executor_2 entities respectively.

The graphical representation shown in the above Figure 2, can be translated into the sketched notation included in Table 5. This notation is to be computationally interpretable by a workflow engine.

The interpretation of this notation is the following. The whole "*Application_A*" is mapped as the content of a *Compound_Task('identifier', 'process_name')* expression. The key word *Process_Defintion()* is used to describe each one of the activities that conform the application as a whole. Inside this *Process_Definition()* section, the constituent process steps are specified using the parameters of the *Activity ('identifier', 'step_name', 'BPRole')* key word expression.

Each one separately, the *Compound_Task(..)* and the *Activity(..)* expressions, must specify its *input* and *output* information using the key constructs *Input_Group('identifier')* and *Output_Group(' identifier'),* and inside each of them, the actual data to be exchanged between the process steps is specified with

input('input_object_identifier','type') and *Output('output_object_identifier', 'type')* constructs.

Table 5. Sketch of a Textual Notation

```
Compound_Task( ct_1, Application_A){
 Input_Group( ig_1) { Input( io_1, 'd' ); }
 Output_Group( og_1 ) { Output( oo_1, 'd' ) ; }
Process_Definition() {
  Activity( act_1, Activity_1, Executor_1 ) {
   Input_Group( ig_1 ) { Input( io_1, 'd' ); }
   Output_Group( og_1 ) { Output( oo_1, 'd' ); }
  }
  Activity( act_2, Activity_2, Executor_2 ) {
   Input_Group( ig_1 ) { Input( io_1, 'd' ); }
   Output_Group( og_1 ) { Output( oo_1, 'd' ); }
  }
Dependency (){
   dep( 1, ct_1.ig_1.io_1, act_1.ig_1.io_1 );
   dep( 2, act_1.og_1.oo_1, act_2.ig_1.io_1 );
      dep( 3, act_2.og_1.oo_1, ct_1.og_1.oo_1 );
}
    }
```

The section *Dependency ()* is used to describe the dependencies among all steps that conform the workflow process. Each individual dependency is specified using the expression *dep(..)* which have three arguments: an identifier of the dependency, the identifier of the source of the information and the identification of the target of the information.

In case of an event oriented workflow process, the event oriented control dependencies could be declared with a syntax similar to the one used when declaring *Input_Group* and *Output_Group*. Following it is a simplified sketch (see Table 6) of the key expressions to declare the intention to produce or to consume workflow events.

A workflow entity specifies its intention to receive a workflow event of type '*event_type*' through the key word *Subscriber(..)* and inside the *NotificationRule* key word are to be specified the associated conditions and control actions to be taken (following the suggested CIMOSA rules as mentioned in section 3.

Table 6. Sketch of a Textual Notation

```
Subscriber( event_type) {
NotificationRule ( id, 'CIMOSA_rules');}

Publisher( 'event_type') {
EventNotice ('reference_to_event_data');
RegraDeExposição ( id, 'trigger_conditions');}
```

5. CONCLUSIONS AND STANDARDIZATION WORK

This paper illustrates that there is a clear mapping between the UML profile for EDOC and the DSTC's workflow process definition metamodel that helps to avoid the misunderstanding between business analysts (workflow modelers) and software developers. At the same time that it is possible to build from the DSTC's workflow process definition metamodel, a workflow definition metamodel with a clear separation of the data and event oriented control flow dependencies consistent with the UML profile for EDOC systems. And that the CIMOSA behavioral rules are a plausible way to orient the work to further specify the event oriented control dependencies in a workflow process definition based in the suggested workflow definition metamodel.

Furthermore, the approach proposed in this paper is coherent with the current process that goes on in the Object Management Group (OMG) to standardize the specification of a process definition. The OMG has issued (January 2003) a Request for Proposal (Cohete, 2003) that solicits submissions that specify a business process definition metamodel, which is platform independent with respect to specific business process definition languages. This metamodel will define an abstract language for specification of executable business processes that execute within an enterprise (with or without human involvement) and may collaborate between otherwise-independent business processes executing in different business units or enterprises.

The specification developed in response to that RFP (Cohete, 2003) is expected to achieve the following:

A common metamodel to unify the diverse business process definition graphical and textual notations that exist in the industry.

A metamodel that complements existing UML metamodels so that business processes specifications can be part of complete system specifications to assure consistency and completeness

The ability to integrate process models for workflow management processes, automated business processes, and collaborations between business units.

Adoption of this specification is expected to improve communication between modelers, including between business and software modelers, provide flexible selection of tools and execution environments, and promote the development of more specialized tools for the analysis and design of processes.

With reference to the behavioral rules as the CIMOSA ones mentioned in the above section 3 of this paper there is no generally accepted approach for defining or representing business rules (Hendryx, 2002b). In this sense the work on modeling business rules is in the early stages. The OMG has issued (September 2002) a Request for Information (RFI) for Business Rules in Models (Hendryx, 2002a) and more recently (June 2003) a Request for Proposals (RFP) for Business Semantics of Business Rules (Hendryx, 2003).These works and the relationship of business rules to business processes should be considered in the development of a business process definition metamodel.

REFERENCES

Boldt, Juergen., (2000). "UML Extensions for Workflow Process Definition. Request for Proposal", OMG Document Number: bom/2000-12-11. Available at: ftp://ftp.omg.org/pub/docs/bom/00-12-11.pdf

CIMOSA. "A Primer on key concepts, purpose and business values". Available at: http://cimosa.cnt.pl/Docs/Primer/primer0.htm .

Cohete, Manfred., (2003). "Business Process Definition Metamodel Request for Proposal," OMG Document: bei/2003-01-06. Available at: ftp://ftp.omg.org/pub/docs/bei/03-01-06.pdf

Duddy, Keaton. (2001). "Workflow Process Definition Metamodel submitted by The Cooperative Research Center for Enterprise Distributed Systems Technology (DSTC)", OMG Document Number: cem/2001-12-01. Available at: ftp://ftp.omg.org/pub/docs/cem/01-12-01.pdf

Heaton, Linda., (2001). "Unified Modeling Language Specification" v1.4, OMG Document Number: formal/01-09-67 to formal/01-09-80. Available at ftp://ftp.omg.org/pub/docs/formal/01-09-67.pdf

Heaton, Linda., (2002). "UML Profile for Enterprise Distributed Object Computing", OMG Document Number: ptc/2002-02-05. Available at: ftp://ftp.omg.org/pub/docs/ptc/02-02-05.pdf

Hendryx, Stan., (2002a). "Business Rules in Models Request for Information". OMG Document Number: ad/2002-09-13. Available at: ftp://ftp.omg.org/pub/docs/ad/02-09-13.pdf

Hendryx, Stan., (2002b). "Defining Business Rules. What are they really". OMG Document Number: ad/2002-10-03. Available at: ftp://ftp.omg.org/pub/docs/ad/02-10-03.pdf

Hendryx, Stan., (2003). "Business Semantics of Business Rules Request for Proposal". OMG Document Number: br/2003-06-03. Available at: ftp://ftp.omg.org/pub/docs/br/03-06-03.pdf

Hollingsworth, David., (1999). "Workflow Management Coalition White Paper-Events". Available at: http://www.aiim.org/wfmc/standards/docs

OMG, "The Object Management Group". [http://www.omg.org]

WfMC, (1999). "Workflow Management Coalition. Terminology and Glossary". Document Number WfMC-TC-1011, v3.0 (Feb 99).

Chapter 26

XML-BASED E-CONTRACTING

Michael Merz

Abstract: Under the research projects eLEGAL and OCTANE (funded by the IST programme of the European Commission), European companies have joined forces to develop a variety of tools and templates that are required to support forming and closing electronic contracts over the Internet without any media breaks. This article informs on some of the projects results.

To explore practical business needs, scenarios are taken from different industries: on one side, construction companies and subcontractors enter a joint-venture to accomplish a construction project. In this setting, the e-contracting software is required to let partners compose contracts in an ad-hoc manner out of a clause database. A second scenario shows how bilateral agreements on data exchange relationships, data formats, or rules for information exchange can be made by using e-contracting tools.

Key words: Legal issues, electronic contracting, XML Signature, XML Schema

1. INTRODUCTION

Past decades have witnessed a continuous reduction of cost and media breaks in general office procedures. In many cases, this increase of efficiency has been achieved inside organizations, while inter-organizational processes could only be improved mainly at the very general level of using email. Today, it can be recognized that there is still much room for improvement – specifically in the area of contract editing, negotiation, signing, and fulfillment:

While on one side, processing and exchange of contract data lags behind the formalization of electronic documents used in industrial supply chain applications (orders, call-offs, delivery notes, inventory status, etc.), means for the next step of formalization – and therefore automation – are given already today (XML, including XML Signatures, certification authorities, collaborative document management systems, etc.).

Research on e-contracting started already in the early 80ies by with a focus on formal contract semantics [1]. Over the 90ies, formalization of business rules and their mark-up started (e.g. [2, 3, 4, 5]) and also security aspects have been explored, e.g. [6, 7]. A first XML-based contract editor was presented by [8] as a result of the COSMOS project (funded under the European ESPRIT programme).

This paper takes a rather practical position and explores business advantages, software applications and future business models that are related to e-contracting.

The presented Contract Editor Software can be downloaded from: http://www.ponton-consulting.de, where it is branded as "Ponton X/E".

2. WHAT IS E-CONTRACTING?

e-Contracting is the process of setting-up, negotiating, signing, and maintaining contracts purely in electronic form.

e-Contracting may either require electronic signatures, or it may take place implicitly by getting a receipt after a retail transaction, it may involve two or many parties and it may consist of a single contract or a set of them – in any case, e-Contracting leads to a legal contractual situation.

E-Contracting affects very different aspects of business life: It is not only an economic act but also a collaboration of two or more parties. It starts with the selection of business partners over the negotiation of positions and modalities and does not even end with the final signing of a document. In many business relations, a contract is not engraved in stone – it is rather the starting point for further refinements, amendments or extensions as the business relation develops its own dynamics.

E-Contracting can thus be viewed from several perspectives: First, a legal viewpoint, second, a business viewpoint and, third, from a document management viewpoint:

1. *Legal Viewpoint.* Here, the legal situation abstracts away from its materialization. A contract might exist in written form, as an oral agreement, or in electronic form – the legal situation is always the same. For lawyers, contracts are abstract concepts that describe this situation and govern the behavior of the parties involved. In practice, many lawyers are occupied with setting-up contract documents that should describe the legal situation as precise as possible.
2. *Business Viewpoint.* In many business environments, a contract is an instrument that is integrated into a long-term relationship: Suppliers and customers agree on contracts for the delivery of a certain quantity of a product at a certain quality, price, and date. These figures will be referred

to and recycled for a long-term period, e.g., in supply-chain relations, supply contracts can be closed for a time span of several years. On the other hand, contract are rather snapshots of the current state of a business agreements. They may be amended, extended, changed or cancelled – all this just requires the agreement of the parties involved and the contract is quickly adjusted to the new situation. From a business viewpoint, the re-use and later change of contracts is key.

3. *Document Management Viewpoint.* Contracts involve more than one party and they are electronic documents, at least over their agreement phase. Therefore, online collaboration is the main facilitator to achieve an agreed and consistent electronic document. Since electronic signing is supported by law today, not only negotiation, but also the joint signing of a contract takes place across the Internet. If the contract is represented in a structured, well-defined form, it may also be interpreted and re-used by other applications such as Enterprise Resource Planning (ERP) systems, Viewers, or archiving tools.

3. THE CONTRACT EDITOR – A TECHNOLOGICAL FRAMEWORK FOR ELECTRONIC CONTRACTS

Business innovation and information & communication technologies (ICT) are often so closely intertwined that it would be a hen-egg loop to determine the actual driver of the innovation process. For the following set of business cases, we will provide as little technology input as necessary. However, to set up the technical environment, at least certain XML-related technologies should be introduced beforehand:

3.1 XML-based legal documents

Many documents in application areas such as supply chain management, electronic publishing, visualisation of database content are based on XML today. However, there is no use of XML in the case of legal documents – only very early standardisation approaches like LegalXML [9] or LexML [10] can be found in the research field. We assume cultural reasons are the main obstacle: Contracts are usually negotiated, exchanged and signed by non-technical persons, therefore tools like word processors and manual exchange by e-mail remain state of the art even today.

On the other hand, contracts are syntactically and semantically highly structured documents. They consist of very specialised clauses, and in highly formalised cases they can even be expressed in terms of a set of parameters – as for example in many form-based standard contracts (car sale, lease, etc.).

For these reasons, it is very likely that contracts will be expressed based on XML in the future. The first approach to achieve this is to define a standard grammar for contractual documents (formally expressed as an XML Schema). This might be a horizontal standard for the representation of contracts in general, or it might be a vertical one for specific applications such as EDI framework contracts, purchase orders, or reinsurance contracts, for example.

3.2 The XML Signature standard

Usually, electronic signatures are created using public key technologies as implemented, e.g., by algorithms such as RSA or DSA [6]. In both cases, the signature is represented by an encrypted hash value derived from a binary document. Usually this is a file of a format like MS Word, PDF, XML, ASCII text, etc. The benefit is that all document types are treated the same way – as a contiguous chunk of bits and bytes. Although this simplifies the process of signing, it complicates signature management at the same time: Signatures and certificates have to be kept as files separate from the original document. If there are several parties signing the same document, the number of managed files consequently increases. In practise, document management systems (DMS) are required in order to maintain document consistency.

While "classic" approaches for electronic signature support signing at the physical level, the XML signature standard allows signing of logical substructures of a document. Thus, an XML document can contain XML structures that are considered relevant for the signature as well as other XML structures that not signed, i.e. they are not within the scope of legal consideration.

Moreover, XML Signature allows embedding of the signature itself into the signed document. This can be done in three different ways: Either the signature becomes a part of the signed document itself (embedded signature). Here, the XML Signature standard requires from technology providers to explicitly exclude any embedded XML Signature sub-elements from the rest of the signed part of a document. The second approach is to embed the signed part of the document into the XML Signature structure. This can be done by copying the signed part from the original document into the signature document (embedding signature). Finally, the signature may be allocated "next to" the signed part, e.g., in a sibling XML path or in a separate document (detached signature).

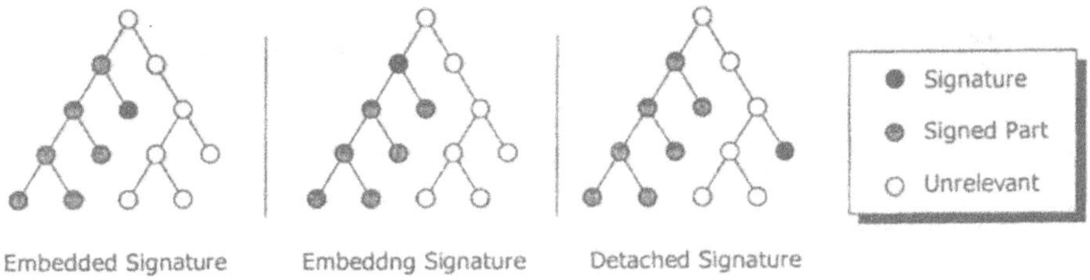

Figure 1: Different relationships of XML Signatures and the signed content within the XML document tree

The signature itself is represented as a standardised XML structure, following the W3C standard (www.w3.org/Signature). Such a <Signature> element not only contains the signature value, but also other data including a reference (URI) to the signed part of the document. In more advanced cases this might even be a list of references that are all signed at once. Also, electronic certificates could be embedded into the XML Signature structure to support signature verification.

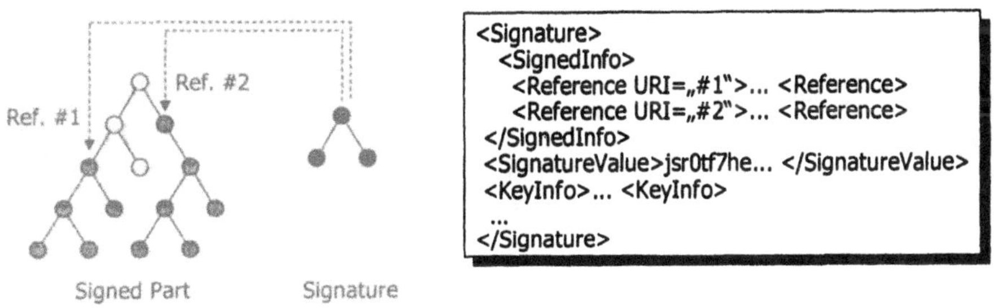

Figure 2: Signing several parts of a document "at one strike"

XML Signature thus helps represent signatures in a much more appropriate form compared to the classic approach. The signing algorithm, however, remains the same: Instead of processing a binary document, XML Signature compliant tools concatenate the value of elements and attributes into a canonical form and apply the signing algorithm (RSA or DSA) to the result. Signature verification is consequently based on the same canonicalized string.

The following figures show how a signed XML document looks like:

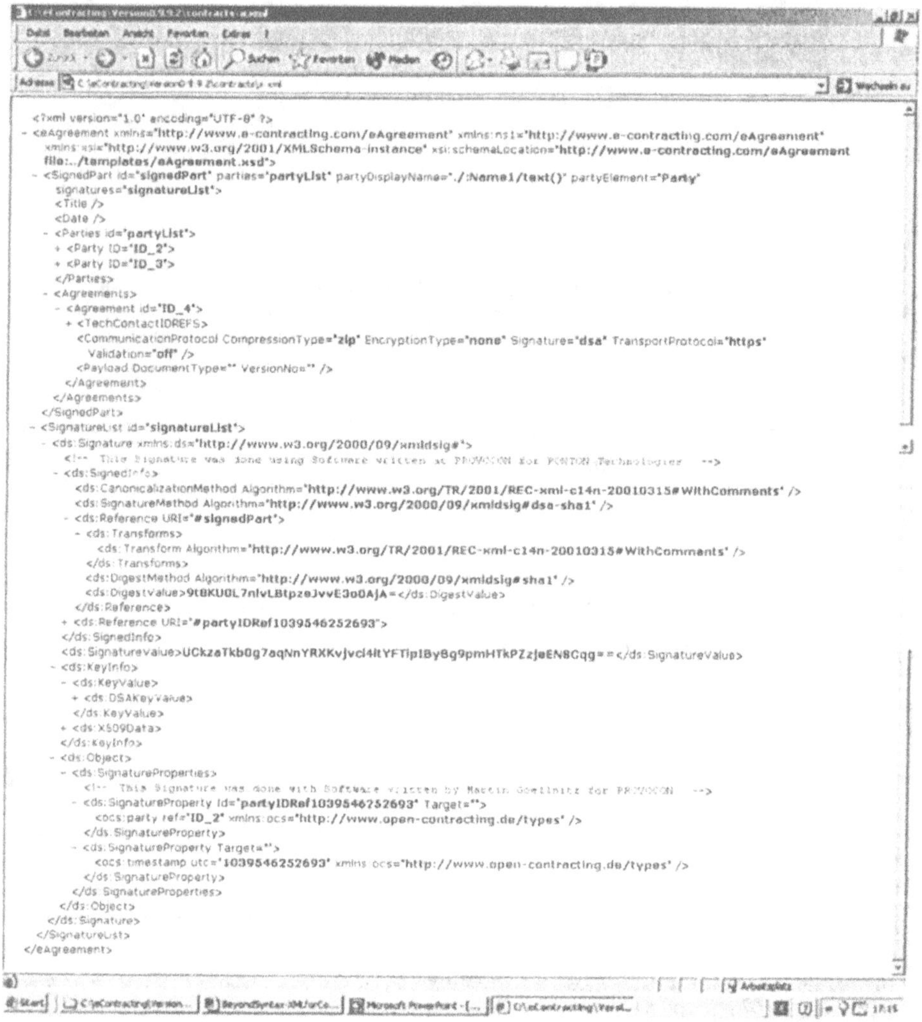

Figure 3: An XML document containing an electronic contract with a detached XML Signature

Although this summary might seem a bit technical, we will see further down in the "business process" section how these can be streamlined.

3.3 Contract Management Software

Nobody is willing to understand the binary document format of a word processor. In the same way, nobody is willing to edit an e-contract as an XML document as displayed in figure 3. For this reason, users require a contract editor that facilitates contract processing, specifically creating contract templates, exchanging contract data, filling out contract forms, signing them and verifying them afterwards. In many business cases e-contracting specifically pays off because contract documents can be directly re-used as input to operational IT systems.

The requirements to contract management systems are therefore:

- *User interface*: Conceal syntactic XML details from the user.
- Support users or consultants in *defining structure and layout* of a contract.
- Support the users in easily *composing contract clauses.*
- In case of *standard contracts*, support users in entering parameters for the clauses.
- Assure syntactic and semantic correctness of the contract as far as possible.
- Support the signing and signature verification process.
- Allow for additional processing options like transformation into other data formats, integration of application-level functions (e.g. selection from an online catalogue), and access to a document management system that provides functions for document storage, user rights management, and archiving of documents.

3.4 Contributions from the OCTANE and eLEGAL projects

As a result of two projects – eLEGAL and OCTANE – funded by the IST programme of the European Commission, a contract editor software was developed.

The Contract Editor can be divided into two different parts: First, the *base software* can be used to edit any XML document, with optional support of a customisable document layout. The base software provides an interface to *plug-in* additional functions, such as an electronic signing tool, an XSL-based HTML converter, remote access to a document management system, and e-mail support to send the XML document directly to a business partner. This part of the software was developed for the *OCTANE* project.

One of the most complex plug-ins is the *Contract Wizard.* This tool requires a certain set of well-known XML substructures in the contract consisting of nested <Clause> elements. The purpose of the Contract Wizard is let the user select clause types and clauses from a database and insert then into the contract. Moreover, clauses can be checked for consistency using rule definitions. This part was developed for the *eLEGAL* project.

3.4.1 The Contract Editor

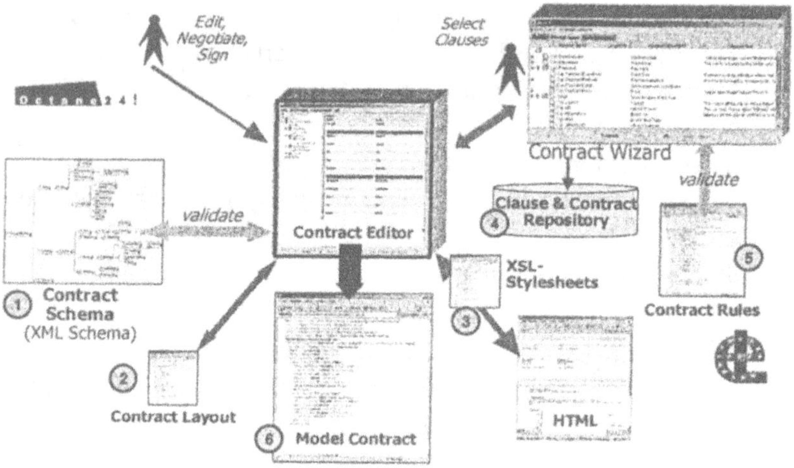

Figure 4: Components of the Contract Editor

Figure 4 illustrates the different document types and data sources used in creating an electronic contract:

1. An *XML Schema* is used to define the structure of the contract. Based on this specification, the Contract Editor does not allow any document structure to be created which does not comply with the underlying XML Schema.
2. By default, no *contract layout* is needed. In this case, all data elements are displayed in canonical order and in a vertical flow.
3. Another option is to develop *mapping definitions* for converters. Under OCTANE, a HTML converter was developed based on XSL transformations. This may be extended by MS Word converters or other transformations in the future.
4. The *Contract Wizard* uses a database to retrieve clause types and clause texts. The data model for clauses and clause types is illustrated further down.
5. To structure database access, *contract rules* can be defined by the creator of a contract template. These rules define which clause types are to be used for a specific contract type, whether the clauses are mandatory or optional, and which logical constraints apply to them.
6. Finally, model contract can be created based on the given tools by legal experts. They can be published or used as templates for actual contracts.

In short, the Contract Editor deals with syntactic aspects of XML-based contracts, while the Contract Wizard deals with semantic aspects of the content.

The following two figures illustrate the use of the Contract Editor for a reporting task between printers and their customers (Inventory Stock Report). The first figure displays the document without a layout definition, whereas the second illustrates the application of a layout definition:

The screenshot in figure 5 shows the default user interface to edit a Schema-based XML document. A layout definition is applied in figure 6. Here, the user interface is structured much better, and the data elements are displayed with German labels.

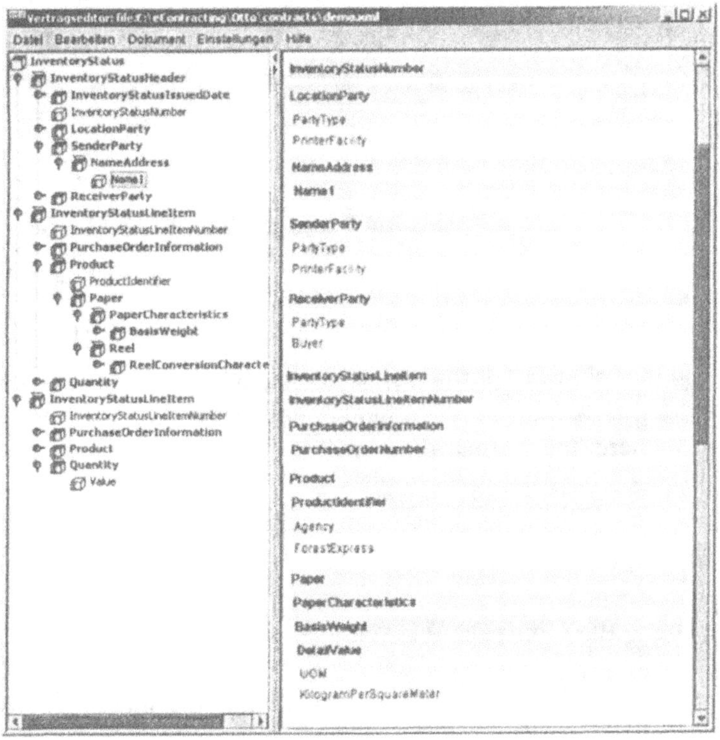

Figure 5: Inventory Status report displayed in canonical form

One of the most interesting features of the Contract Editor is the signing of contracts based on the XML Signature standard. In principle, the Contract Editor allows you to sign any part of an XML document and place the signature in any other part. To define these locations, several settings have to be provided at the level of the contract template:

- Which part of the document is to be signed?
- Which part will carry the signature?

- Will the contract be signed by more than one party? If so,
    - o Which element contains the list of parties?
    - o Which element represents a party?
    - o Which element contains the name of the party, as it should be displayed so that the user can choose the party to sign for?

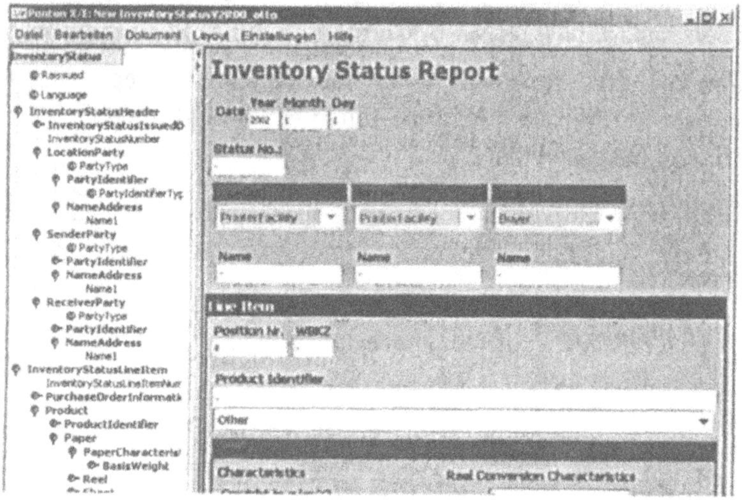

Figure 6: Inventory Status report applying a layout definition with German localisation

The signing plug-in of the software expects some well-known XML attributes defined within the document. These attributes indicate the part to be signed, where the signature(s) will be stored, and where the party descriptions are located.

This way, any XML document can be extended with minimal effort to accept XML Signatures.

3.4.2 Using the Contract Wizard

The Contract Wizard is a plug-in that allows contract clauses to be composed for a given legal context. This context is defined by the language, the applied legislation and the domain (e.g. a purchasing contract, a lease, a consortium agreement, or company statutes, just to name a few).

To use the Contract Wizard, one precondition must be met by the XML Schema for the underlying e-contract: It has to support a nested structure of text clauses, as in the following example:

```
<Clauses>
    <Clause ClauseID="XXX" ClauseTypeID="XXX">
        <ClauseTitle>XXX</ClauseTitle>
        <ClauseText>abcde pqr stuvwx yz</ClauseText>
```

```
    </Clause>
    <Clause ClauseID=" XXX" ClauseTypeID="XXX">
        <ClauseTitle>XXX</ClauseTitle>
        <ClauseText>abcde pqr stuvwx yz</ClauseText>
        <Clause ClauseID=" XXX" ClauseTypeID="XXX">
            <ClauseTitle>XXX</ClauseTitle>
            <ClauseText>   abcde   fghij   klmno   pqr   ...
            </ClauseText>
        </Clause>
    </Clause>
    ...
</Clauses>
```

This is the target structure the Contract Wizard operates on. figure 7 shows an End User License Agreement (EULA) as another example. Here, a set of clauses has already been composed giving a deeply structured tree of <Clause> elements. This can be seen on the left hand side in the layout tree and in the content window showing the top-level clause "Restriction of Use" with subclauses in the subordinated window.

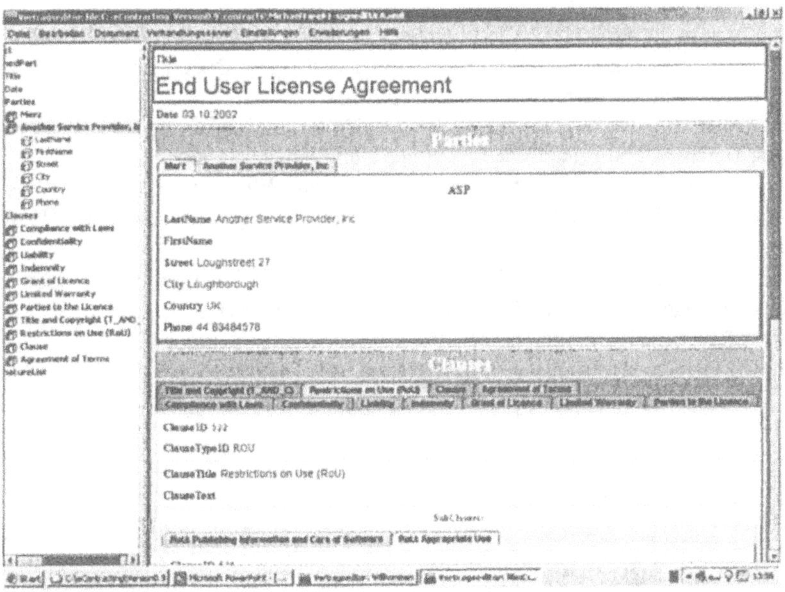

Figure 7: An electronic contract with a nested structure of <Clause> elements

4. CONCLUSION

In this paper some of the concepts and the developments performed within the framework of both eLEGAL and OCTANE Projects have been presented. By means of the eContracting scheme and tools developed within the eLEGAL Project, such development lead to an environment able to establish a sound and legally valid business relationship fostering a new business model for B2B scenarios and consultants seeking to broaden their business activities capturing the opportunities provided by the available ICT tools and infrastructures.

REFERENCES

1. Lee, R.M – A Logical Calculus for Describing Financial Contracts, PhD Dissertation, Department of Decision Sciences, The Wharton School, University of Pennsylvania, 1980
2. Zoran Milosevic, David Arnold, Luke O'Connor, Inter-enterprise Contract Architecture For Open Distributed Systems: Security Requirements, WET ICE'96 Workshop on Enterprise Security, Stanford, USA, June 1996.
3. Michael Schroeder, Gerd Wagner, eds.: Proceedings of the International Workshop on Rule Markup Languages for Business Rules on the Semantic Web, 14 June 2002, Sardinia, Italy, CEUR Workshops, http://sunsite.informatik.rwth-aachen.de/Publications/CEUR-WS/
4. Rule Markup Initiative Web site, http://www.dfki.uni-kl.de/ruleml/
5. Laurence L. Leff: Automated Reasoning with Legal XML Documents; *Conference '00*, Month 1-2, 2000, City, State.
6. Bruce Shneier: Applied Cryptography, Second Edition, John Wiley & Sons, 1996
7. Michael Gisler, Harald Häuschen, Yvonne Jöhri, Arion Meier, Otto Müller, Bernd Schopp, Katarina Stanoevska: Requirements on Secure Electronic Contracts, Report No. mcm-institute-1999-01, Institute for Media and Communication Management, University St. Gallen, 1999
8. M. Merz, F. Griffel et al, Supporting Electronic Commerce Transactions with Contracting Services, International Journal of Co-operative Information Systems, 1998, p.16.
9. Web site of the Legal XML standardisation group: www.legalxml.org
10. Web site of the Lex ML research group: http://www.lexml.de/mission_english.htm

Chapter 27

ICT SUPPORT FOR EVOLVING HARMONIZATION OF INTERNATIONAL ALLIANCES

Ronald M. Lee , Elizabeth Dominguez Campillo

Abstract: International alliances are agreements between multiple countries to cooperate on trade, or other forms of economic activity, for mutual benefit. This paper focuses on how information and communication technologies (ICT) can support transaction efficiency and effective controls in the context of evolving international alliances. We concentrate on the potential for electronic procedures to manage the document flows automatically, in a fashion analogous to workflow systems within organizations. The key challenge, however, is how to support the evolution of these documentary procedures to match the dynamic structure of international alliances.

1. INTRODUCTION

There is a saying that no man is an island – we are social animals and need to participate in social networks. The same, it seems, also applies to nations. More and more, nations are interlinking via various forms of international alliances. By international alliance we mean agreements between multiple countries to cooperate on trade, or other forms of economic activity, for mutual benefit. Examples include the World Trade Organization (WTO) which is the administrative organization that implements the General Agreement of Tariffs and Trade (GATT), European Union (EU), North American Free Trade Alliance (NAFTA), MERCOSUR (Argentina, Brazil, Paraguay, Uruguay), CARICOM (Caribbean countries), and many others. (Here we are not specifically concerned about military alliances, though they may share some aspects.) In some cases, such as the European Union, there may even take on the character of a government, with a president, and parliament. The purposes of these international alliances are varied, but typically include efforts to reduce transaction costs among their members and to achieve economies of scale in their markets.

1.1 HETEROGENOUS MEMBERSHIP; NEED FOR HARMONIZATION

International alliances are inspired by good intentions among the initial members. Often they are motivated in response to some outside pressure or threat. But the implementation often becomes bogged down as the members realize the extent of their various differences on detailed, but yet relevant points. The goal, ideally, is a harmonious solution that allows the parties to inter-operate, yet enables each to maintain their distinctive mode of operation. This is of course not always possible, and compromises need to be hammered out. This can take time, especially if some of the parties are reluctant.

1.2 EVOLUTIONARY

Working out these various harmonious solutions and compromises can be slow and arduous – often taking years to complete. In the meantime, conditions may change requiring further revisions. These changes in policy can be:

a. monotonic -- adding new features without affecting existing features of the policy

b. obsolescent -- adding new features that make other features ineffective or obsolete

c. non-monotonic -- adding new features to the policy that actually contradict previous policies.

(these are the kinds most difficult to handle)

1.3 POTENTIAL FOR TECHNOLOGY SUPPORT:

The heterogeneity of international alliances is mainly due to the differences in social and economic values of the member countries. These are determined by various types of domestic political processes. However, the participation in international alliances can create issues that need to be addressed by these domestic processes. While the political process itself is outside the scope of this research, we want to address how information and communication technology (ICT) can facilitate the deployment and evolution of international alliances, especially in their more procedural aspects. Here we introduce the term "protocol" to encompass procedures, processes, documents and associated rules agreed upon by the international alliance.

1.3.1 NORMAL OPERATIONS

The normal operation of documentary procedures for an international alliance includes the following features:

* computerized implementation of international alliance protocols
* facilitate learning and adoption of international alliance protocols
* deployment -- making international alliance protocols readily available and useable to all appropriate international parties
* secure -- controlling access, authentication as appropriate so that the international alliance protocols are not abused or used fraudulently
* monitoring usage and performance of international alliance protocols

1.3.2 CHANGE, EVOLUTION MANAGEMENT

The dynamic nature of international alliances also imposes the following additional requirements:

* detecting need for change (based on monitoring)
* helping to organize change proposals
* help to analyze, determine potential consequences and costs of changes

2. EXAMPLE INTERNATIONAL ALLIANCE: CARICOM

The Caribbean Community and Common Market (CARICOM) provides a typical example of an international alliance. It is a regional alliance to provide economic integration, hence reduction of transaction costs and economies of scale, for countries in the Caribbean region. It was established in 1973 under the Treaty of Chaguaramas. CARICOM replaced the former Caribbean Free Trade Association (CARIFTA). As is seen in various other international alliances, CARICOM's primary mandate is to provide a framework for regional, political and economic integration. Also common in other international alliances is the heterogenous status of membership, which presently has four levels: full members of both the community and common market; members of the community only; associate members; observers. Members of the Caribbean Community and Common Market include: Antigua & Barbuda, Barbados, Belize, Dominica, Grenada, Guyana, Haiti, Jamaica, Montserrat, St. Kitts & Nevis, St. Lucia, St. Vincent and the Grenadines, Suriname, Trinidad and Tobago. The Bahamas is a member of the Community, but not the Common Market. Associate members are British Virgin Islands and the Turks and Caicos. Observer nations include Anguilla, The Dominican Republic, The Netherlands Antilles, Puerto Rico and Venezuela. The objectives of CARICOM include the free movement of goods, services, persons, and capital throughout the region. These policies, and the current status of specific items is clearly stated in their Web site (www.caricom.org).

3. CHALLENGE: FLEXIBLE, ADAPTIVE BUREAUCRACIES

In the theory of organizations (e.g. Weber 1956/1978), the term 'bureaucracy' refers to management based on uniform, explicit rules, rather than case-by-case discretionary decisions. In this technical sense, international alliances are bureaucracies. Note: bureaucracies no longer need to be of the brick-and-mortar variety – they may also exist virtually. For instance, the North American Free Trade Alliance, NAFTA, is a virtual organization. A common complaint about bureaucracies is that they are inefficient. Requests are routed through numerous offices, each of which may have a backlog, leading to an accumulation of delays. Non-routine requests are sometimes misrouted, creating additional delays. Errors in the request cause backtracking and still further delays. Other complaints are about the effectiveness of the system. In order to keep the complexity manageable, bureaucratic rules may be simplistic, covering only standard cases. Exceptional requests are either rejected or diverted to a discretionary authority. When there are many exceptions, a change in regulations may be called for. However, because of the complexity already present, this is often difficult to do. Change is either resisted or made minimally.

On a broader scale, the very existence of a bureaucracy is sometimes criticized. Bureaucracies are often erected to perform a particular governmental or social function, and are thus protected from competitive challenges. They may grow to such size that the sheer momentum of their economic weight relative to the society carries them forward. An important aspect of this problem is a failure to effectively manage complexity. Bureaucracies are "sticky upwards". That is, they seem to grow more easily than they shrink. Nearly three decades ago, Elgin and Bushnell (1977, p. 337) complained that "we have rushed to create bureaucracies of such extreme levels of scale, complexity, and interdependence that they now begin to exceed our capacity to comprehend and manage them." Today, despite all the remarkable advances that we have seen in information and communication technologies, these complaints seem just as appropriate. Furthermore, this is certainly not because bureaucracies have ignored these advances. Indeed, bureaucracies are among the most aggressive consumers of ICT. Certainly, ICT has done much to improve bureaucratic efficiency. On the other hand, we claim, it has not helped at all in the management of bureaucratic complexity. Indeed, it has aggravated the problem by obscuring bureaucratic rules and procedures in the form of computer codes.

(For instance, many of us have experienced the frustrations of trying to rectify a billing problem that is computerized.)

This is both a representation problem and a modeling problem. It is a representation problem in that computerized rules and procedures need to be made more accessible and directly controllable by management. It is a modeling problem in that improved methods of systems analysis are needed to more effectively analyze and design complex bureaucratic systems. This is not, we argue, merely a matter of refining present techniques. There is a need for more fundamental revisions in our approach to such problems (Stamper, 1979; Lee, 1985; Kimbrough and Lee, 1986). Bureaucratic systems are not merely information processing systems. They are systems of organizational and social control. They convey more than data; they convey orders, commands, obligations, contracts, permissions, licenses, vouchers, receipts, prohibitions, waivers, verifications, etc. etc. etc. These are performative transactions in that by their communication they change the nature of social relationships within the organization.

These problems are well-recognized in large rationalized organizations and governmental agencies. However, we argue that these kinds of problems are most salient for international alliances, which are an evolving dialectical balance among multiple national interests.

4. APPLICATIONS SOFTWARE AND ORGANIZATIONAL CHANGE

It is a commonplace observation that organizations, to survive, must adapt to changes in their environment. Those that do not are forced out of business, if they are companies in a competitive market; have their budgets canceled, in the case of government bureaucracies; or are overthrown, in the case of governments themselves. Just how an organization should be designed to accommodate change is, of course, a much more difficult matter, and has been the subject of many volumes of organizational theory. One aspect of this general problem seems to have been neglected, namely, the effect of information technology on the organization's ability to adapt and change.

Certainly, there are numerous clear cases where the installation of an information system adds to the organization's flexibility. For instance, the installation of a centralized database may allow data to be accessed and combined in a variety of ways that would have been practically impossible when that data was recorded in paper files scattered throughout the company. The flexibility of a given computer application obviously depends on the foresight of its designers. To this end, programming students are generally taught to seek the most general definition of the problems they are given so that the resulting program can handle not only the immediate problem but also variants of it that might arise.

This strategy has obvious limitations. In seeking to find a generalized solution, the programmer may waste undue amounts of time on conditions that will never arise. He/she must therefore make a choice as to how much flexibility to encode into the program logic. We refer to the level of flexibility chosen as the 'designed flexibility' of the system. Selecting the appropriate level of designed flexibility is however difficult and, almost certainly, new requirements will later arise that were not planned for originally, so that the program must be modified. This is where the problem arises. Anyone who has written even small programs will know that it is much easier to incorporate a given feature in the program logic in its original writing rather than try to add this feature afterwards. This difficulty rises exponentially with the complexity of the original program or system.

To summarize, the basic problem with current application systems is that they are 'brittle'; i.e. they cannot easily be reformed to adapt to changing circumstances. This brittleness has profoundly disturbing consequences as more and more organizations, ranging from small and medium size companies to immense governmental agencies, convert their information processing to computer software. The immediate gains of increased efficiency, speed of processing, rapid access to centralized data files, etc., are clear (or the investment would not be justified). However, there may be a long term, possibly devastating hidden cost as the organization finds its ability to adapt and respond to new environmental conditions hampered by its inability to modify its information systems accordingly.

Concerns such as these have been major motivations in the development of object oriented programming languages (including Java) and, component-based architectures.

5. SOFTCODED ELECTRONIC PROCEDURES

As illustrated by the example of CARICOM, international alliances typically deal with the free movement of goods, services, persons, and capital among the countries that are party to the alliance. One may view this as a transaction simplification process, reducing the bureaucratic overhead of these kinds of exchanges. However, this process of simplification tends to be a gradual one, over an extended period of time.

To give a sense of the scope of complexity that is involved, consider the procedure in Figure 1, which is for the import/export of goods through a seaport. This is a typical procedure, not involving added complexities of dangerous goods, animals, foodstuffs, or terrorist controls.

A point that we want to emphasize here is that the reduction of bureaucratic requirements can be complemented with the introduction of

ICT solutions, which can make the existing bureaucratic overhead much less costly and time consuming.

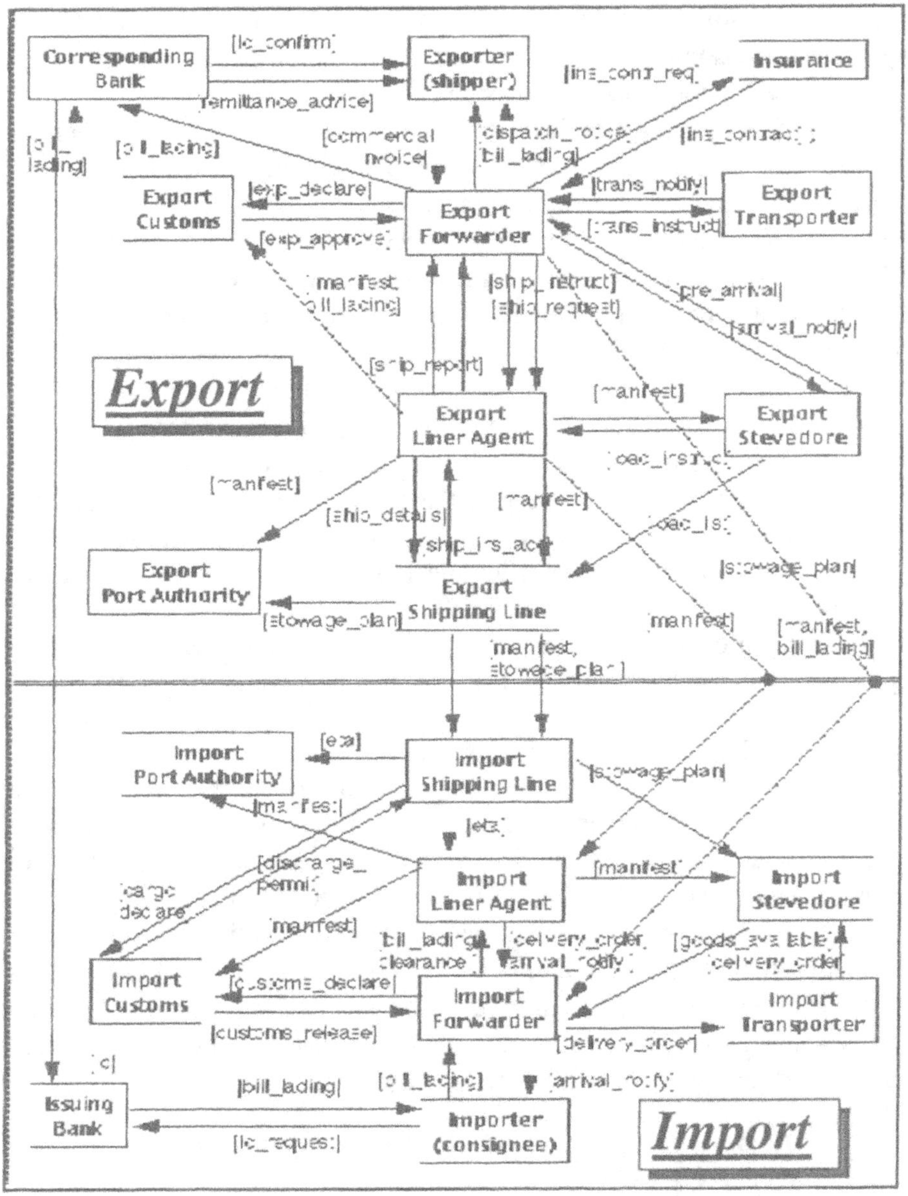

Figure 1. Typical International Trade Transaction

We might illustrate this point with a comparison to a standardization of keyboard layout for Latin alphabet languages. The US and UK have a keyboard standard called QWERTY, which refers to the keys in the left side of the top alphabetical row. The French prefer a layout that is AZERTY. During the time of Salazar, Portuguese adopted a standard layout HCESAR. At one time, the choice of a standard keyboard layout was a point of serious debate. Eventually, however, computers became flexible enough to support any variety of keyboard layouts, simply by changing software. Thus, the advance of technology made the problem disappear.

In a similar way, we believe that software solutions can also help to eliminate negotiation obstacles for international alliances. That is, instead of everyone agreeing to a single standardized solution, there may be a harmonized approach that allows existing national standards to inter-operate, with a minimum of adaptation. In many cases, software has already been developed to automate international procedures at a national level. However, it is likely that the procedures have been 'hard coded' into the application software – that is, the sequence of procedural steps is expressed in the program code. A key aspect of the architecture presented here is that the international procedures are 'soft coded', in a declarative, rule-based form. Application software then needs to be made generic, not committed to a particular version of the procedure. Instead, the application will interpret the procedure from an external specification (analogous to the idea of 'table-driven' software).

Within a national context, this has an immediate benefit of expediting the deployment of changes to procedures. Rather than sending out new versions of manual instructions (for non-automated procedures) or new versions of software (hard-coded programs), one may simply distribute a revised representation of the procedure, or have it accessed from a central repository – as illustrated in Figure 2. Even more significant is that such procedures can be analyzed and managed using computational tools. For example, analytical techniques can be applied to check for formal correctness (boundedness, etc.), as well as fraud potential and other audit controls. Further, soft-coding allows for the representation of generic models that are parameterized for specific circumstances. Additionally, soft-coding enables the navigation, synthesis and negotiation of procedures from different trading sectors or regulatory environments.

6. REPRESENTATION OF DOCUMENTARY PROCEDURES

We found Petri Nets to be the best formal representation for procedures that offer both a graphical representation and a formal basis for the verification of various properties of these procedures. The main advantage of the Petri Net formalism, in addition to its capability to graphically model both concurrency and choice, is that it offers various kinds of both formal and informal analysis methods, which make Petri Nets especially suitable for modeling "Discrete Dynamic Systems" (van der Aalst, 1992). We have extended the Petri Net representation to include the exchange of documents (as document places) in what we call Documentary Petri Nets (DPN's). This DPN representation is explained in detail in Lee and Bons (1996). Figure 3 presents an example Documentary Petri Net, as modeled using the graphic design environment InterProcs (see www.euridis.nl/?view=Interprocs). This system produces executable electronic procedures that are decomposed by role to execute autonomously in each of the separate organizations involved in the transaction.

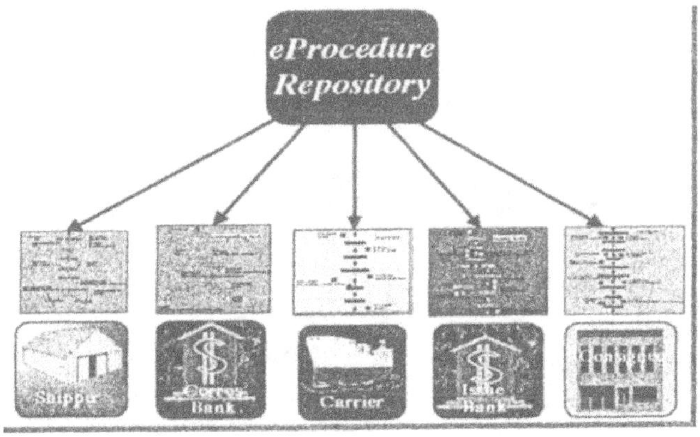

Figure 2. Parties Downloading from eProcedure Repository

These observations highlight the focus on aspects of multi-lateral harmonization as a priority for future research. A technical solution to enable multi-lateral harmonization that we describe is the agent-based approach called the Messenger Model. Further refinement of this technology focuses on the constraint resolution mechanism for Procedure Constraint Grammars (PCG's) (Lee, 2001).

Another research direction relates to normative modeling of documentary procedures: given the goals and control concerns of the various member countries for a particular type of transaction, what constitutes a 'good' procedure, which balances the excess transaction costs of over-control against the risks of under-control? This direction of future research will pursue development of automated verification tools that may be used to check whether a proposed documentary procedure conforms to specified control requirements. An advantage of using Petri nets as the basis of the DPN formalism is that various kinds of automated verification algorithms and tools already exist, for instance to detect possible dead-lock situations. However, our modeling experience has shown that these are of limited usefulness in practical situations – these kinds of problems seldom arise, and if they do, are usually quite obvious. The more serious control problems have to do with fraud and collusion, which require a deeper semantic understanding of what the procedure is actually doing. An approach to this is the notion of "audit patterns", originally developed by Lee (1991), and elaborated in the dissertation work of Chen (1992) and Bons (1997). Further research opportunities exist in this direction.

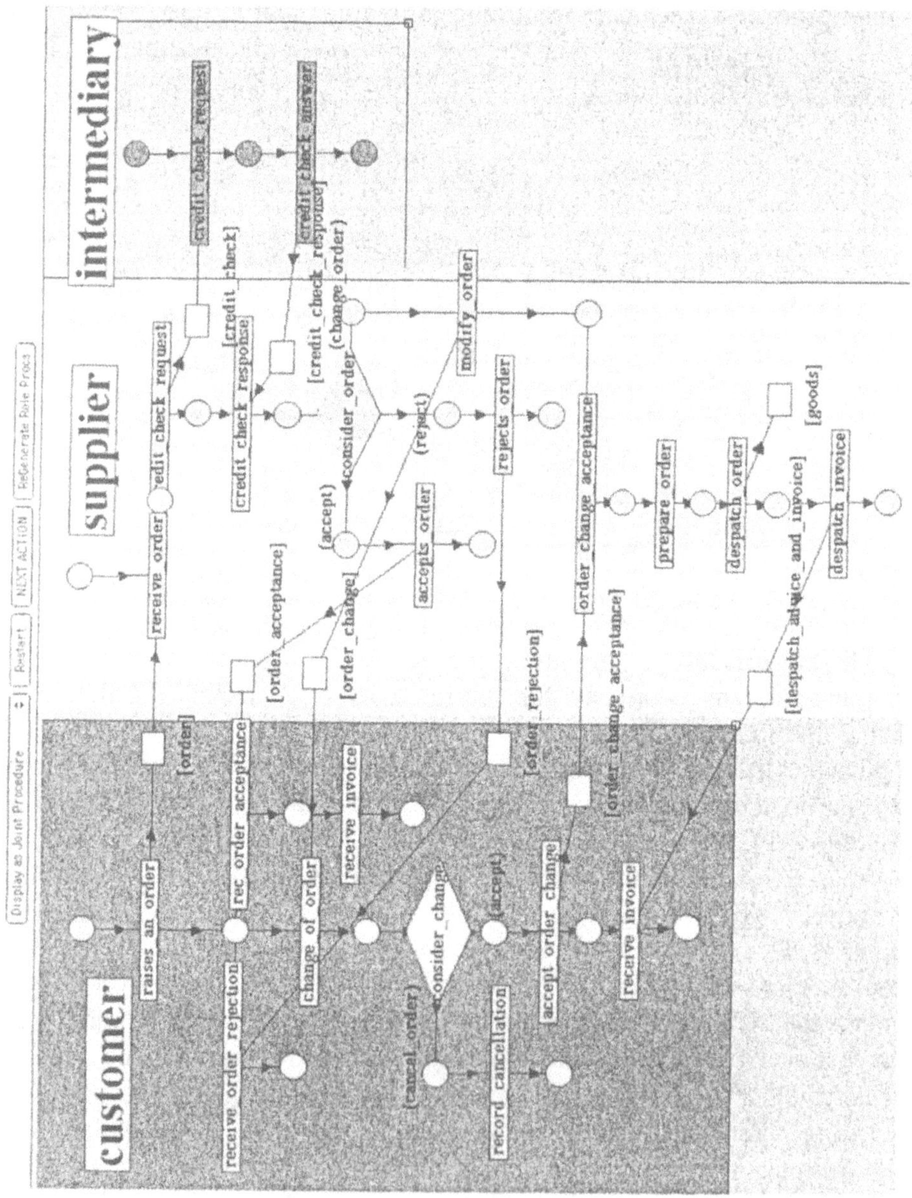

Figure 3. InterProcs Display of Order Process as DPN

REFERENCES

Aalst, W.M.P. van der "Timed coloured Petri Nets and their application to logistics", *PhD thesis Eindhoven University of Technology*, 1992

Austin, J.L. "How to DO things with words", *Harvard University Press*, Cambridge, MA, 1962

Bons, Roger W.H., *Designing Trustworthy Trade Procedures for Open Electronic Commerce*, PhD Dissertation, Euridis and Faculty of Business, Erasmus University, September, 1997.CARICOM, see www.caricom.org.

Chen, Kuo-Tay. *Schematic Evaluation of Internal Accounting Control Systems*, University of Texas at Austin, PhD Dissertation, 1992.

Chen, KuoTay and Lee, Ronald. "Schematic Evaluation of Internal Accounting Control Systems", Euridis Research Monograph (RM 92.10.01), October, 1992.

Dewitz, S. K. and Lee, R. M. "Legal Procedures as Formal Conversations: Contracting on a Performative Network", *Proceedings of International Conference on Information Systems*, Boston, December 1989, pp. 53 – 65.

Elgin, D.S. and Bushell, R.A. 1977. "The Limits to Complexity: Are Bureaucracies Becoming Unmanageable?", **The Futurist**, December.

Genrich, H.J. and Lautenbach, K. "System Modelling with High-level Petri Nets", *Theoretical Computer Science*, Vol. 13 pp 109-136, New York: North-Holland, 1981

Genrich, H.J. and Lautenbach, K. "The Analysis of Distributed Systems by Means of Predicate/Transition Nets", *Semantics of Concurrent Computation: Lecture Notes in Computer Science*, Kahn G. (Ed.), Vol. 70 pp 123-146, Springer Verlag, 1979

ICC, "The Uniform Customs and Practices for Documentary Credit Procedures", *International Chamber of Commerce publication 500,* Paris, France, January 1994

Kimbrough, S.O. and Lee, R.M. "On Illocutionary Logic as a Telecommunications Language", *Proceedings of the International Conference on Information Systems* (San Diego; December, 1986)

Lee, R.M. 1985. "Bureaucracy as Artificial Intelligence", in Humphreys, P.(ed.) *Knowledge Representation for Decision Support,* Proceedings of IFIP WG 8.3 Working Conference (Durham, England; July, 1984), North-Holland.

Lee, R.M. "Auditing as Pattern Recognition: Automated Analysis of Documentary Procedures", Working Paper, Department of Management Sciences and Information Systems, University of Texas at Austin, August 1991.

Lee, R.M. "Automated Generation of Electronic Procedures: Procedure Constraint Grammars" *Decision Support Systems*, Special Issue on Formal Models for Electronic Commerce, eds Kimbrough, S. and Tan, Y-H, 2001, pp. 291-308.

Lee, R.M. and Bons, R.W.H, "Soft-Coded Trade Procedures for Open-EDI" *International Journal of Electronic Commerce*, Vol 1, No 1, 1996, pp. 27-49.

Lee, R.M., Bons, R.W.H, Wagenaar, R.W. "Pattern-Directed Auditing of Inter-organisational Trade Procedures" Proceedings of the 1st IFIP Conference on eCommerce, eBusiness, and eGovernment, Zurich, Switzerland, 4-5 October 2001.

Peterson, J. L. "Petri Net Theory and the Modeling of Systems", *Prentice-Hall*, 1981

Petri, C.A. "Kommunikation mit Automaten", *PhD thesis University of Bonn*, Germany, 1962.

Searle, J. "Speech Acts: An Essay in the Philosophy of Language", *Cambridge University Press*, London, 1969

Stamper, R. 1979. "LEGOL as a Tool for the Study of Bureaucracy", *Proceedings of the IFIP TC8 WG 8.2 Conference on the Interaction of Information Systems and the Organization.*

Weber, M. 1956/1978. *Economy and Society.* Berkeley, California: University of California Press, translated from *Wirtschaft und Gelsellschaft.* Tuegingen: J.C.B. Mohr, 1956.

Chapter 28

MODELING FRAMEWORK FOR E-BUSINESS SYSTEMS

Murali Mohan Narasipuram

Abstract: Use of internet technologies has expanded rapidly from the initial profit-oriented commercial systems to wide ranging business systems including administrative, governmental, non-profit oriented businesses, etc., which include intra-organizational and inter-organizational systems (IOS). All these systems are broadly referred as e-Business (e-Biz) Systems in this work. e-Biz systems are developed based on strategic alliances among the participating business systems to exploit the IT(information technology)-driven synergies. Emphasis in the traditional systems development methodologies is on process models, data models, event models, etc. of the business system under study where as in an e-Business system the emphasis is on modeling the strategic alliances between business systems and the integration architecture, and thus necessitating the need to model the strategic dependencies and relationships. Current e-Biz system development tools assume that the changes that are being caused by the technology to business systems' goals and objectives, and nature and scope, are well understood, analyzed and documented ready for implementation by the developer. This assumption is not true. There is a requirement for an e-Biz system modeling methodology which facilitates modeling the strategic relationships, goals and objectives of partners of an e-Business system, and which helps convert the conceptual strategic model to an implementation-oriented model. In this paper, an effort is made to develop such a methodology and demonstrate it using a case study.

Key words: eBusiness, Modeling, inter-organizational systems, system development, methodologies, I*framework

1. E-BUSINESS SYSTEMS

Internet technologies are affecting the way we conduct business in almost all walks of life these days. Over the last decade, several traditional business systems have been reengineered and/or redesigned to exploit the benefits from Internet technologies. However, the open and global nature of Internet necessitates not only reengineering the existing business processes of a business system, but also redefining the very nature and scope of the business being conducted. The emphasis in profit making has evolved from reducing costs and increasing productivity and process efficiency to increasing the revenues, developing new products/services, expanding the markets, and to making the processes add more value to the system. This change in the nature and scope of business systems has affected and has been effected by the developments in the domain of Information Technology (IT).

Traditional business systems have looked at two types of IT-driven requirements, viz. mandatory requirements necessitated by market-adoption (ex. Automatic Teller Machines in Banking Sector) or statutory legislations (ex. security protocols), and voluntary requirements driven by business process automation decisions. In this traditional perspective, requirements necessitated by the IT-driven strategic alliances with other organizations are not given priority. The emphasis is on development of isolated computer-based business systems to help meet the sectoral market requirements. As the IT applications in the business systems have grown, the computer-based business systems have grown larger, but yet isolated, unwieldy in terms of facilitating synergies through strategic alliances. Development of such large systems has given raise to development methodologies such as Structured Analysis and Design methods. These methods essentially concentrate on modeling the process, data, network and interface aspects of the business systems to help improve the efficiency of the current processes within the broad umbrella of current corporate goals and objectives. The expected life of these systems is considered to be at least about 5 years, with the development effort needed ranging from 1 person year for a small system to over 10 person years for a large system.

Advent of Internet technologies has brought significant changes to the basic paradigm of business systems. With the open and global nature of the Internet, businesses find it easy to interact on both ends of their business chain, viz. suppliers and customer. Several small 'intermediary' businesses have evolved challenging the traditional big businesses. The intermediaries specialize in specific tasks and add more value making them more cost-effective than the traditional systems. These systems are organic and dynamic and respond to the changes in the environment very quickly, thus making the expected life of these computer-based systems short (as short as 3 months in case of some dotcom companies) and necessitating rapid systems development. They attempt to find and exploit a gap in the value chain of inter-organizational systems and develop business systems quickly to fill the gap by providing value added services and information products in a synergic manner. In this paper, these systems are broadly referred as e-Business (e-Biz) systems.

Evolving e-Business system development tools such as Microsoft's DNA, Sun's Enterprise Javabeans and Java Platform for the Enterprise, IBM's San Francisco, SAP's R/3, etc. assume that the changes that are being caused by the technology to business systems' goals and objectives, and nature and scope, are well understood, analyzed and documented ready for implementation by the developer; however it is not the case as the issues involved are complex and more importantly, dynamic in nature. There is a gap, which needs to be filled by an e-Biz system modeling methodology, which facilitates modeling the strategic alliances and also conversion to a development-oriented model.

2. KEY ISSUES IN E-BUSINESS SYSTEM MODELING

In the study conducted in the 1998 Bled Electronic Commerce Conference on 'Electronic Commerce in the Information Society', a group of 56 academic, business and government leaders from 19 countries on five continents identified organizational change and process improvement/modeling as the top two most important research issues in International Electronic Commerce out of the 174 issues that were raised (Vogel 98). Regarding organizational change, the comments of the group "were oriented around the changes in structure and process that accompany the introduction of the technology", whereas in the case of process improvement/modeling, issues raised included business process modeling, modeling changing value chains (e.g., virtual value chains), business process simulation techniques and new ways of combining parts of the value chain (e.g. through online auctions, etc.), etc.

During the 1960s and 1970s, development of computer-business systems was more programming-oriented. A single person or a small group of persons used to study, analyze, design and develop systems without much documentation. A clear disadvantage felt of this approach was that maintenance cost was exorbitantly high. Many variants of structured analysis and design methods were developed in 1970s and 1980s to help alleviate this problem. Most of the application 'silos' referred earlier are developed using this approach. These approaches tend to spend substantially more time in understanding and documenting the system requirements specifications compared to the earlier programming-oriented approach. This has substantially reduced the maintenance cost, and hence the overall system cost over the system's life. However, an implicit assumption in this approach is that the requirements are going to be stable over a reasonable amount of time in order to make the initial investment made in requirement elicitation and specification is cost-effective.

E-biz systems, as explained in earlier sections, are dynamic in nature and are expected to respond very quickly to the changes in their user and technology environment. More importantly, user group of an e-Biz system is far more heterogeneous compared to the user groups in traditional systems. Customization of user interfaces and even database structures to suit the individual customers is a hallmark of the contemporary e-Biz systems such as On-line newspapers, Electronic clipping services, etc. Not only the user groups are heterogeneous, but their requirements are dynamic and evolving. In this environment, the structured analysis and design approaches have major disadvantage due to their implicit assumption.

Object-oriented (OO) approach to computer-based business systems development helps in providing continuum in the analysis, design and development process in terms of objects – their structure and behavior, thus facilitating to respond more quickly and effectively to the user requirements changes. However, there is no specific improvement provided in OO approach during the phase of user requirement elicitation and documentation in the e-Biz system modeling. Moreover, the basic premises of e-Biz systems is that they develop IT-driven strategic alliances with other business systems to exploit synergies in tapping the core competencies of competing and cooperating business systems to collapse the supply chains. These alliances are strategic in nature in supporting the core business of the e-Biz system under concern, whereas the operational model of the e-Biz system includes on the operational models of the partner systems also. In other words, while modeling the e-Biz system, it is essential to model the strategic requirements of the system in addition to the operational requirements.

3. E-BIZ SYSTEMS DEVELOPMENT

The system development tools over the last four decades are evolving more in the direction of providing 'components' with higher (larger) 'business granularity'. Some of the developments in the last two years include e-business development environments such as Microsoft's DNA, Sun's Enterprise Javabeans and Java Platform for the Enterprise, IBM's San Francisco, SAP's R/3, etc. These development tools assume that the changes that are being caused by the technology to business systems' goals and objectives, and nature and scope, are well understood, analyzed and documented ready for implementation by the developer; however it is not the case as the issues involved are complex and more importantly, dynamic in nature. The approach suggested in the earlier section to model the strategic alliances in an e-Biz system helps to meet this gap.

The component based development (CBD) defines a business system software at five levels of abstraction, i.e. granularity, viz. software language class (i.e. the actual code), distributed component (ex. business function such as validating credit card, etc.), business component (ex. Web-based Order manager, etc.), business component system (BCS) (ex. Web-based flower mart), and federation of business component systems (ex. Flower mart, FedEx, Visa, and HSBC). Though CBD envisages the five levels, it is not necessary that every business system needs all the levels. Major strengths of CBD approach are scalability, ability to build solutions to business systems in new domains efficiently, ability to adapt existing solutions to new problems efficiently and ability to integrate and evolve business system solution sets. As one of the major characteristics of e-Biz systems is to strike new strategic partnerships and evolve into business systems with new products and markets, CBD meets this requirement of integrating and evolving new business system solution sets quickly and efficiently. As CBD is scaleable, it can be applied to smaller as well as larger systems.

The essence of the emerging e-Business models is establishing strategic alliances, driven by information technology (IT), with other businesses to reduce the cost of production of the `services and products, to increase the market share and revenues, and to improve customer services. Business systems differ in size (mom and pop operations to conglomerates), nature of business (low diversity - possibly single – product to multiple products), range of services offered, extent of external alliances, mode of internal organization, level of automation, etc. CBD approach helps to generate solutions quickly to this wide range of systems using the component libraries. In addition to component libraries, it is envisageable to develop BC libraries and BCS libraries as the e-Biz systems get more standardized. In other words, it is only the strategic alliances which will define the new business systems whereas individual business systems (or component systems) are defined by operational models supported by BC or BCS libraries. In such an environment, system development takes very short time so that the business systems can go on line very quickly to capture the market.

4. CONCEPTUAL MODELING OF E-BIZ SYSTEMS

The current e-Business system development environments are based on the functional requirements of the business system. These functional requirements are, in fact, an outcome of the strategic alliances and the e-Business model adopted by the business enterprise. As explained in the earlier sections, the strategic alliances may and in most cases will affect the corporate goals and objectives of the enterprise. Hence, it is imperative that we model the strategic dependencies and relationships of an enterprise involved in e-Business systems based on the strategic alliances made in the supply chain and the e-Business model adopted.

There are several efforts in applying conceptual modeling techniques to understand and design business processes. These efforts have modeling formalizations using basic concepts such as entities, activities, assertions, and time. In order to capture the strategic objectives of e-Biz systems such as intentions, dependencies and relationships, a higher order modeling paradigm is required. Yu has proposed I* framework with a specialized ontology with additional concepts to more fully support the types of knowledge and reasoning involved business systems modeling. The framework has been described in terms of requirements engineering, business process reengineering and software process modeling [Yu 1993] [Yu 1994]. This framework argues that organizations consist of actors who perform tasks. Actors depend on one another to perform the tasks. We need to understand not only what and how tasks are performed but also why, the intentions and motivations, in order to model the strategic nature of business process

However, the I*framework emphasizes on the dependencies and tasks performed within a business enterprise whereas e-Biz systems are inter-organizational in nature. It is proposed to extend the framework to capture not only the individual actor goals, but also the organizational goals. In some cases, the collection of the goals of actors within a system may result the overall goals of the system, but not necessarily in all cases. So it is proposed that in addition to the actors of an e-Biz system, the other participating e-Biz systems also are identified as actors with dependencies either on the individual actors of the e-Biz system and/or on the e-Biz system as whole. It is also proposed to define new constructs to capture the collective goals of higher (than actors) order concepts such as the e-Biz system components, subsystems, etc. It is envisaged such an approach helps convert the conceptual model into an implementation-oriented model.

5. CASE STUDY

In this section, we will demonstrate the use of the conceptual modeling techniques to model the strategic alliances in an eBiz system, viz. on-line supermarket store. This case is a typical example for B2C eBiz environment. The aim of this on-line store is to bring a selection of food and groceries products to the shoppers through online to genuinely provide a one-stop shopping solution. This eBiz system taps the benefits of connecting the traditional value chain & supply chain models into a strategic alliance as discussed in the following paragraphs. Corporate level Actor Dependency model of this eBiz system is given in Fig 1.

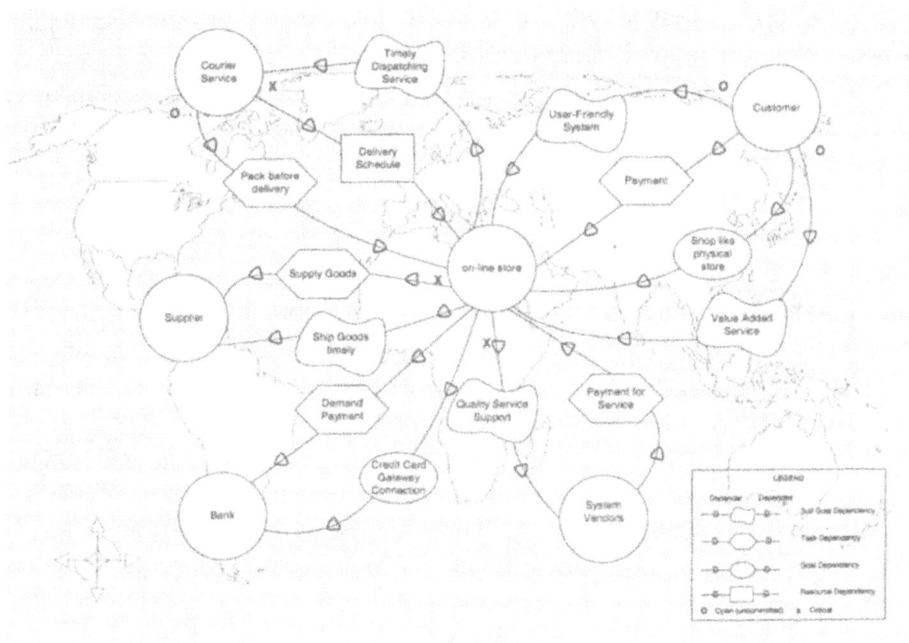

Fig 1: Corporate level Actor Dependency model

Strategic Alliances

Within the value chain and supply chain of the store, there exist opportunities for alliance. In fact, the development of strategic alliances is to support the store's competitive strategies that would eventually add value in terms of efficient operations and improving customers' services.

In the upstream activities of supply chain, the store relies on the flow of goods/materials from the suppliers. The online store purchases some groceries such as snack, milk, flour, drinks, toilet tissue, foods bags and etc. from major brands supplies. To avoid disruption of supplies, it has partnered with some local dealers to ensure the constant offering of snack, drinks, candies and etc. Such alliance contributes a lot of advantages to the store such as lower purchase costs, reduced inventory, and enhanced efficiency of logistics as well as increased sales and lower marketing costs. The store also made strategic investment by establishing its own brands selling groceries, toilet tissues, snack, daily consumable products and soft drinks. The benefit is direct sales by disintermediating the middlemen. Besides, the store could build up its own brand loyalty from the customers. On the part of distribution, although the store has established an in-house delivery team, they have outsourced some delivery activities to several transportation companies so as to ensure efficient, reliable and prompt delivery services.

We refer to these intermediaries as Courier Services. The store has also made strategic alliance with a bank to process its payment collection. Customers buying goods from the net may use Credit Card, EPS or Personal Check for payment. The bank handles all the payment collection, electronic payment transfer and Credit Card processing on behalf of the store for its consumers and suppliers, thus, eliminating its tasks in cash management. The store strategically teamed up with canned food manufacturers to demonstrate "Fresh Cooking Ideas" everyday on the net. The cooking ideas are provided free of charge to customers enhancing its value-added services for its shoppers. Based on the value chain description, the components of the eBiz system are described and modeled using the I*framework as following.

Procurement System: Under procurement system, Purchasing Department (PD) is the main actor to be involved. PD would like to have a user-friendly system, which can help to minimize workload and manual procedures, and it depends on the Information System Department (ISD) to furnish such a system based on PD requirements. In addition, it has to interact with Warehouse Department (WD) to exchange information on purchase order information and stocks received information. Suppliers receive payment after an order is shipped.

Payment System: Accounting Department (AD) will settle supplier invoices based on the stock received information provided by PD. In fact, PD obtains stock received information from the WD. ISD's duties are to develop and maintain a user-friendly information system. In addition, AD critically depends on the accuracy and timeliness of the reports generated by the system for payment. On the other hand, suppliers would like to receive payment timely, which in fact depends on the efficiency and effectiveness of AD. If the system cannot provide timely and accurate payment information, the whole chain of actions will suffer.

Distribution System: In the distribution model, the main actors are Customer, Courier and Warehouse Department (WD). WD requires timely and accurate information from the information system for preparing delivery schedule. On the other hand, Courier requires delivery schedule to arrange trucks to dispatch the goods. In addition. Courier expects all goods are properly packed before loading to the trucks; however, there may be cases that WD does not have enough time to pack all the orders. In this case, Courier will assist WD to pack the goods. The information system plays an important role in this model. The whole value chain critically depends on ISD.

Mercantile System: Customer would like to compare what the virtual store offers with that available in the physical store; therefore, synchronization of the products/services and their prices is an important task when both physical and virtual stores exist. There may be situations that prices or offers cannot be synchronized due to some reasons; however, this basic rule should not be violated too often. User-friendly operation interface is a key critical successful factor of e-storefront system. It allows novice customers to operate and feel at ease when they are purchasing online. While orders are received online from the web, it transfers the order information to Warehouse Department (WD) for dispatching. WD will then pick and pack the goods for dispatching. The dispatching service is in fact provided by Courier, which is omitted in this model for simplicity.

Revenue System: Customers can pay online via credit card or pay by cheques or cash on delivery. When customers paid the goods online via credit card, the system will interface to the card center to verify the validity of the card and to record the transaction. The bank will then pay to the store after receiving the supporting document (electronically) from the store. The turnaround time should be short to maintain a healthy cash flow. Customers demand user-friendly interface, online refund and payment history records.

eCRM System: Buying online is mainly a personal task. Vendors would not be able to provide direct services to customers who are shopping around. eBiz systems incorporate the concept of eCRM into their web-based functions, i.e. serving the customer electronically. For example, feedback or suggestion loops built into the eStore help to identify customers' preferences and suggestions. In addition, customization can be done by profile administration. Most of the successful eStores, like Amazon.com, require user registration in order to provide personalized services, to identify their need s and to perform Online Analytical Processing (OLAP) to understand and manipulate customers' buying habits / behaviors. In building an eCRM system, industrial practices, customers group preferences, marketing trends and innovative idea are all of great importance.

6. CONCLUSIONS

Strategic alliances are very important in the realm of eBiz systems. Systems development methodologies should incorporate concepts and techniques to conceptualize and model these strategic alliances while developing eBiz systems. This paper illustrates a case of virtual supermarket store and its strategic alliances modeled using the I*framework.

REFERENCES

1. Farbey et al, A taxonomy of 15 applications: the benefits evaluation ladder, ERS, 1995, 4, 41-50.
2. Vogel, D. W. White Paper on 'Research Issues in International Electronic Commerce', Results from the June 8, 1998 Special Session of the 11th annual Bled, Slovenia Electronic Commerce Conference on 'Electronic Commerce in the Information Society', Bled, Slovenia.

3. Yu, Eric S. K., John Mylopoulos. An Actor Dependency Model Of Organizational Work – With application To Business Process Reengineering. University Of Toronto. 1993
4. Yu, Eric S. K., John Mylopoulos. From E-R To "A-R" – Modeling Strategic Actor Relationships For Business Process Reengineering. University Of Toronto. 1994

SECTION 9
Modeling and Construction of e-Services

Chapter 29

REFERENCE MODELS FOR ADVANCED E-SERVICES

Chris A. Vissers, Marc M. Lankhorst, Robert J. Slagter

Abstract: Reference models (RMs) capitalize on the experience that key functions and relationships determine a system's main design structure which has to be established before other design details can be settled. As such RMs can play an important role in designing complex (distributed) systems, in allocating design tasks to cooperating design teams and in facilitating their communication. These roles are also eminent in standardization. This paper discusses the need for precisely defined basic architectural concepts to construct RMs, building on experience with designing the OSI-RM. We apply these concepts in the design of a number of RMs for networked applications that provide advanced e-services.

Keywords: Reference model, architecture, basic architectural concept, service, protocol, networked application.

1. INTRODUCTION

Suppose you contract an architect to design your house. You discuss the size of the house and how it is situated on the premises. How rooms, corridors, staircases, windows, bathrooms, balconies, doors, a roof, etc. will be measured and put together. You agree on a master plan on the basis of which the architect will produce detailed specifications.

How come that you communicate so efficiently about that master plan? We suggest it is because you share a common frame of reference: you both know the concepts of premises, situation, measure, room, balcony, staircase, etc. You know their functions and their possible relations. Mentally, you both use a *reference model* of a house. This reference model (RM) defines a structure of major functions and how they are related while applying commonly known concepts. It provides an abstract design, ignoring many details that will be filled in later such as construction details, materials to be used and colors to be applied.

Likewise, the design of information systems will be more effective and efficient if a suitable RM as a blueprint for information systems is taken as a starting point. Suppose you want to design a system to support electronic transactions. Then you have to answer questions such as: What services may users expect from the system? What should be the building blocks of the system? How should they interact? It would then be quite helpful if you also have a RM that shows building blocks for managing transactions, for providing financial and logistic services, etc., and that shows what information these blocks exchange.

We introduce RMs for a broad range of networked applications: structures of functional entities, implemented by software, that offer *services* to distributed users while exchanging information via a network. Examples are applications that enable users to trade by exchanging orders, bills, and payments via an underlying network, or applications that allow users to simultaneously access design drawings despite the fact that they are miles apart, or applications that allow users to search through remote video libraries and download a video. The majority of such services fall in one, or are a combination of the following categories:

- collaborative services;
- transaction services;
- content services.

Networked applications can become quite complex if you consider all the details of functional definitions, message formats, programming schemas, programming code, operating system calls, etc. RMs help you to master this complexity by focusing first on the high level system design, suppressing detailed design issues until this high level design has been settled. In fact you follow the strategy of the architect of your house.

2. REFERENCE MODELS, THEIR NATURE AND PURPOSE

A reference model provides people with a common reference to an object, e.g. a distributed system, as a basis for their common understanding, discussion, and further action. They are about models in the sense that they describe essential aspects of systems while abstracting from details not considered essential for the pursued goal. Typically the goal is to focus on what systems should do, rather than how they can be constructed and operate at the implementation level. We define a RM as a structure, or organization, of related functional entities that defines only globally the key functions and key relationships of these entities. This means that RMs are incomplete system designs in two respects:

1. The high abstraction level implies that functions and relationships are defined only functionally and as implementation independently as possible, leaving freedom for the individual manufacturer to choose his own implementation strategy.
2. Key functions and relationships are defined only globally, leaving freedom for design teams to complete the functional design by adding design details that at the RM level are not considered key.

The latter requires some extra explanation since the term "key" is usually intuitively, rather then explicitly applied. By a *key* we mean a characteristic that largely determines the function of an entity, system or relation and consequently has a large impact on the structure and further design of a system. It means that extending the RM to a complete design by refining the key functions can be done while preserving the structure and relationships (interfaces) of these functions.

For example, when it is key that a service is connection oriented, it will appear that this key requirement largely determines the nature of the service and the protocol that implements it. The same applies when it is key that a service is reliable. At the protocol level it will appear that protocol data unit (PDU) numbering is a derived key functional element to achieve both connection orientation and reliability which completely dominates the structure of the protocol.

The development of RMs is a design activity that should follow qualitative design principles such as: do not link what is independent (orthogonality), introduce functions in their most general form and avoid slightly diverging alternatives (generality), do not introduce what is immaterial (parsimony), and do not restrict what is inherent (propriety).

A RM serves several purposes, such as to act as a basis for:

- understanding the essential user requirements and derived (key) properties of the real world system (a networked application in our case) that has to be developed;
- formulating these properties in an initial high level design and preserving the conformance between the service provided and the protocol implementing this service;
- communication between the users of the real world system and the RM architect to communicate requirements and document their agreements;
- communication between the RM architect and the designers of systems;
- communication between different design teams that may work together on a system design.

There exists an extensive literature about RMs, reference architectures, and design patterns, their purpose, their form, the language in which they should be expressed, etc. (see e.g. [2] , [7], [16], [17]). A well-known example is the RM for Open Systems Interconnection (OSI-RM) [11] which globally defines the services and protocols of networks, and applications on top of these networks (*Figure 1*).

Other well-established examples are the RM for Open Distributed Processing (RM-ODP) [13], the Workflow RM of the Workflow Management Coalition (WfMC) [9], and the Object Management Group's Object Management Architecture RM (OMA-RM) [14].

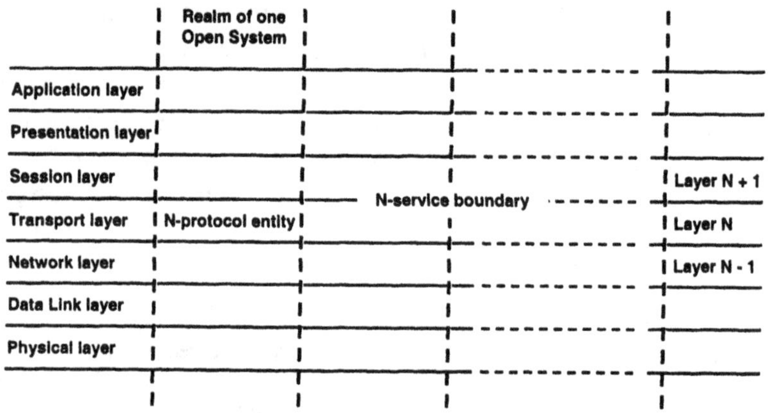

Figure 1. Original illustration of the RM for Open Systems Interconnection.

3. BASIC ARCHITECTURAL CONCEPTS AGAINST A HISTORICAL BACKGROUND

To serve its purpose as a common reference, a RM must be "constructed from" sound abstract and basic architectural modeling. This implies that such concepts should model properties considered essential for real world systems, that are precisely defined at the appropriate abstraction level, and that can effectively be applied as basic building blocks to construct a high level design. They should also be well understood and commonly supported. Here is the "Achilles heel" of many RMs. Experience with the definition of the OSI-RM, and its related services and protocols, has learned that whole crowds of experienced designers were perfectly willing to believe in the most bizarre and absurdly defined concepts as soon as some form of abstraction comes into play. Correspondingly it is astonishing to observe what confusion can be introduced in block diagrams while people are believing that the mere use of it would ensure clarity and precision. Since sound basic architectural concepts, apparently, are not so easily established, we define below some architectural concepts that we will use for our RMs for networked applications. To underline that their definition is not so trivial we contrast them with some bizarre interpretations that have circulated in the OSI world.

In this section we use an intuitive graphical notation; in the sections describing our reference models we use a more formal notation.

3.1 Service

We define a *service* as the (possible) behavior of a system as it can be observed and experienced by its users. The service concept is of prime importance since it defines precisely what benefit a system provides to its users. Actually it defines a system's "reason d'être".

In the ISO-OSI standards meeting in November 1981 in Berlin there was a big fight between those who wanted to have separate service standards and those who rejected this idea and only wanted to consider a service definition as an informative addendum to a protocol standard. The argument was that OSI aimed only at systems interconnection and that only interconnection needed to be defined in a conformance testable way (by monitoring the PDU exchanges). Service standards were considered not to be conformance testable since they focus on user interaction (which apparently was considered irrelevant) and would require testing the common behavior of products of different manufacturers. The argument, of course, was nonsense, since one (i.e. ISO itself) can proof the conformance of the protocol entity specification against the service specification and then conformance test the protocol entity implementation against the protocol entity specification. Fortunately the advocates of service standards won the battle, however at a certain price: the service standard could not be published independently from the protocol standard.

A service defines a system as a black box, i.e. it provides the most simple but complete definition of the observable behavior while obscuring how the system is internally constructed. See *Figure 1*.

In the early OSI documents a service was defined as "the functions of a layer, while using the functions of all layers below. See *Figure* . Since the functions in an N-layer are the N-protocol entities, this definition in fact confronts the users with the whole complexity of all protocol entities below an N-service boundary. It took until the end of 1982 before this definition was revised, following the work of the Formal Description Techniques (FDT) Group.

3.2 Interaction Point or Service Access Point

In order to make use of a service, users have to interact with it. The fact that a user can interact with a service can be formalized by the *interaction point* (IP) concept, in OSI called the *service access point* (SAP) concept. The formal semantics of an IP is basically the identification of the functional entities that can interact. For an IP these entities are the user and the service (provider). See *Figure 1*.

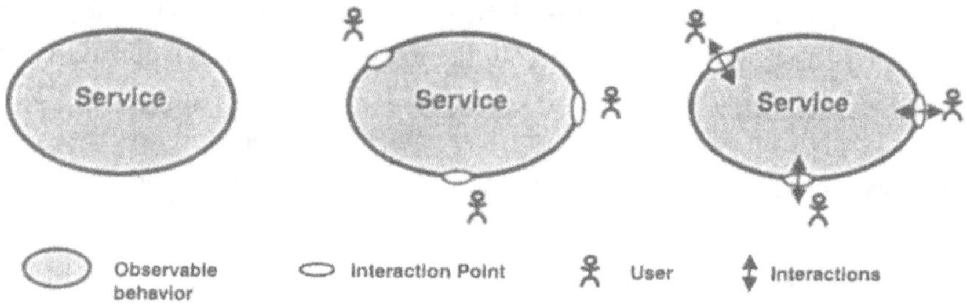

Figure 1. Service as a black box, service with IPs and users,service with IPs and user/service interactions.

Associated with IPs are the concepts of *addresses* and *names*. Generally an address is considered as a physical or logical location, while a name is considered as the identification of a specific user, the location of which is left undefined.

3.3 Interaction Primitive or Service Primitive

The interaction of a user with a service is always in terms of one or more "units of interaction", called interaction primitives, or simply *interactions* (Is). See *Figure 1*. In OSI they are called *service primitives* (SPs). The semantics of an I is basically a unit of common activity, which in principle is the same as a unit of common behavior, at an IP and resulting (with a certain probability) in some data at some moment. Interacting parties may provide different contributions to the I, and may have different use of and constraints on the results. At the appropriate abstraction level only the existence of the I and its attributes (the IP, the contributions, the results, the probability and the time moment) need to be defined.

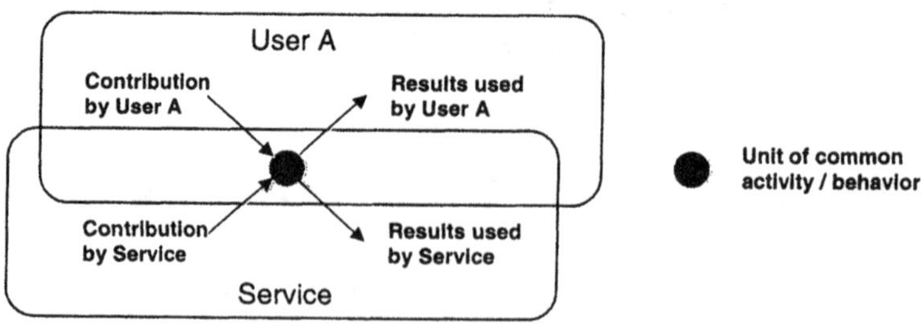

Figure 2. An interaction as an unit of common behavior of two entities providing different contributions and using different results.

Figure 2 shows a graphical way of illustrating a common activity, where a surface indicates activity and a surface overlap indicates common activity. The I can be defined by parameters and operations on them.

During the design and documentation of the OSI Transport Service, the design team (of which the first author was a member) did not want to view the abstract notion that a service primitive can be considered as a unit of common activity of two adjacent protocol entities. Instead, service primitives were only understood as one-directional message transfers, an interpretation many designers still have. Considering the establishment of a SP as a unit of common activity would have allowed to define the connection endpoint identifiers (CEIs) as a parameter of the SP whose establishment can be left to (protocol) implementation. By not appreciating this advanced design concept, CEIs were left out of the definition of SPs resulting in a politically agreed remark in the text of the standard [12].

3.4 Abstract Interface

An *abstract interface* (AI) defines the possible Is at one interaction point and their causal and parameter relationships. Whereas the IP is an abstract (logical) location, the AI defines the common behavior at that location.

There is quite some confusion between the notions of an AI and a *real interface* (RI) where many designers seem to only understand the latter. Where the AI is at the top architectural level, such as at the RM level, a RI is always at the level of a real implementation. In fact the same RI may be used in the implementation of a variety of AIs. The notion of AI never played a dominant role in OSI.

3.5 Service Design

A service can be designed by defining the IPs, the possible Is at the IPs, and all their causal and parameter relationships. Alternatively this can be designed by first defining the AIs at different IPs and then adding the causal and parameter relationships between the Is at the different AIs. The latter makes use of a constraint-oriented specification style [22], in which local and remote constraints on the service's behavior are separated. This helps in understanding complex systems, in deriving protocol entities from a service, and in proving their conformance.

In order to show the causal relationships of interactions, time sequence diagrams are often used. Although quite illustrative for simple situations, the two dimensional drawing scheme is not suitable for more complex relationships. For that purpose we better take resort to an architectural specification language such as AMBER [5].

Continuing the discussion under 3.1: In OSI one did not want to speak of Service Specifications in line with Protocol Specifications but only of Service Definitions. The reason to speak about Definitions was that they were not supposed to be implemented. So they were not specifications meant as prescriptions for implementation. This again was a misunderstanding since a service is implemented by implementing the protocol. A Service is just one step in abstraction level (i.e. formulating the essential!) closer to the user. In fact also an OSI Protocol Specification is not implementable directly, since only the OSI Protocol Entity is, and its specification has to be derived first from the protocol standard by the designer (see also the discussion in Section 3.6).

OSI service standards were defined by English text and illustrated by block diagrams and time sequence diagrams. An example of the latter is shown in *Figure 3a.* To draw such diagrams, so called OSI Service Conventions were prescribed according to *Figure 3b.* Originally these conventions prescribed straight angled lines, as shown in *Figure 3c,* where even the magnitude of the angle had the meaning to indicate "transfer speed". This idea was probably inspired by the advantageous use of rulers to draw such lines. Its absurdness could only, after heavy debates, be convincingly demonstrated by showing the effect of Expedited Data which can overhaul Normal Data. While the conventions would prescribe a line with a larger positive angle for higher speed data transfer it actually would result in a line with even a negative angle.

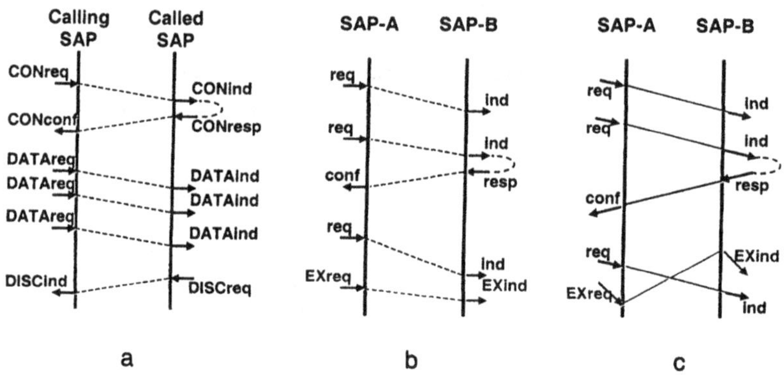

Figure 3. The OSI connection oriented service (a), OSI service conventions (b) and its early predecessor (c).

3.6 Service as a composition of Protocol Entities

In OSI it was said that a protocol renders a service. If so, then there should be a direct relation between a service, and the protocol that implements it. This can be easily obtained by defining that the composition of the protocol entities renders the service. This is illustrated in *Figure 4*.

From the above it follows that the IPs of the protocol that are accessible by the users are the same as the IPs (SAPs) of the service, which means that the protocol entities that have these IPs (SAPs) are also involved in the execution of Is at these IPs (SPs at these SAPs). Formally one can say that the behavior of the composition of protocol entities, as observable at the IPs, should conform to the service.

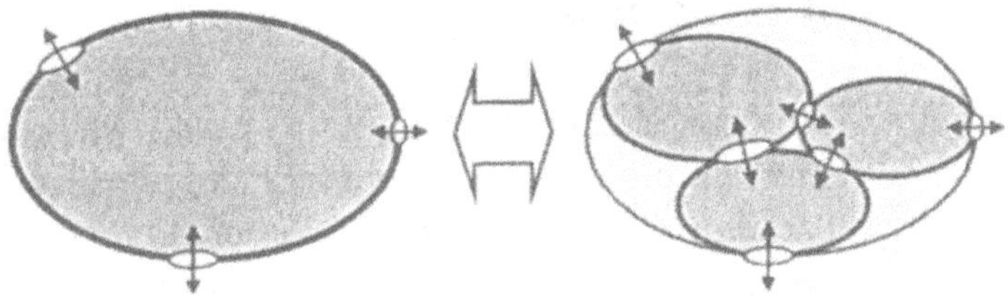

Figure 4. A service as a composition of a(n arbitrary) structure of protocol entities.

Consistent with this approach, an OSI (N)-protocol should be defined as the composition of the (N)-peer protocol entities in the (N)-layer and the underlying (N-1)-service of that layer, where the latter can be seen as a particular entity (protocol entity according to *Figure 4*), functionally quite different from the peer (N)-protocol entities and interconnected to them via (N-1)-SAPs. See also *Figure 5*. In OSI, however, the attitude was as if the (N)-peer protocol could be defined rather independent of the underlying (N-1)-service. This was also reflected in the task assignments to the various subcommittees: the Session Group did notdefine the Session Protocol and the Transfer Service, given the Session Service, but defined the Session Service and the Session Protocol given the OSI-RM.

Pre-occupied with the focus on the interconnection of open systems, OSI defined a protocol only as the relationship between peer-to-peer protocol entities (i.e. entities in the same layer). This led to the funny consequence that OSI protocol specifications never defined the explicit relation between SPs and PDUs. This in spite of the fact that concepts such as segmentation and reassembly, concatenation and separation, blocking and de-blocking, splitting and recombination, and multiplexing and de-multiplexing were defined in principle. Consequently, a conformance relationship between a service standard and a protocol standard rendering this service could never been proved. Certain OSI officials were even quite surprised that the FDT group wanted to formally specify both the (N)-service and the (N)-protocol and proof their conformance, questioning why one should make two different specifications of the same standard.

Saying that a service is a composition of protocol entities one could easily be tempted to also say that a protocol is a decomposition of a service. And frankly this use of words often occurs. However, the reader should be warned that a protocol *cannot* be derived by simply decomposing a service. The basic reason for this is that in a decomposition internal structure is revealed that requires additional design choices. These choices are also incurred by implementation concerns. Consequently a protocol is generally much more complex than a service. We will not further elaborate on this since it is beyond the scope of this paper.

3.7 Layered Protocols

The notion of protocol as a decomposition of a service, as introduced above, allows to define a protocol with an arbitrary structure. I.e. a structure with an arbitrary, problem dominated composition of protocol entities like we do hereafter in sections 5, 6 and 7. Frequently, like in the OSI-RM, we find layered protocol structures. As mentioned in our example above we achieve this by defining an (N)-protocol as the composition of the peer (N)-protocol entities and an underlying (N-1)-service, where the latter can be understood as an (N)-protocol entity with a specific nature. This allows to define the (N-1)-service again as a composition of peer (N-1)-protocol entities and an underlying (N-2)-service, as shown in *Figure 5.*

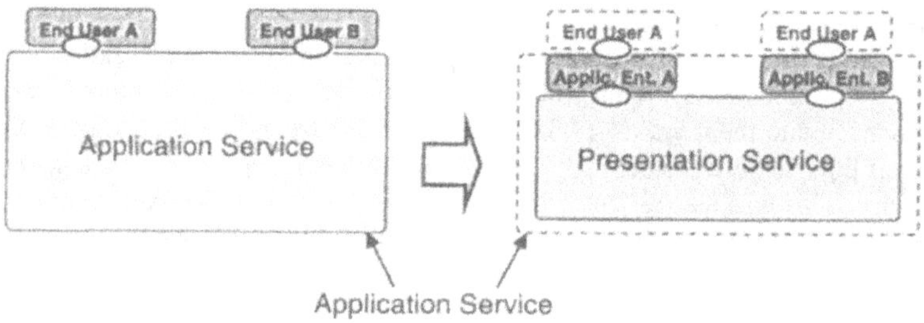

Figure 5. The OSI-RM as nested services interconnected by layers of protocol entities.

In so doing, a layered structure results that consists of a set of *nested* services interconnected by layers of protocol entities. In fact, the OSI-RM consists of 8 (not 7) nested services, from top to bottom starting with the Applications Service and ending with the Medium Service, where the latter is not further decomposed.

3.8 A Protocol Entity Considered as a Service

Considering a service as a composition of protocol entities poses the question how such entities should be specified. The answer is simply by specifying them in the same way as a service is specified. This view allows one to consider a protocol entity as a service in its own right, however at a lower abstraction level, implying that a protocol entity can again be considered as a composition of lower level (protocol) entities. This view is consistent with the view to consider an (N-1)-service as a composition of (N-1)-protocol entities and a (N-2)-service.

This in fact gives us the basis of a top-down design methodology. It proves the importance of the service concept as a black box that can also be seen as a composition of lower level (protocol, or if you wish service) entities.

4. MODELING E-SERVICES

Networked applications are top-level protocol entities that directly interact with end users, corresponding to the OSI-RM application layer entities. They provide application services, often called *e-services,* while using network services. Inside application layer entities we do not necessarily have again a layered (sub)structure, but generally have a specific structure determined by the nature of the service to be provided. Some networked applications, for example, exhibit a centralized control structure as shown in the collaborative services of Section 5. Others may be structured according to phases in a process as is the case for the transaction services described in Section 6, or have a pipeline structure like the content services of Section 7. We illustrate the quality of our decompositions with examples that show interface-preserving refinements of some of these key functions.

Table 1 summarizes some characteristics of collaboration, transaction, and content services, the basic service categories we discuss. It shows that these categories are quite distinct.

Table 1. E-service characteristics.

	Examples	Data volumes	Procedures	Media
Collaborative services	Video conferencing Shared whiteboard	medium – high	Some control	Multi
Transaction services	Auction Workflow Management	low	Rigid	Single
Content services	Digital video library	high	Loose	Multi

The modeling approach we take evolved from the architectural concepts we described in the previous sections. The language AMBER [5] was designed as a specification language for business processes, with strong roots in design and description techniques for telematics services and protocols. The RSD modeling language [19] augments AMBER with concepts for describing networked enterprises. In the following sections, we use (self evident parts of) RSD to describe our models. These languages themselves cannot be discussed as they are beyond the scope of this paper.

In the design of networked applications the people and organizations as users of these applications play important roles. In the modeling process we therefore start with identifying the necessary roles and next derive the associated key functions from these roles. Our RMs, therefore, consist of two types of sub-models.

The first sub-model is the *role model* (*Figure 6*) that defines which roles are involved in delivering and using the service defined by the RM. A role can be considered an abstract carrier of behavior (like end users, or service provider in the OSI model) and is denoted by an octagon. Arrows between roles represent the service delivery or the flow of information, goods or money.

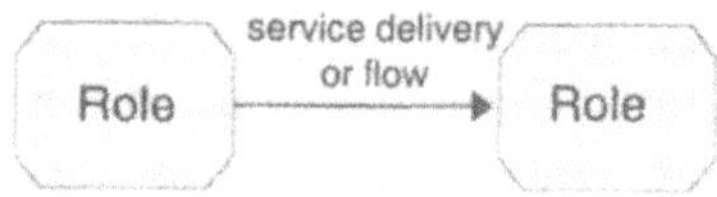

Figure 6. Role model.

The second sub-model is the *function model* (*Figure 7*) that defines the functions that together realize the behavior of a service. The responsibility for performing a function is associated with one or more roles. An actor that fulfils a role must carry out the associated functions or delegate this to another actor.

A function can be further detailed as a composition of sub-functions. Arrows between functions denote again the flow of information, goods or money. Flows may split and join, depicted by diamonds and squares, respectively; they enter and exit functions via triangular input and output interfaces. The thickness of the arrows is used to distinguish between different types of flows. Primary flows, e.g. the content delivered by a Service, may be depicted by a thick arrow; secondary flows, e.g. control flows, may be denoted by a thin arrow.

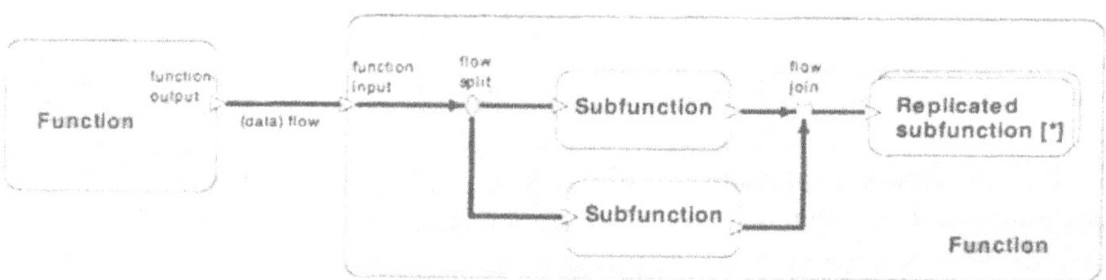

Figure 7. Function model.

Figure 8 shows how functions and flows correspond to protocol entities and interactions discussed in Section 3. Interactions may be refined to one or more flows that indicate the direction of the exchange of information, goods or money. The interaction between PE1 and PE2 in *Figure 8*, for example, is split into two flows between Function 1 and Function 2. Flows may be augmented with text.

The RMs described in the sections 5, 6 and 7 are described in much more detail in [3]. Here, we zoom in on parts of the RMs that are illustrative for our approach.

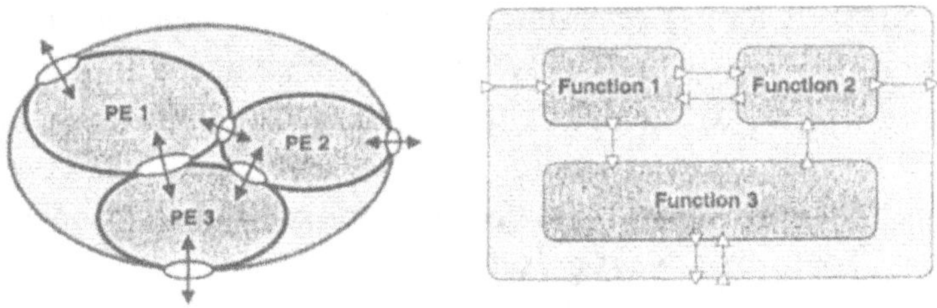

Figure 8. Protocol Entities and Interactions versus Functions and Flows,

5. COLLABORATIVE SERVICES

Collaborative services are designed to support groups of interacting people in their cooperative tasks. Examples of such cooperative settings are teachers tele-educating students, physicians tele-consulting each other, tele-conferencing board members, and engineers collaboratively working on designs. Collaborative services typically allow people to invite others to a virtual meeting, to communicate, share documents, share agendas, divide and together carry out work, etc. In this paper we use the term *conference* as a unit of abstraction, denoting a group of cooperating people that complete a common task while supported by collaborative services.

The prime use of collaborative services is to bridge distance and time between geographically dispersed collaborating people by allowing them to communicate and access data at different moments in time. Ensuring that users have consistent views of their collaboration and shared information is an important aspect of this service. A shared whiteboard service, for example, must provide mechanisms to ensure that all users can view the same information, can see all updates, and can know who may change the information. Typically, collaborative services exert a modest control over the users in order to organize their cooperation. One user may play the role of chairman and employ procedures to give the floor to others, interrupt them, etc.

5.1 A Reference Model for Collaborative Services

In a collaborative setting each user (participant in a conference) has access to collaboration support functions. These functions use a network for interconnection. See *Figure 9.*

The clustering of collaborative support functions into functional building blocks in our RM is based on the separation of concerns principle. Literature on how people cooperate (e.g., [1, 6, 8]) indicates that cooperative settings differ (in abstract terms) regarding four aspects: 1) the set of people that cooperate, 2) the set of tools they use to communicate and access shared information objects, 3) the set of rules they apply, and 4) the mechanism they apply to start cooperation (denoted as conference enabling). On the other hand, functions to start and end cooperation and to manage the set of cooperating people are always present.

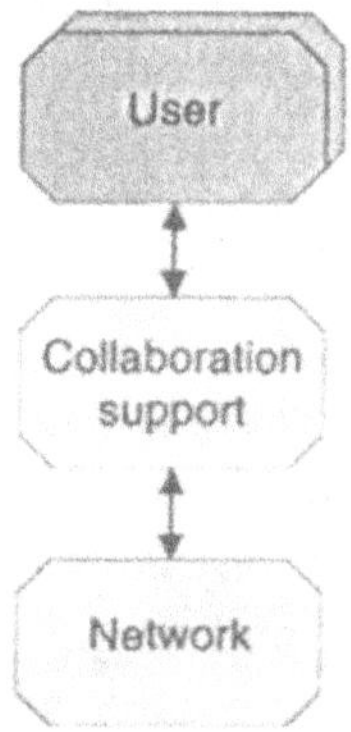

Figure 9. Roles in a collaborative setting.

Based on this we define the following key functions, as indicated in *Figure 10*: conference tools, to communicate and access shared information; coordination, to define and enact rules; conference enabling, to bring people together for cooperation; and conference management, to provide start and stop conferences and to manage the set of cooperating people. Conference management also keeps track of the conference tools that are in use and specifies what coordination policy applies.

Conference tools allow participants in a conference to communicate or collaborate using shared information objects. An audio conferencing tool and a shared whiteboard are examples of conference tools. Coordination defines and enacts the rules that may apply during a conference: the access rights for using collaborative services. The main function of Conference Enabling is to bring people together for cooperation, by providing awareness about other users who can be invited to a conference, or about other ongoing conferences that can be joined.

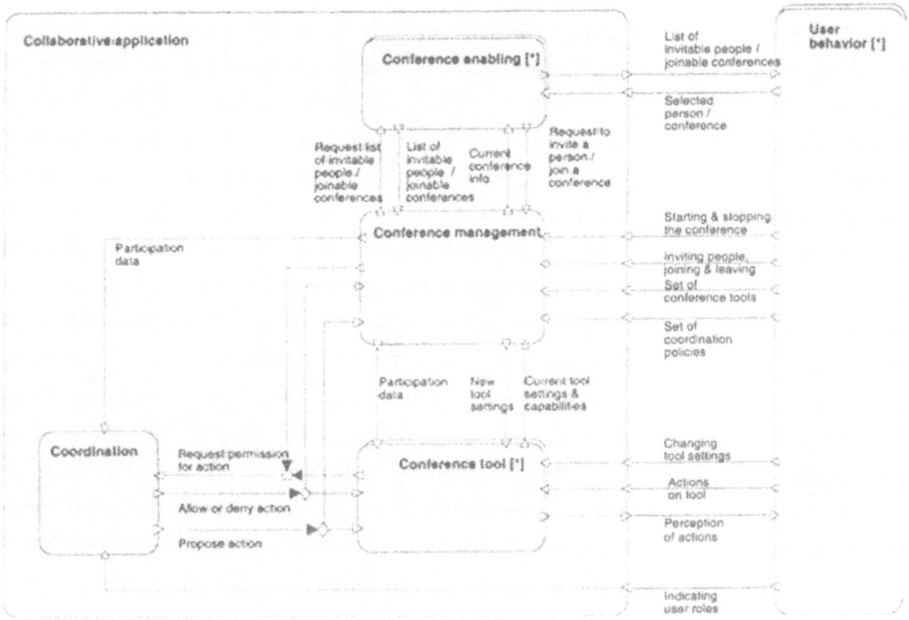

Figure 10. Key functions realizing a collaborative service.

5.2 Conference Management

As an example of interface preserving decomposition, we decompose the Conference Management function (*Figure 10*) that is responsible for providing services to users to manage an online conference. For a complete description we refer to [3] and [18]. These services allow users to:

- start and end conferences;
- join and leave conferences;
- invite people to a conference;
- select the tools to use in a conference;
- select the coordination policy that applies during the conference.

The specification of conference management sub-functions in *Figure 11* adds detail (i.e. lower level functional entities) to the high-level description in *Figure 10*. It shows how the specified interactions are provided by sub-functions. For instance, users can interact with conference management to indicate which tools should be active in a conference. Subsequently, conference management can adjust the conference tools settings, taking into account the tool capabilities and current tool settings.

The conference management function is furthermore responsible for providing information regarding the conference and its participants to other collaborative functions.

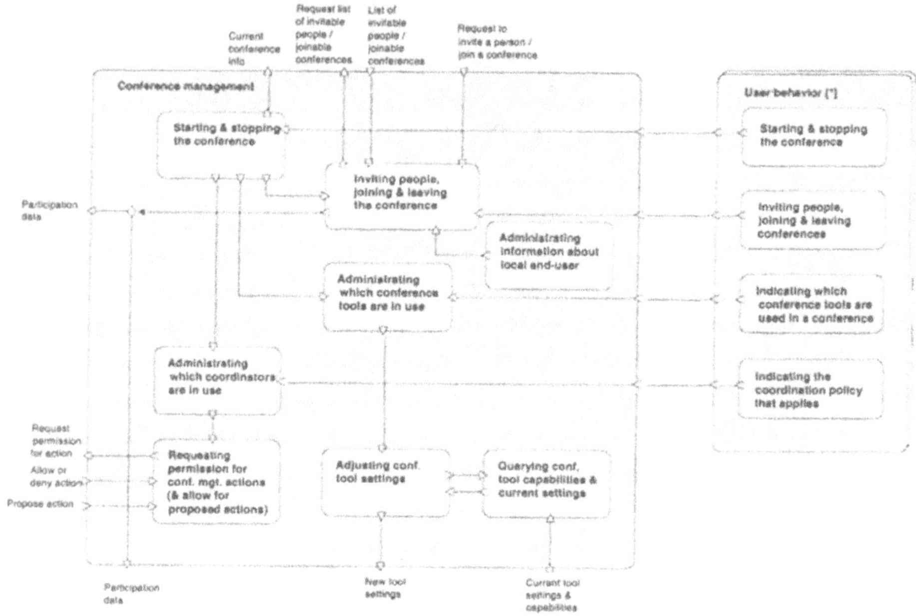

Figure 11. Conference Management sub-functions.

6. TRANSACTION SERVICES

Transaction services support formal, and often legally traceable, transactions between parties by organizing the exchange of pre-defined messages in pre-defined orders. Their prime use is to exert legal, commercial and financial commitments, and parties may be held liable when they fail to follow them. A procurement may be organized, for example, as a request for a product, leading to the indication of a price and followed by the acceptance or rejection of the offer. Companies trade according to predefined procedures such as: "you pay first, then I deliver", or "I deliver, you pay in two installments", or "you order, I ship unless you cancel", etc. The procedures may be dictated by law or may have been agreed before in negotiations, etc.

Typically, transaction services do not convey massive volumes of data per transaction; rather they exchange simple messages like orders, reservations, bills, payments, etc.

Commercial transactions occur over and over in the course of a business day, and many different types of, so called, e-commerce services have emerged recently including online marketing, searching, ordering, payment, and after-sales support [20, 21]. Here, we will focus on e-commerce that takes place around *electronic marketplaces* in a business-to-business setting.

6.1 A Reference Model for E-Marketplaces

Figure 12 shows the basic roles involved in an e-marketplace and their relations [10]. The primary user roles involved in any electronic trading system are those of buyer and seller. Another essential role is the one of marketplace operator as a front-end service provider. The latter is responsible for offering a large spectrum of services to the buyer and seller. All of these services are meant to support the fulfillment of the distinct phases of a business transaction process.

Almost always, the provisioning of some parts of these services is delegated or subcontracted to back-end service providers (e.g. providers of logistics services, financial services, foreign trade services, communication services, trust services). The services can be delivered either directly to the users (like in the case of logistic services), or through the front-end service provider, e.g. the marketplace operator (as in the case of trust services). Of course each marketplace has its own particular role, flow structure, and degree of complexity, and therefore various scenarios can be imagined to describe a commerce system.

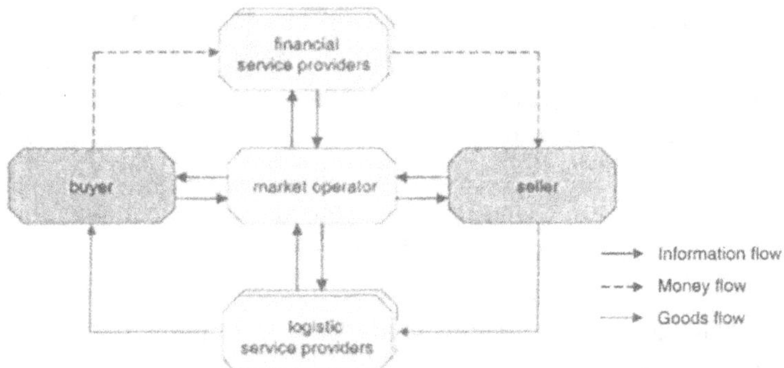

Figure 12. Roles involved in an electronic marketplace.

Let us now zoom in on the services provided by the market operator. The central organizing principle of our RM (see *Figure 13*) is to group those functions that occur within the same time frame and need the same information. Any commercial transaction has three main phases: information, negotiation and settlement. We therefore define three key functions of the marketplace: Information Management, Negotiation and Agreement Management, and Settlement and Fulfillment Management. Note that we distinguish between the functionality of the two user roles: buyers and sellers.

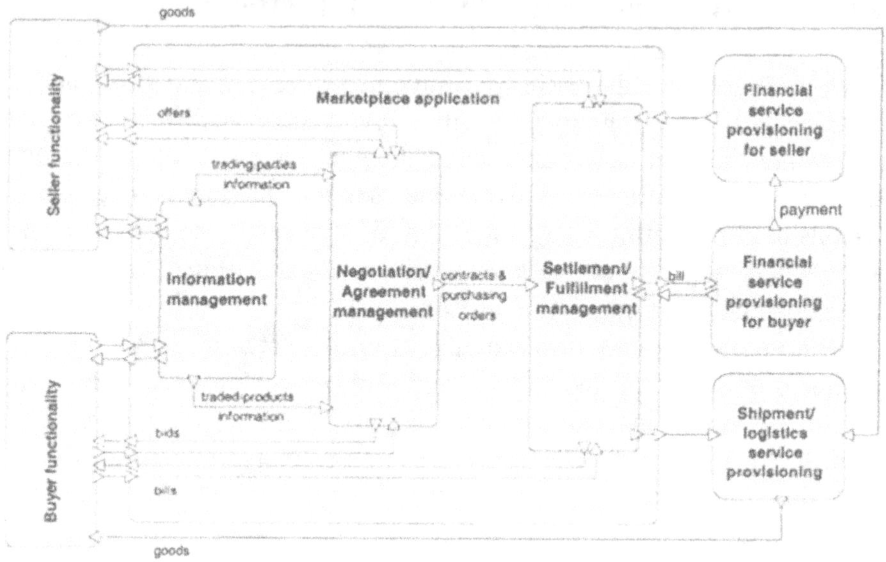

Figure 13. Key functions in an e-marketplace.

Information Management deals with the acquisition, storage, maintenance, provision, and presentation of information to buyers and sellers. This function covers information areas such as general user information and profiling, user offerings and user demands information, administration and maintenance of electronic catalogues, search engines in catalogues, marketing & advertising. This function is also responsible for the provision of information regarding the selected products and the trading parties to the function Negotiation and Agreement Management.

Once a user has decided to perform a transaction, the Negotiation and Agreement Management function takes care of the negotiation process, provides support for pricing mechanisms (auctioning, bidding, bartering, exchange, etc.), and for the issuing and distribution of electronic contracts and purchasing orders to the Settlement and Fulfillment Management function.

The latter operates whenever an agreement over a commercial transaction between two (or several) parties has been reached and has resulted in an electronic contract or order. This function provides support for the transfer of goods and money between the trading parties, and thus triggers the financial and logistic completion of the transaction. Some of the tasks that are covered by this function are invoicing, billing, and payment, tracing, and tracking orders, coupling to the back-office (including legacy systems), shipment facilitation or logistic services, etc.

6.2 Negotiation and Agreement Management

As an example of interface preserving decomposition, we now zoom in on the Negotiation and Agreement Management function. Again, we use functional decomposition according to the phases in the transaction as the guiding principle in the RM. First, the seller and the buyer engage in a negotiation process in the marketplace. This negotiation follows certain procedures through which the price and other conditions are settled. Second, once the price has been set, the negotiation ends, resulting in an agreement. This also entails issuing a contract (or the purchasing order) describing the terms of the transaction (products, amounts, prices, trading parties, guaranties, penalties, etc.). In our RM, this commercial functionality is embedded in the function Negotiation and Agreement Management (see *Figure 14*). This consists of a number of sub-functions.

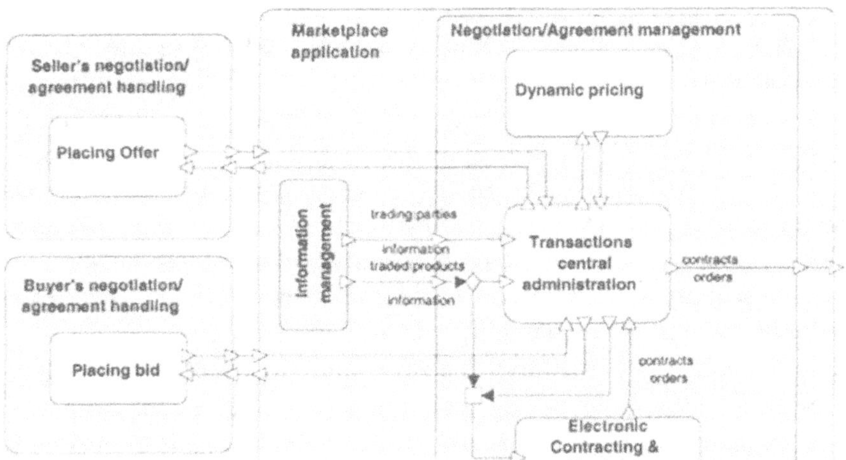

Figure 14. Negotiation and Agreement Management sub-functions.

The Transactions Central Administration is responsible for the information regarding the negotiation procedure, for controlling this procedure, and for commanding, in case of agreement, the issuing of an electronic contract (or purchase order). The inputs of this function are the information about the trading parties and traded products received from the information management functions, and, in case of dynamic pricing (such as an auction) the bids and offers received from the Placing Bid function and Placing Offer function of the buyer(s), and of the seller(s) respectively. In the case of dynamic pricing, these inputs are forwarded to the Dynamic Pricing function where they are processed according to the rules of the pricing algorithm. The results are sent back to the trading parties, through the Transactions Central Administration function, and, eventually, a new iteration can start.

When the price has been set, the Transaction Central Administration provides all the necessary information to the Electronic Contracting & Electronic Purchase-Ordering function. This initiates the issuing of an electronic contract (or purchase order). This function is also responsible for the storage of these documents, which is in fact the most important output of the transaction central administration, and of the negotiation/agreement management itself. They will be forwarded to the next main function of the marketplace, the Settlement and Fulfillment Management.

7. CONTENT SERVICES

Content services allow people to access and manipulate electronic content such as: accessing digital video libraries, viewing video programs, mixing parts of different video programs to compose a new program. The primary use of content services is to give users access to large amounts of data. Here, we will mostly discuss digital video disclosure because video is in some respects the most demanding form of content manipulation. To disclose video, we need a set of services for content production, a web enabled distribution channel and a method to consume it. The particular focus will be on video handling, data-interoperability, storage of high volume content, semi-structured metadata and structured data for services like metering, accounting, billing and payment ([15, pp. VI and 9]).

The main engineering problem of a multimedia system is to deal with the fundamentally analogue nature of real sounds and images, and to extract the meaningful information from the theoretically infinite amount of information they contain. An important practical consideration is that for this reason multimedia tends to be expensive. One has to process large amounts of data while keeping track of financially important information like usage. In content engineering, properties of data and the way data is represented and stored are important architectural issues.

7.1 A Reference Model for Content Services

In the content supply chain from media producer to digital content consumer, we identify four different roles to be played as shown in *Figure 15*.

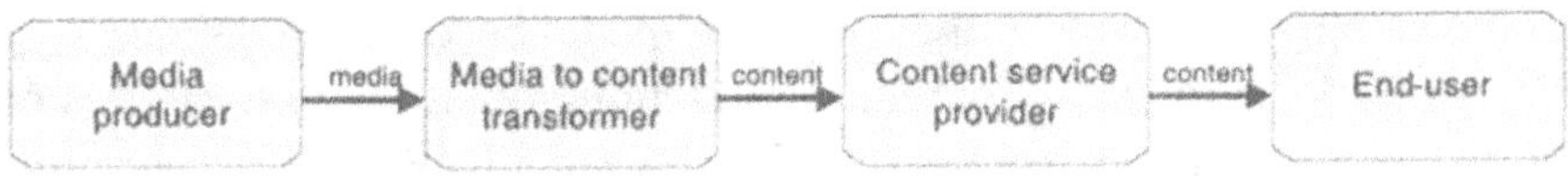

Figure 15. Roles in the content supply chain.

These roles perform consecutive steps in the 'content pipeline', from the production to the consumption of the content. This, therefore, determines the main structure of our RM. We decompose the content supply chain along the same boundaries as in *Figure 15*, resulting in the key functions of *Figure 16*.

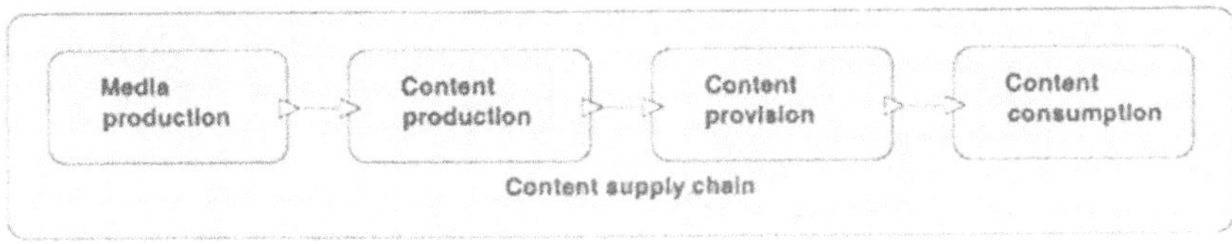

Figure 16. Key functions of the content supply chain.

The function Media Production is responsible for producing media assets (movies, music, pictures, texts, etc.), and metadata for these assets (e.g. bibliographical data, background data).

Content Production transforms media assets and associated metadata into digital content, by digitizing the media (if not digitally produced), and/or transcoding it into various digital formats. Additionally, it adds metadata for easy searching, data management, and business purposes, metadata and watermarks for digital rights management, and web pages, links to other pages and assets, advertisements etc.

Content Provision makes content available to end-users. It enforces digital rights, obtains payment from end-users, and pays media producers for the use of their content. Finally, Content Consumption provides functionality for searching and retrieving of content, paying for content, authentication for digital rights management, and possibly enjoying the content.

7.2 Content Production

As an example of interface preserving decomposition, we further refine the Content Production function. The full RM can be found in [15]. Content production is typically done by service industries for both the media industry and for content providers. The organizations involved are the organizations that provide the technical infrastructure for media production (which may or may not be owned by that media producer). The boundaries between media production and content production are not entirely clear cut, and the boundaries may blur even further if digital media become the default.

In decomposing the Content Production function, we use another important organizing principle in RMs: the distinction between primary functions, i.e., dealing with the content itself, and secondary functions, which control and monitor these primary functions. This leads to the decomposition shown in *Figure 17*.

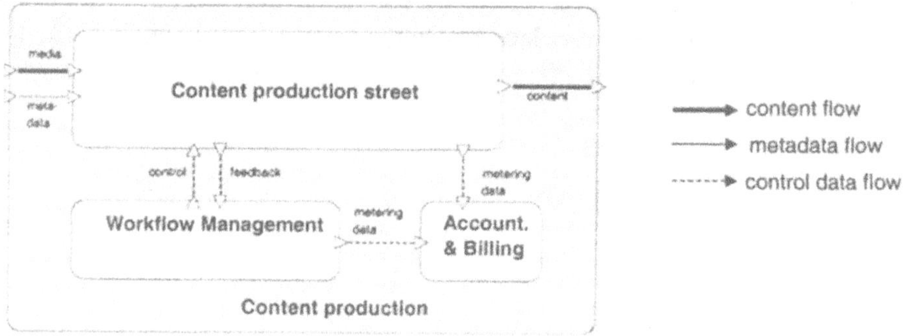

Figure 17. Content production sub-functions.

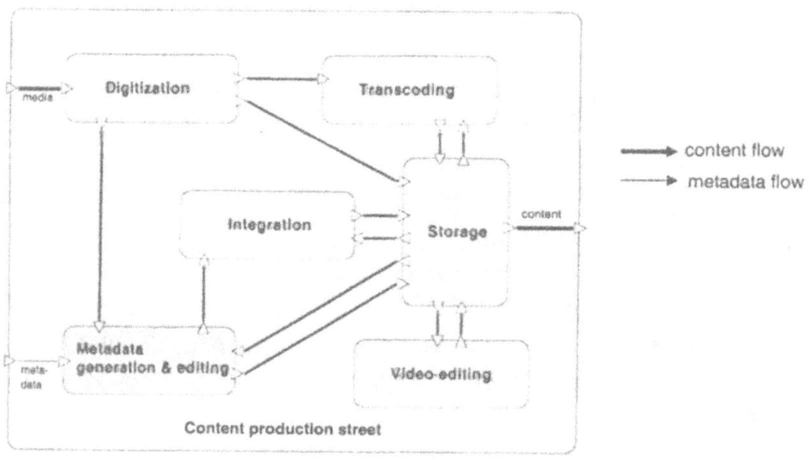

Figure 18. An example content production street.

The Content Production Street is the "factory" in which analog or digital media assets and the corresponding metadata are transformed into content. The Workflow Management function does job scheduling, job monitoring and resource allocation inside the production street. The Accounting and Billing function keeps track of resource consumption. The core functions of the content production process are contained in the example decomposition of the Content Production Street, shown in *Figure 18*. The sub-structure of this function is very specific to the content production domain, but its general organizing principle is that of a data-centric architecture, which is very well suited to a data-intensive application such as content production.

8. CONCLUSIONS

We have taken collaborative, transaction, and content services as categories of basic e-services from which, in principle, more complex e-services can be composed. By focusing on the most important functional aspects of a service we have shown that it is possible to define a high level, main design structure for the networked application as a composition of key functions and their relationships that supports this service and that can act as a RM. The structure of this composition is very much determined by the nature of the service to be provided. This main design structure has to be established before other design details can be added. The architectural quality of the compositions has been illustrated by showing interface preserving decompositions of some key functions as examples of adding design details.

The approach shows that RMs can play a significant role in designing complex systems, in allocating design tasks to cooperating design teams and in facilitating their communication. As such, RMs can also play a vital role in standardization activities.

To serve its purpose as a common reference, we think it is of great importance that a RM and its derived functions is built on sound architectural modeling concepts that are not only generic, complete, and consistent but also realistic abstractions of real world system properties, are well understood and commonly supported. In that respect we strongly support Brooks' statement, "conceptual integrity is the most important consideration in system design" [4]. Therefore we have outlined some concepts we consider essential for architecting RMs and contrasted them against some historical misjudgements.

Since e-services that are distributed across the Internet are relatively advanced developments, the body of knowledge regarding their architecture and design is still very much under development. This paper aims to contribute to these developments.

ACKNOWLEDGMENTS

Much of the work on basic architectural concepts, to which this paper refers, has been done at the University of Twente. We would like to acknowledge Marten van Sinderen, Luis Ferreira Pires, and Dick Quartel for their important contributions to this work. The material on the example reference models originates from the GigaPort program (http://www.gigaport.nl), a large Dutch initiative on future Internet applications and networks, funded by the Dutch Ministry of Economic Affairs. We would like to acknowledge Frank Biemans, Rogier Brussee, Maria-Eugenia Iacob, Paul Porskamp, and Henk Jonkers for their contribution to these reference models.

REFERENCES

1. Ackerman, M. (2000). The Intellectual Challenge of CSCW: The Gap Between Social Requirements and Technical Feasibility. *Human-Computer Interaction*, Vol. 15 (2-3), pp. 181-205.
2. Bass, L., P. Clements, and R. Kazman (1998). *Software Architecture in Practice.* SEI Series in Software Engineering, Addison Wesley Longman, Reading, Massachusetts.
3. Biemans, F.P.M., R. Brussee, M.E. Iacob, H. Jonkers, M.M. Lankhorst, P.A.P. Porskamp, and R.J. Slagter (2002). *Reference Models for Networked Applications: Blueprints for systems that support commerce, collaboration and content deployment via the Internet.* TI/RS/2001/069, Telematica Instituut, Enschede, The Netherlands.
4. Brooks, F.P. (1995). *The Mythical Man-Month: Essays on Software Engineering*, Addison-Wesley, Reading, Massachusetts.
5. Eertink, H., W.P.M. Janssen, P.H.W.M. Oude Luttighuis, W.B. Teeuw, and C.A. Vissers (1999). A Business Process Design Language. In: Wing, J. M., Woodcock, J., & Davies, J. (Eds.), *FM'99 – Formal methods: World Congress on Formal Methods in the Development of Computer Systems*, Toulouse, France, September 1999. LNCS 1708. Berlin; Heidelberg: Springer, pp. 76-95.
6. Ellis, C.A., S.J. Gibbs and G. Rein (1991). Groupware: some issues and experiences. In: *Communications of the ACM*, 34, 1 (Jan. 1991), pp. 39 - 58.
7. Gamma, E., Helm, R., Johnson, R., & Vlissides, J. (1995). *Design Patterns: Elements of Reusable Object-oriented Software.* Addison-Wesley.
8. Hofte, G.H. ter (1998). *Working Apart Together: Foundations for Component Groupware.* Telematica Instituut, Enschede, The Netherlands, ISBN 90-75176-14-7. https://doc.telin.nl/dscgi/ds.py/Get/File-7623/
9. Hollingsworth, D. (1995). *The Workflow Reference Model*, TC00-1003 Issue 1.1. Workflow Management Coalition.
10. Iacob, M.E., and A. Smit (2001). *Electronic Markets: classification and reference model*, TI/RS/2001/021, Telematica Instituut, Enschede, The Netherlands. https://doc.telin.nl/dscgi/ds.py/Get/File-15266/
11. ISO – International Organisation for Standardisation (1984). *Basic Reference Model for Open Systems Interconnection*, ISO 7498. ISO.

12. ISO – International Organisation for Standardisation (1986). *Transport Service Definition*, ISO 8072. ISO.
13. ISO – International Organisation for Standardisation (1995). *Reference Model for Open Distributed Processing*, ISO/IEC DIS 10746-1. ISO.
14. Object Management Group (1997). *A Discussion of the Object Management Architecture*. Object Management Group, Inc.
15. Porskamp, P.A.P., and A. Tokmakoff (2000). *Content Engineering Architectures*, TI/RS/2000/055, Telematica Instituut, Enschede, The Netherlands. https://extranet.telin.nl/docuserver/dscgi/ds.py/Get/File-9813.
16. Rechtin, E., and M.W. Maier (1997). *The Art of Systems Architecting*, CRC Press, Boca Raton, Florida.
17. Shaw, M., and D. Garlan (1995). Formulations and Formalisms in Software Architecture. J. van Leeuwen (ed.), *Computer Science Today: Recent Trends and Developments*. Springer-Verlag LNCS Vol 1000, pp. 307-323.
18. Slagter, R.J. (2004). *Cooperating People and Systems: Guiding the Design of Component Groupware*. Telematica Instituut, Enschede, the Netherlands. (forthcoming).
19. Steen, M.W.A., M.M. Lankhorst, and R.G. van de Wetering (2002). Modelling Networked Enterprises. In Duddy, K. & Aagedal, J.Ø. (Eds.), *6th IEEE International Enterprise Distributed Object Computing Conference (EDOC 2002)*, Lausanne, 17-20 September 2002., pp. 109-119.
20. Technology Strategy Team (1999). E-commerce: the Third Wave (2nd ed.). White paper. US: EC Cubed, Inc.
21. Timmers, P. (1998). Business Models for Electronic Markets. *EM - Electronic Markets*, 8 (2), 3-8.
22. Vissers, C.A., G. Scollo, and M. van Sinderen (1988). Architecture and Specification Style in Formal Descriptions of Distributed Systems. In: Aggarwal, S. and Sabnani, K. (eds.) *Proc. IFIP 6.1 on Protocol Specification,Ttesting and Verification*, Atlantic City, pp. 189-204.

Chapter 30

MAPPING "ENTERPRISE BUSINESS ARCHITECTURE" TO "INFORMATION SYSTEMS FRAMEWORK"

Atsushi Yamaguchi, Motoyuki Suzuki, Masanori Kataoka
Hitachi INS Software

Abstract: Globalization accelerates worldwide Business Modularization, and Business Modularization cannot be achieved without the help of Internet. In this paper, we will clarify the role of Internet in Business Modularization, and will discuss the systematic methodology to leverage Enterprise Business Architecture, which is placed at the origin of Information Systems Development Lifecycle.

Key words: Enterprise Business Architecture, Business Modularization, Information Systems Framework

1. INTRODUCTION

1.1 Business Modularization

Fast-paced globalization of business has forced enterprises to categorize their business elements, the process which is referred to as "Business Modularization". Enterprises concentrate their efforts on their core business and sell their core products and services worldwide. On the contrary as for the non-core business, enterprises procure them from or outsource them to their international partners.

Examples of Business Modules are shown in Table 1.

Table 1. Examples of Business Module

Module	Example
People	Human Organization, Service Package
Things	Products, Parts
Money	Financial Package
Knowledge/Information	Knowledgeware, Information Contents

1.2 Outline of Methodology

In this paper, we will place "Enterprise Business Architecture" at the origin of Information Systems Development Lifecycle, and will discuss the systematic techniques to leverage this architecture in system design phase.

Figure 1 illustrates the outline of the methodology.

There are two achievements that are Enterprise Business Architecture in business analysis phase and Information Systems Framework in system design phase. Our challenge is to bridge the gap between Enterprise Business Architecture and Information Systems Framework. Information Systems Framework is build by leveraging Enterprise Business Architecture.

The outline of methodology to build Enterprise Business Architecture, the outline of methodology is:

(1) To clarify the Competitive Strategy.

(2) To make Business Model Transformation with the above strategy.

The result of transformation is important for Information System Framework.

The outline of methodology to build Information Systems Framework, the outline of methodology is:

(1) To build IT system with PIM (Platform Independent Model)

(2) To transform PIM to PSM (Platform Specific Model)

2. ENTERPRISE BUSINESS ARCHITECTURE

Enterprise Business Architecture consists of Business Competitive Strategy and Business Systems Architecture. In addition, Business Systems Architecture consists of Business Module and Inter-Module Structural Operator.

Business Module is the basic element of business such as "People", "Things", "Money" and "Knowledge/Information". Every Business System contains Business Modules. Well-balanced structure of "People", "Things" and "Money" is important, and the structure should be well armed with

"Knowledge/Information". Relationship between Business System and Business Modules is shown in *Figure 2*.

Business Module is referred to as *Module* in this paper.

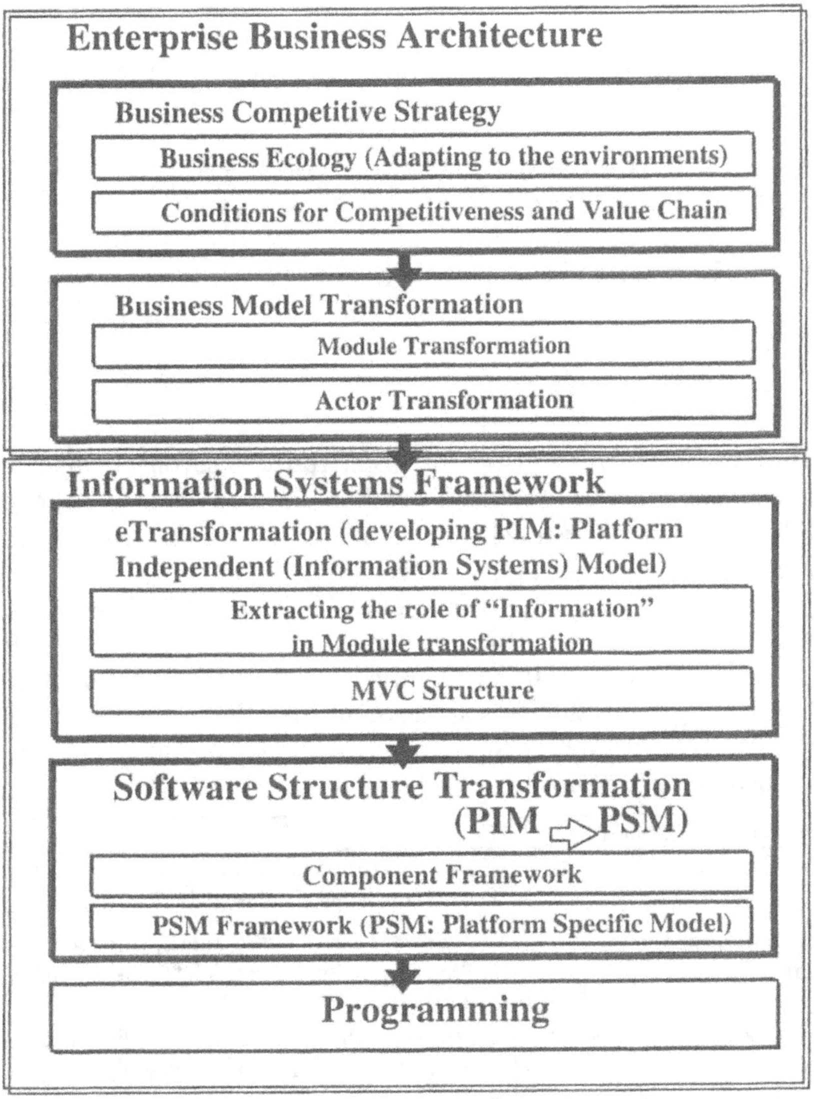

Figure 1. Methodology

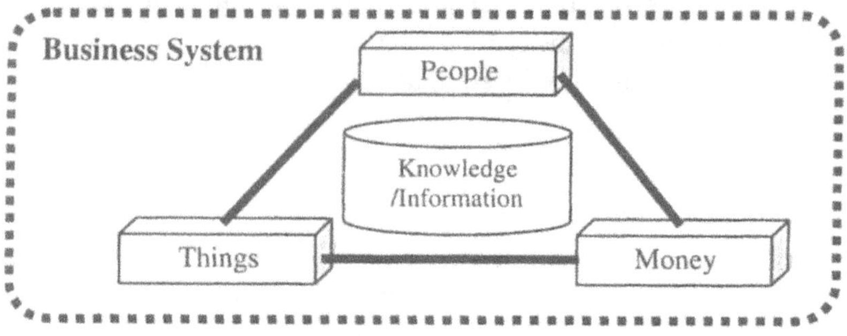

Figure 2. Relationship between Business System and Business Modules

2.1 Business Competitive Strategy

2.1.1 Business Ecology

Business Ecology is strategy for an enterprise to become competitive and to adapt itself to the environment, which is based on the idea that enterprise as a living being survives with suppliers, customers/consumers, and social environments. Factors that influence the enterprises are diversity, interaction, reciprocal action and natural selection. Globalization of recent years is strengthening the effect of these Influential factors. Environmental elements are Social Environment, Customer/Consumer Environment and Value Chain with the Suppliers as shown in Table 2.

Table 2. Environmental Elements

Environmental Elements	Example
Social Environment	National Organization, Politics, Geographical Feature, Economy, Culture, Labor
Customer/Consumer Environment	Value for the Customer/Consumer, Differentiation of Product Quality, Information "Contents" added to the Products
Value Chain with the Suppliers	Cost of Materials/Parts/Products, Quality/Function of Products, Selection/Fostering of Suppliers

2.1.2 Conditions for Competitiveness

The following are the 4 conditions to become competitive:

(1) Zoning, Maintaining territory

(2) Adaptation and Evolution in the Environment
(3) Flexibility of the organization to be adaptive
(4) Survival of the fittest, Natural selection

Parameters to become competitive are Resource Capacity, Scale, Internal Control, Scarcity, Variety of supply source, Market Structure and Transition as shown in Table 3.

Table 3. Parameters to become competitive

Parameters for competitiveness	Example
Resource Capacity	People, Things, Money, Knowledge/Information
Scale	Expansion, Reduction
Internal Control	Concentration, Distribution
Scarcity	Place, Position, Resource, Knowledge
Variety of supply source	Vendors
Market Structure	Symbiosis, Competition
Transition	Transition to a new domain

The following 3 sequences of procedures are important to map out the strategy to achieve competitiveness:

(A) To clarify external factors.
(B) To change internal factors depending on external factors.
(C) Design/Modify the Business Systems Architecture in order to change internal factors.

Examples of external factors are product quality, product price, amount of supply, scarcity, brand, CRM(Customer Relationship Management), competitors and market domain. Examples of internal factors are people (especially leaders), production facilities, production engineering, development (innovation) capacity, SCM(Supply Chain Management) and partners. Examples of module operations in the Business Systems Architecture to change internal factors are Module Transformation and Actor Transformation.

2.2 Business Systems Architecture

2.2.1 Characteristics of a module

Business Systems Architecture consists of *Module* and Inter-Module Structural Operator. It can be said that *Module* is the characterized box as shown in *Figure 3*.

External specification of *Module* consists of knowledge that can be expressed explicitly by language or sentence, in other words, a "protocol".

Meanwhile Internal structure of *Module* consists of subjective or physically acquired knowledge such as thoughts/views that are sometimes hard to express by language or sentence, in other words, "process".

It is important to keep the external specification of a *Module* stable and to keep the internal structure of a *Module* flexible to adapt to and survive in free competition environment.

Thus, advantages of Modularization are as follows:

(1) Modules are adaptive to rapid environmental change.
(2) Modules are divide jobs into independent parallel jobs.
(3) Only strong *Modules* with good internal structure survive the free competition between modules.

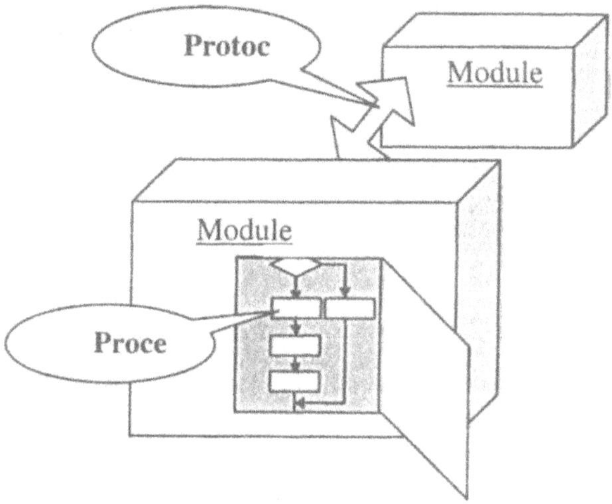

Figure 3. Characteristics of a Module

2.2.2 Inter-Module Structural Operator

Inter-Module Structural Operator illustrates how the connections between the modules are established. The formats of Inter-Module Structural Operators are Set, Hierarchy, Chain, Hub and Web. The Inter-Module Structural Operators are explained in Table 4.

While Web is important to energize the relationship among *Modules*, Hub is important to coordinate *Modules*. Many links are established between variety of enterprises and people to form a Web, and a Hub works as the central place to coordinate *Modules*. In the real world, we usually utilize both Hub and Web.

2.2.3 Module metrics

Module metrics consists of Module Strength, Module Coupling, Module Granularity and Degree of Encapsulation [2]. Module metrics are important to evaluate successfulness of Modularization.

Module Strength is a metric to evaluate module stability and reliability. When a module is evaluated in the perspective of module strength, it can be said that a module with high Functional Strength has high module strength and is highly independent, and a module with high Set Strength has low module strength and is highly dependent. Informational Strength positions itself in between Set Strength and Functional Strength, and it indicates that modules are related to some shared information. Module Strength is expected to be strong.

Module Coupling is a metric to evaluate dependency between *Modules*.

Table 6 illustrates Module Coupling metrics. As to Module Coupling, weaker coupling means self-organized and greater independence. Control coupling with complete dependence has no independence. *Modules* should have high independence and weak coupling.

Module Granularity is a metric to evaluate module size or module complexity. In case management resource is people, Module Granularity indicates the size of an organization unit or an enterprise, e.g. 10 people, 100 people, 1000 people and 10000 people. In case of things, Module Granularity is size or complexity of parts or products, e.g. m, km, g. In case of money, Module Granularities are amount of money for investment, development, etc, e.g. $1, $100 and $10,000. It is not necessarily appropriate to suggest that large *Module* or small *Module* is better.

Degree of Encapsulation is a metric to evaluate the degree of module encapsulation or capacity to provide good external interface. The measures are as follows:

(A) Single function/Multi Function

(B) With/Without memory

- With/without internal state
- With/without internal data

(C) Dependent/independent on platform

Higher degree of encapsulation is expected for *Modules* to be highly independent.

Table 4. Inter-Module Structural Operator

Operator	Figure	Description
Set	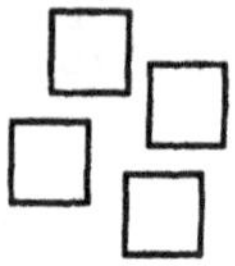	Set of Modules without Relation. e.g.: Parts stocked in warehouse.
Hierarchy	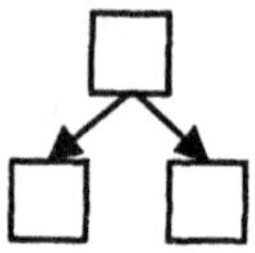	Hierarchy of Modules. e.g.: Hierarchical structure of physical parts.
Chain	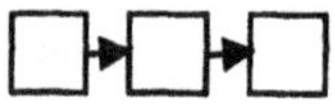	Modules are chained to form a flow. e.g.: SCM (Supply Chain Management)
Hub	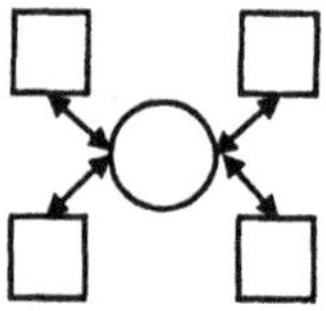	Modules are coordinated through a central place. The place could be physical place or abstract coordinator. cf. "Chain" could be treated as a kind of "Hub".
Web	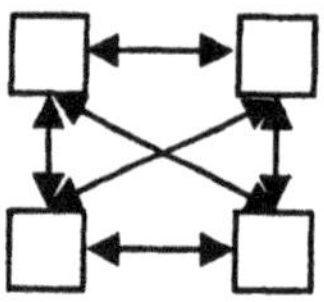	Free relationship among modules. cf. "Web" is frequently used in open connection with extra-enterprise systems.

2.3 Business Transformation Operations

It is required for Enterprises to change their internal factor of Business Systems Architecture to adapt themselves to environmental (external factor) changes. Business model transformation is such an adaptation process that includes following module transformation operations.

(1) Augmenting new *Modules*.
(2) Excluding existing *Modules*.
(3) Dividing *Modules*.
(4) Integrating *Modules*.
(5) Substituting *Modules*. (Actor Transformation)
(6) Porting *Modules*. (to be common modules)
(7) Creating a new hierarchy by extracting common parts from multiple *Modules*.

Table 5 explains the Module Strength of each module.

Table 5. Module Strength Metrics

Strength		Weak ⟵	⟶	Strong
Modules		Set Strength	Informational Strength	Functional Strength
People	Entity	Conglomerate	Grouped enterprises	Single function company (e.g. venture)
	Process	Set of variety of business processes	Information sharing among processes	Organization of professionals
Things	Entity	Set of independent elements	Set of related elements	Standard (Reusable) parts
	Process	Set of variety of processes	Concentrated management of process information	A single function process
Money	Entity	Set of variety of funds	Funds from market (e.g. stock funds)	A single fund (e.g. parent company, venture capital)
	Process	Set of variety of fund raising	Fund raising from market	A single investor

Table 6. Module Coupling Metrics

Coupling		Strong ⟵		⟶	Weak
Modules		Control Coupling	Sequential Coupling	Common Coupling	Independent
People	Entity	Hierarchical Organization	Team with sequential processes	Team, Group	Independent organization
	Process	Direction, Order	Sequential processes	Information sharing	Independent governance
Things	Entity	Basic frame	Related parts	Common parts	Commercial (open) parts
	Process	Installation to basic frame	Sequential installation process	Shared supply in the group	Modularized process
Money	Entity	Capital governance	Conditional funds	Funds from market	Venture capital
	Process	Direct Management	Investment	Investment	Investment

Actor Transformation is most important among above operations. Examples of Actor Transformations are as follows:

(A) Automation:
- With machine automation
- With computer systems

(B) Handmade

(C) Improvement of internal structure:
- Improvement of production organization/process.
- Upgrading production machine for better performance.
- Upgrading computer system for better performance.

(D) Changing relationship with partners:
- Out-Sourcing
- In-Sourcing

3. INFORMATION SYSTEMS FRAMEWORK

Information Systems Framework has two transformations. One is Information Transformation (eTransformation), and the other is Software Structure Transformation (Software Transformation).

Information Transformation consists of following sequential procedures:

(1) Abstract information from *Modules*.

(2) Allocate information to Information System based on PIM.

Software Structure Transformation consists of a following procedure:

(3) Transform PIM to PSM (Platform Specific Model) with leveraging Component Framework and PSM Framework.

After these transformations, we will develop programs with the help of above PSM and Design Patterns.

3.1 Information Transformation (eTransformation)

3.1.1 Entity (Atom) & Information (Bit)

Every *Module* consists of physical Entity (Atom) and Information (Bit). Essentially in real world, Entity and Information are inseparable. However, digital technology and information technology have enabled the separation of information from Entity. Hereby, Information can be handled independently from the media and information processing efficiency has been largely improved.

On the other hand, because independent handling of Information from Entity has been enabled, it created many security problems such as information without real Entity and synchronization error between Entity and Information.

By Actor Transformation (*Module* Substitution) Processes inside a *Module* are substituted. In this case, "Bit" is split from "Atom", and then, computer processes "Bit". After that "Bit" is synchronized with "Atom" to be re-integrated.

For the re-integration, MVC model that consists of Model, View and Controller is required. *Figure 4* shows MVC model.

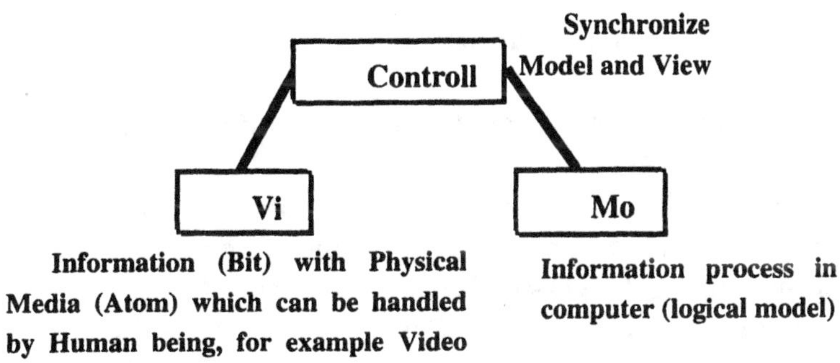

Figure 4. MVC Model

3.1.2 Procedure of Information Transformation

Information Transformation will be done according to the following hierarchies shown in Figure 6

(A) Splitting/Synchronizing Atom and Bit

Splitting Bit from Atom and synchronizing Atom and Bit is the most basic operation of Information Transformation. In case of order form as an example, picking out the described Bit such as product price and buyer's name means splitting Bit (i.e. order Information) from Atom (i.e. order form). MVC model is helpful to install this procedure into Information Systems.

(B) Information Processing

Information Processing is to process, accumulate and transform Bit that is picked out from Atom at above procedure. As to the order form, the typical example is to entry Bit that is picked out from order form into the database.

3-Dimensional Model of Information Processing as describe later is a good model that clarify the semantics of Information Processing. Appendix A illustrates the relation between Competitive Strategy, Module Transformation and the role of Information Processing.

(C) Knowledge Creation

Knowledge Creation is to create a new knowledge based on high-level Information Processing. The example of Knowledge Creation is to analyze customer's purchase pattern by analyzing the customer data and purchase activity history.

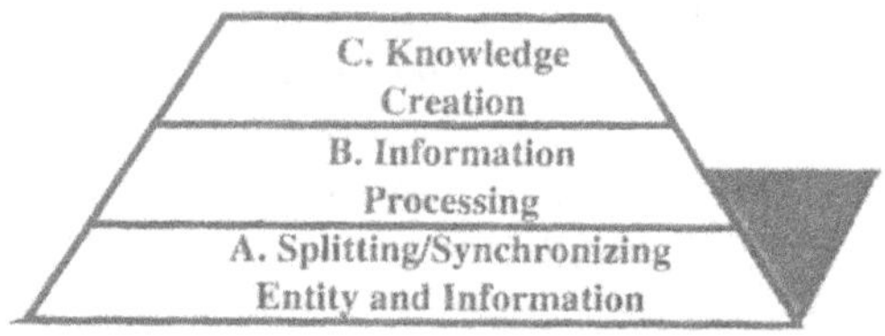

Figure 5. Hierarchy in the Role of Information

3.1.3 3-Dimensional Model of Information Processing

3-Dimensional Model of Information Processing that is used here is helpful to clarify the semantics of Information Processing. 3-Dimensional Model of Information Processing is shown in *Figure 6* consists of Horizontal (Common Use) Axis, Vertical (Value Chain) Axis and Time (Life Cycle) Axis.

Horizontal Axis means sharing variety of resources (physical resources and logical resources) via network. Not owning resources privately but sharing resources increases efficiency in the community.

Vertical (Value Chain) Axis means building value chains by connecting *Modules* and leveraging the value of each *Module*. It enables integration /coordination of skilled specialists worldwide, which increases efficiency in the global business operation.

Time (Lifecycle) Axis means supporting Module Lifecycles to upgrade quality with recording, storing, improving and reusing information.

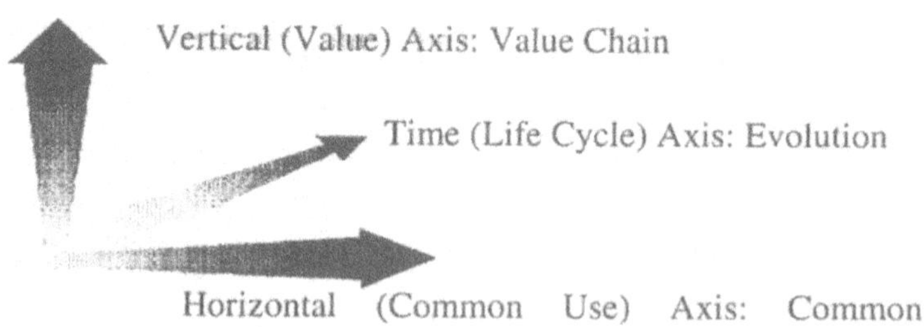

Figure 6. 3-Dimensional Model of Information Processing

3.2 Software Structure Transformation (Software Transformation)

3.2.1 Hierarchical Structure of IT System

Software system can be classified in 3 layers as shown in *Figure 7* – Basic Platform Layer, Managed Middleware Layer, and Application Layer.

Basic Platform Layer and Managed Middleware Layer belong to PSM (Platform Specific Model), and Application Layer belongs to PIM (Platform Independent Model). Basic Platform Layer and Managed Middleware Layer are often simply called "Platform", and Platform will be made of ready-made operating system, packaged software and software reusable components.

3.2.2 Software Structure Transformation in AP Layer

Software Structure Transformation in Application Layer consists of the following procedure:

(1) Build a MVC Model that contains Model, View and Controller. An example of MVC model is shown in *Figure 8* using cFramework structure, which is a product of EC-One, Inc. in Japan [8].

(2) Apply reusable *Module* in each element.

This methodology enables dispersed development module by module independently.

PSM described later is mapped from the PIM that is made hereby.

3.2.3 Software Structure Transformation in BP/MM Layer

The PSM that includes BP/MM Layer is built by transforming PIM, that is, PIM is mapped to PSM. The PSM is a combination of a variety of reusable components, which enables high-efficiency development.

Basic Platform Layer consists of variety of components, and many vendors release the products that fit each component. Appendix B illustrates examples of relation between BP Layer components and ready-made products. Appendix C illustrates examples of the detail of the BP Layer components in Sun Microsystems' Solaris environment.

Many vendors also release products that apply to MM Layer components as well as BP Layer components. Appendix D illustrates examples of relation between MM Layer components and ready-made products.

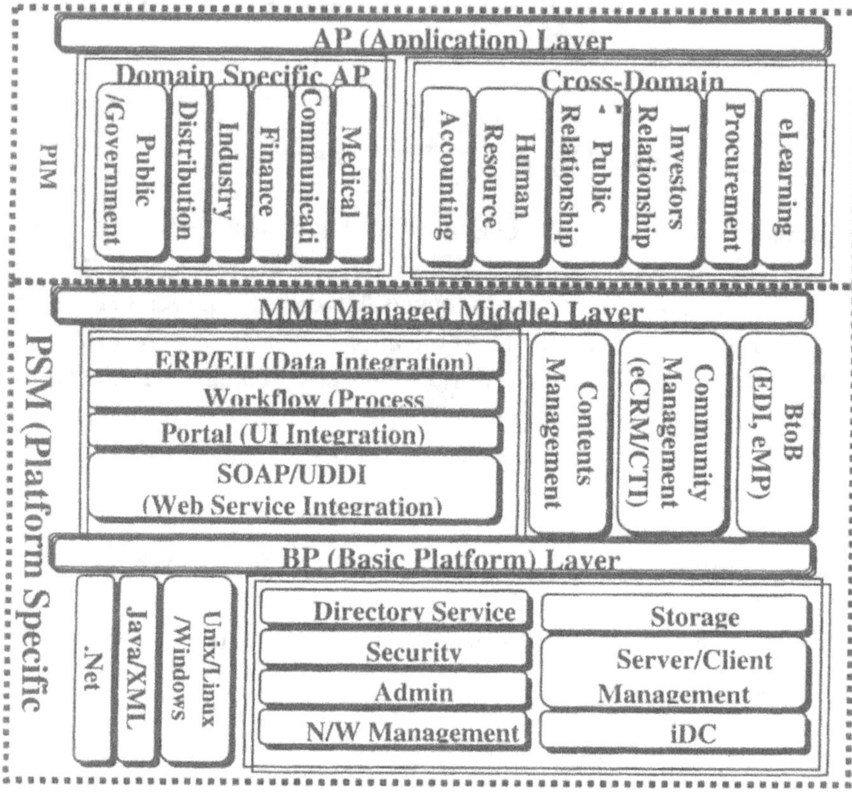

Figure 7. Hierarchical Structure of IT System

4. CONCLUSION

Globalization accelerates Business Disintegration and Modularization. We have discussed on "Enterprise Business Architecture Based on Modularization Concept". We also have presented a systematic methodology to map "Enterprise Business Architecture" to "Information Systems Framework". This mapping can be done as module-transformation, that is Business Module to PIM (Platform Independent Model), PIM to PSM (Platform Specific Model). The transformation is expected to bridge the gap between Business Systems Model and Information Systems Model.

We are enhancing UML (Unified Modeling Language) to describe Enterprise Business Architecture. With this enhancement, the transformation of modules is expected to have good support of automated tools.

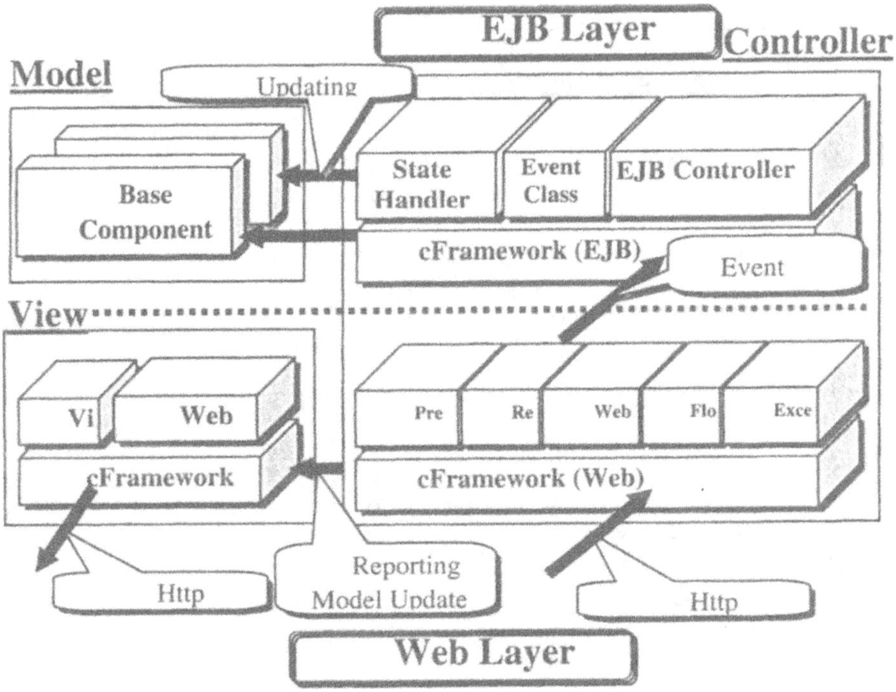

Figure 8. Software Structure Transformation in Application Layer

ACKNOWLEDGEMENT

The authors are extremely grateful to Mr. Koichi Yamano, CEO of Bunasoft Co. Ltd., who gave us a great deal of advices and Ms. Mika Sawada, who initiated this methodology.

REFERENCES

1. Aoki, Masahiko and Haruhiko Ando, *Modularity*. Tokyo: TOYO KEIZAI INC., 2002.
2. Myers, Glenford J, *Reliable Software through Composite Design*. Mason: Charter Publishers, Inc., 1975.
3. Eriksson, Hans-Erik and Magnus Penker, *Business Modeling With UML Business Patterns at Work: Business Patterns at Work*, OMG Press, Wiley Computer Publishing, 2000.
4. Mintzberg, Henry and Ludo Van der Heyden, *Organigraphs: Drawing How Companies Really Works*, Harvard Business Review, September-October, 1999.
5. Negroponte, Nicholas, *Being Digital*, New York:The Alfred A. Knopf, 1995.
6. The Object Management Group, http://www.omg.org/
7. The Object Technology Institute, Inc., http://www.otij.org/
8. EC-One, Inc., http://www.ec-one.com/

Appendix A: Extracting the role of "Information"

#	Competitive Strategy on Business System Model	Module (Actor) Transformation	Role of Information
A	Connecting Business Modules	Substituting the Interface	
A1	Computerizing Information Exchange between Modules	Computerizing Interface between Modules	EDI (Electronic Data Interchange)
A2	Business Modules related to the process from procuring parts to selling products are reorganized to form a "Value Chain"	Interface between Modules are computerized to form a "Value Chain"	SCM (Supply Chain Management)
B	Knowledge/Information Sharing among Business Modules	Formation of upper layer module consisting of Business Modules	
B1	Knowledge/Information sharing among Business Modules	sharing same knowledge /information	Web Portal
B2	Modules related to a specific business category are connected by workflow, enabling the circulation of electronic documents	Modules circulate electronic documents by workflow	Workflow
B3	Customer Information are shared among related Business Modules	Modules share the Customer Information and related processes	CRM (Customer Relationship Management)
C	Substituting Business Modules	Module Substitution (Actor Transformation)	
C1	Out-Sourcing	Business Module is substituted by Business Modules of other parties	EDI with Out-Sourcing Partner
C2	Procure parts and office supply from eMP	Procurement partner is changed from current partner to eMP	eMP (e Market Place), MRO (Maintenance Repair and Order)
C3	Utilize Internet Service (e.g. ticket reservation, purchase)	Services provided by internal organization or by third party are substituted by Web Service	Web Service
C4	Distributed Manufacturing	Manufacturing Organi-zation Module is divi-ded and some portion of it is shifted to distri-buted manufac-turing facility.	EDI with distributed facility

Appendix B: Examples of PSM Models on BP Layer

Vendors / Components	Hitachi	IBM	Microsoft	Sun Microsystems	Oracle	Other
OS	HI-UX	AIX	Windows	Solaris		Linux
Web Browser			IE			NetScape Opera
Web Server	Hitachi Web Server	IBM HTTP Server	IIS	Sun ONE	Oracle9i AS*	Apache Zeus Novell exteNd
RDB	HiRDB	DB2	SQL-Server		ORACLE	PostgreSQL
XML Parser		Web Sphere*	Biztalk	(MSXML)	Sun ONE AS*	XML Parser for Java WebLogic Server* (BEA) Xerces (Apache) The Expat XML Parser
Directory	Hitachi Directory Server		Microsoft	Sun ONE		Novell eDirectory
Security	Real Secure	Tivoli	ISA Server	Sun ONE	Oracle9i AS*	Novell iChain
Storage Mgt.	JP1/ Hi Command	Tivoli		Storage ONE		
System Administration	JP1	Tivoli	App. Center		Oracle9i AS*	

*Bundled with Server

Appendix C: PSM Layout for Solaris

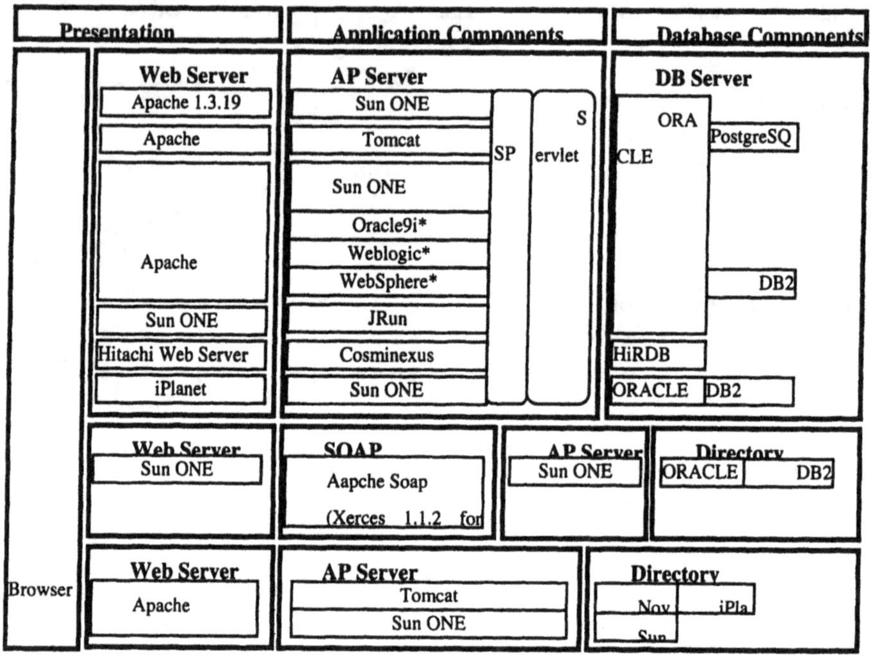

*Application Servers with XML

Appendix D: Examples of PSM Modules on MM Layer

No	Vendors / Components	Hitachi	IBM	Microsoft	Sun Micro systems	Oracle	Novell	Other
1	Portal Server	Cosminexus	Web Sphere	Share Point Portal Server	Sun ONE	Oracle9i AS	Novell Portal Service	Web Logic (Ver7.0)
2	App. Servlet Engine	Cosminexus	Web Sphere	IE	Sun ONE	Oracle9i AS	Novell AS	Tomcat Web Logic JRun
3	SOAP/ UDDI			Microsoft SOAP Toolkit	Sun ONE		Novell Group Wise6	Apache SOAP
4	WF/ Group Ware	Groupmax	Notes/ Domino	SQL-Server				
5	ERP/EII					Oracle App.		SAP/R3P eople Soft
6	SCM/ CPFR	SCPLAN Logility Value Chain Solution						RHYTH MILOG
7	CTI/ CRM							Siebel People Soft 8
8	BtoB							
9	ECM		Content Mgr.	Content Mgt. Server 2002		Oracle9i AS	Novell exteNd Composer	

Chapter 31

A COTS-ORIENTED PROCESS FOR CONSTRUCTING ADAPTABLE E-GOVERNMENT SERVICES

Cornelius Ncube

Abstract: As many governments the world over are engaged in an e-Race to provide their citizens, business communities and public organizations (i.e. service consumers) with electronic public service delivery (EPSD) information systems, the number and type of devices that can be used to access these services is increasing all the time. The mobility of the service consumers over a wide geographical range raises new essential user and system requirements for EPSD systems. Therefore, in order to address the needs of the "mobile citizen", this research proposes an infrastructure for developing EPSD systems that are designed to offer service consumers their individualized information or government services wherever they are, whatever time and whatever their connection point or access device used. In this approach, services are created by applications that are dynamically constructed and configured from a set of interconnected COTS-components with the service instances adapted to the access device used based on the context knowledge of the user, {*time, location, access device*} and the capability profile of the access channel. The systems are designed so as to cope with *dynamic* and *evolving system and user requirements*

Key words: e-government services, Commercial-Off-The-Shelf (COTS) software components, mobile citizen, Electronic Public Service Delivery (EPSD), access/delivery channels, dynamic and evolving system requirements, context knowledge, adaptable services, service consumer

1. INTRODUCTION

Citizens and business organizations of the developed nations currently live and operate in increasingly interconnected societies where the internet is used to get government services 24/7 (*i.e. 24 hours a day, 7 days a week*).

More than 60% of all internet users interact with government sites (Forman 2002). Many national governments the world over are currently engaged in ambitious, tight-scheduled e-Race initiatives. The US federal expenditure on IT exceeded \$48 billion in 2002 and is expected to exceed \$52 billion in 2003.

The main stated goal of all these governments' e-Race is to provide their citizens, business communities and government departments with an efficient and effective electronic service delivery mechanism. However, due to the scale and complexities of these systems, this new frontier will require novel and innovative rethinking of both the structure of the applications and the processes for developing those applications. This research therefore, proposes a flexible infrastructure for *constructing adaptable Electronic Public Service Delivery (EPSD) software systems by integrating multiple Commercial Off-the-Shelf (COTS) software components* that are can be procured *off-the-shelf* in the market. Specifically, this research proposes an infrastructure that dynamically integrates and configure on-demand, multiple COTS software components that are suitable for developing applications that provide e-services offered by the government..

The rest of the chapter is structured as follows: Section 2 provides a "normative" framework for e-government services; section 3 maps out the scale and complexity of providing e-government services. Section 4 gives a description of the proposed COTS-Oriented Process. The overall architecture of the proposed infrastructure is also provided in this section. Section 5 describes proposed profiles for profiling COTS software components and access devices. The chapter concludes with an outline and a vision for key future research directions that are necessary to put the proposed e-Race initiatives on the *e- track!*

2. A NORMATIVE FRAMEWORK FOR E-GOVERNMENT SERVICES STRATEGY

e-Government is the provision of services from government-to-citizen (G2C), government-to-business (G2B) and government-to-government (G2G), (Gouscos et al. 2001, von Hoffman, 1999), using *Electronic Public Service Delivery (EPSD) system over the internet.* In addition to services provided, e-government also brings best commercial practices to key government operations to bring internal efficiency and effectiveness (IEE) to government departments, (Forman, 2002). e-Government services portfolios can be summarized as shown in figure 1, below:

G2C	**G2B**
* Use the web for accessing services such as benefits, loans, recreational sites and educational materials * **Key lines of business**: social services, taxes, recreation * **Key benefits**: will fulfill the vision of one-stop, online access to benefits and services; will also bring modern relationship management tools to improve the quality and efficiency of service delivery.	* Reduce burden on businesses by adopting processes that enable collecting data once for multiple uses and streamlining redundant data * **Key lines of business:** regulation, economic development, trade, asset management, etc. * **Key benefits:** will reduce burden on businesses by adopting processes that reduce redundant data collection.
G2G * Share and integrate central, regional and local data * **Key lines of business**: economic development, public safety, law enforcement, transport, etc. * **Key benefits:** will enable the sharing and integration of central, regional and local data to facilitate better leverage of investments in IT systems (e.g. geographical information systems) and to provide better integration of key government operations such as disaster response.	**IEE** * Adopt commercial best practices in government operations – supply chain management, CRM, HR, work flow management, etc. * **Key lines of business**: supply chain management, HR, finance * **Key benefits:** brings commercial best practices to key government operations particularly supply chain management, financial management, knowledge management, CRM, etc.

Figure 1*: Summary of e-Government portfolios (synthesized from Forman 2002 and Gouscos et. al, 2001). A successful e-government strategy focuses on four citizen-centered groups each providing opportunities to transform delivery to government services:*

3. THE SCALE AND COMPLEXITY OF PROVIDING E-GOVERNMENT SERVICES

The development of electronic public service delivery systems is not easy. The services offered to service consumers (citizens, businesses, and government departments) need to be joined up and delivered through a range of channels and backed-up by a complex back-end software support system and communications network. Traditionally, private citizens and business organizations that need to deal with the government have a limited choice of channels for accessing government services. Using traditional methods, service consumers can only use a call center, write a letter to the government, phone the government department directly or visit the government office as depicted in figure 2 below.

However, with an EPSD system, users should be able to access the same government services *anytime, anywhere,* in the right format irregardless of the access method/device used as shown in Figure 3.

As mobility becomes a major paradigm of the future and lifestyle of tomorrow's citizens and businesses, the infinitely increasing choice of access devices and accompanying demands for instant government information and services are among the most challenges for developing EPSD systems. This "*new frontier*" requires rethinking the structure of government electronic information systems applications and the way these systems need to be designed. Private citizens and business organizations will want to be able to access government services using any access device available or that suddenly becomes the "rage" wherever they are!

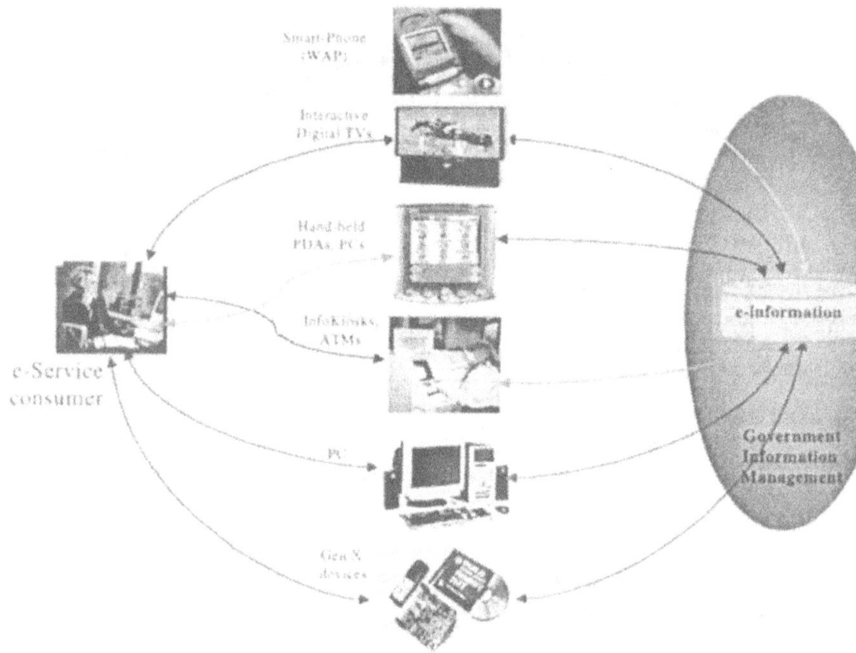

Figure 2: *Traditional access methods. Users can either write a letter to the government, phone a call centre, visit a government officer or government department.*

The rational for EPSD is to envision a government so connected that everyone *–private citizens, business and government people* – can have computerized e-Personal Agents (e-Pa) updating them on almost anything or topics of their choice, wherever they are, whenever they want using any access channel of their choice (von Hoffman, 1999). This will allow private citizens and business organizations the ability to have their entire government on their screen, on the move, anywhere, anytime, whenever they want it. An EPSD system therefore should be designed to:

(a) give citizens access to all their records in an integrated and intuitive fashion, e.g.. *one-stop-shop;*

(b) ensure core consistency across services in the way information is defined, stored and presented, while accommodating uniqueness where appropriate *e.g. private citizen information, business information or government information;*

(c) design services to be independent of delivery channels, i.e. applications should be designed so as to cope with the latest changes on the delivery channels, *e.g. using the object-oriented paradigm*

(d) facilitate use by clients with special needs, *e.g. e-accessibility capabilities, text or screen magnifiers.*

A flexible infrastructure that aims to achieve the above objectives is described next.

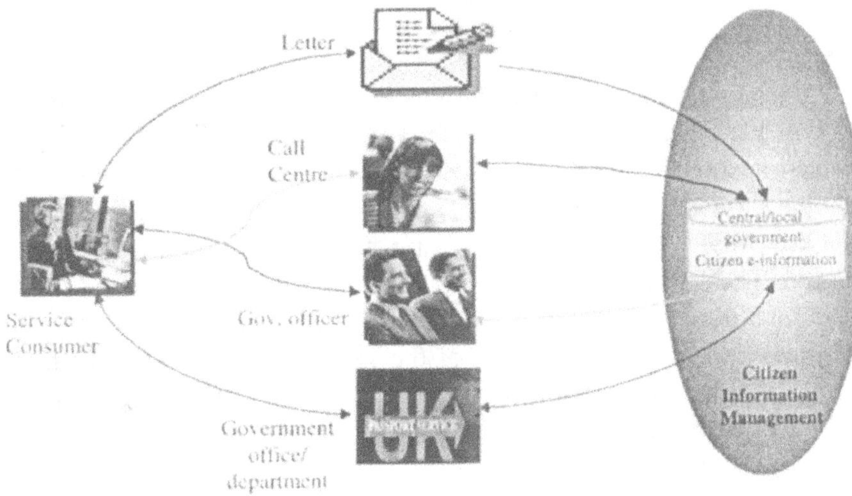

Figure 3: Electronic devices/channels for accessing government e-services

4. A COTS-ORIENTED PROCESS FOR ADAPTIVE E-GOVERNMENT SERVICES

As mentioned above, the mobility of service consumers over a wide geographical range raises new essential user and system requirements for EPSD systems. The number and types of devices that can be used to access government services is increasing significantly all the time (Marvie et.al 1998, Capra et. al. 2000). The use of access devices such as smart mobile phones, PDAs, public inforkiosks and PCs over a large-scale internet and WAP network means that EPSD systems should be designed such that are uniformly accessed. Furthermore, the information content and the system configuration should be relevant to the consumer's context such as *location, time and access device*. Therefore, a framework for building EPSD systems in such a way that a ***single service instance is adapted to the access device based on the user's context*** envisioned.

This research proposes a COTS-Oriented process for building adaptive e-services using COTS-software components. In this approach, an application is dynamically constructed, configured and then executed from a set of interconnected COTS-components with the service instances transformed, formatted, adapted and then routed to the access device based on the context of the user. Depending on the capabilities of the access device used, a single service instance will be created and adapted at the point-of-need, to the relevant access device capabilities, thereby providing the same service information whatever the user's connection point, access device and context.

To achieve this, a Service Oriented Middleware (SOM) infrastructure that will handle all the complexities is required. The SOM would provide service-transformation services and would be able to split, reformat and combine services to be delivered to any access channel. For this, predefined rules for forwarding reformatted services to the appropriate device are required. The service delivery is based on various conditions and the SOM utilize *situation rules* that are defined in the *rules inference engine*, *transformation rules* that are part of the *service transformation engine* and *routing rules* that are part of the intelligent *service routing engine*. Figure 4 below depicts the structure of the envisioned SOM.

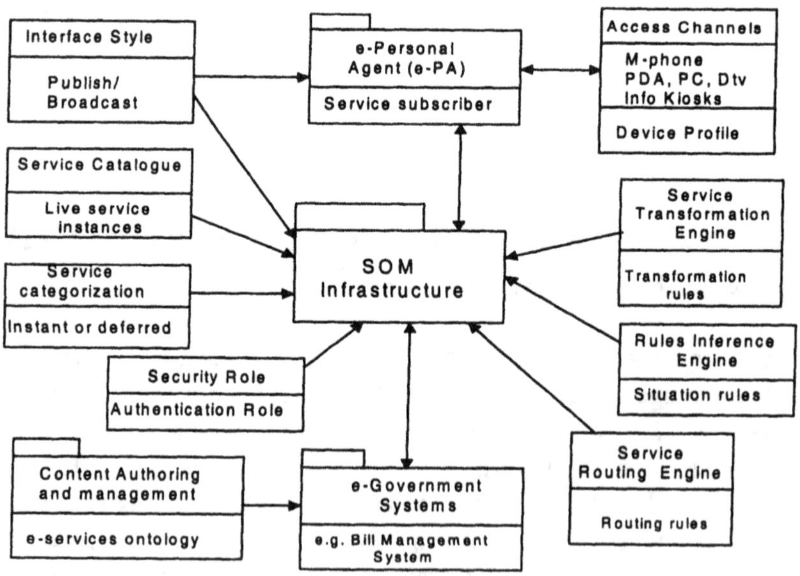

Figure 4: The Service Oriented Middleware (SOM) infrastructure

In developing adaptable EPSD systems, various concepts and technologies need to be considered. As mentioned above, this research envisions a concept whereby a single application is dynamically configured from COTS components stored in repositories provides services that are adapted to an access device/channel depending on the user's context. To achieve this, the following process components are required to be developed:

a) a meta-language for COTS-component and access device profiling,
b) a Service Oriented Middleware infrastructure that is depicted in Figure 4 which utilizes various engines and uses the meta-language to match service instances to access devices, and
c) a process for creating and adapting service instances to access devices.

These new elements are described next and Figure 5 below depicts the overall architecture of the proposed framework.

5. PROFILING COTS SOFTWARE COMPONENTS AND ACCESS DEVICES

The concept of profiling helps us to understand the capabilities of COTS software components and access devices in specific operational environment (Voas 2000). In the context of the proposed COTS-Oriented adaptive e-services, profiling involves developing:

(a) a description of the known capabilities and characteristics of the COTS software components under consideration;
(b) a description of the known capabilities of currently available access devices;
(c) a description of the applications created using the COTS software components;
(d) a description of the service instances that are instantiations of the application developed using the COTS components.

For this, three main profiles – *the abstract, concrete* and *instance profiles* - which are described later in this section, need to be defined. However, to guide the definition of these profiles, the following questions need to be answered:

(a) what will be the formal semantics of both the abstract, concrete and instance profiles that can support effective service provision by integrating, configuring and adapting service instances?
(b) what characteristics of profiles can be determined during service instance creation?
(c) how can the proposed Service Oriented Middleware (SOM) best use the profiles to effectively create service instances, match, adapt and route them to the access device used based on the user context?
(d) how can profiles be integrated with the SOM so that it can use them to provide effective service provision?

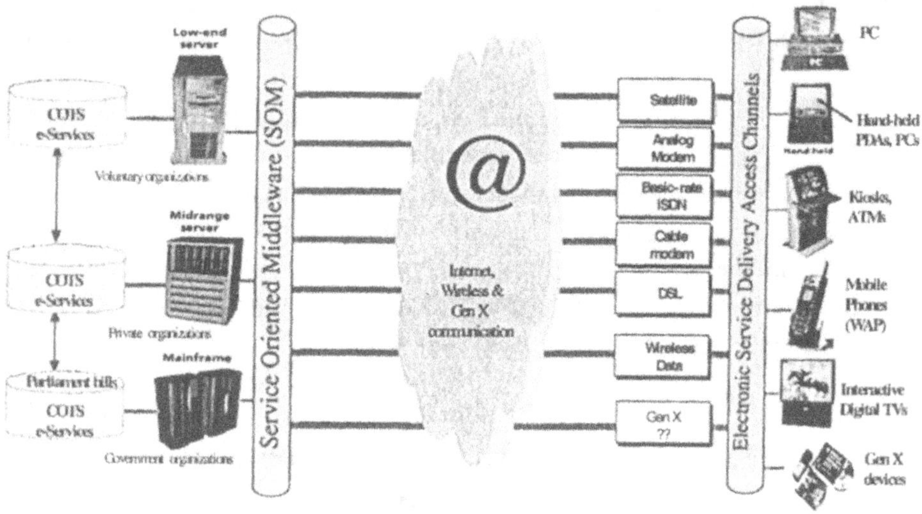

Figure 5: Architecture for the COTS-Oriented adaptive application infrastructure framework

To answer these questions, the proposed SOM needs to reason about the characteristics of the COTS-component and the characteristics of the access devices such as interoperability, assumptions, interface and interaction requirements, (Capra et. al 2000, Marvie et. al 1998). For the SOM to reason about these characteristics, a Profile Description Language (PDL) that adequately captures these characteristics is required. The PDL would be a meta-language that offers syntax and semantics for expressing the characteristics of both COTS software components and access devices. The SOM will speak the language of the PDL, reason about the COTS-components and select the ones that meet a consumer's service request. When the SOM receives a consumer request for a service, it uses its rules inference engine and situation rules to dynamically construct and configure an application from stored COTS software components, instantiate the constructed application to create service instances based on the user context knowledge, adapt and route the service instances to relevant access devices. This will be a matching process between the requested service and the characteristics of the COTS component on one hand and the characteristics of the access channel used on the other, performed by the SOM. The results will be matched, adapted and then routed to the appropriate device.

To cope with the evolution of both the COTS-components and access devices, the PDL will be independent of specific characteristics of COTS-components and access devices. It will be a meta-description language that defines the SOM's view of the COTS-component, service instance and access device. XML-based formats such as Portable Application Description (PAD) or Open Software Description (OSD) can be used for developing the PDL meta-language. PAD is an XML-based format for describing downloadable applications and allows creating an application's description only once. OSD is also an XML-based format used by W3C to describe software components and their dependencies in a distributed environment.. This will be very useful as the proposed process constructs the EPSD systems by integrating multiple COTS software components that are stored in distributed repositories. In order for the SOM to be able to create and adapt service instances, the following 4 key profiles are required:

(a) **COTS-component Abstract Profile** that is an abstract description of the component's architecture and its functional and non-functional properties. It also describes the component's properties such as interconnection mechanisms, published interfaces, performance, reliability, security, i.e. the overall Quality of Service provided by the component (Wang & McClean 1999).

(b) **Application Concrete Profile** that is a concrete description of a set of interconnected COTS components that have been dynamically configured to create adaptive services. COTS component attributes such connector types, interface types, provided interfaces, required interface, etc. (Ning 1999) are used to create the concrete application profile.

(c) **Access Device Concrete Profile** that is a concrete description of the access device's characteristics and capability features. Typical device features that will be described are: mobility , i.e. smart mobile phone, PDA; fixed, i.e. inforkiosk, PC; display screen, i.e. interactive digital TV, desktop PC, mobile phone. The SOM will use device profile characteristics to match and adapt service instances.

(d) **Service Instance Profile** that is an instantiation or the execution of the application concrete profile to create service instances that are uniquely adapted and routed by the SOM to the access device used based on the user's context. The service instances are created based on the context knowledge which is defined as a set {*time, access device, user location*}. The SOM first uses the *rules inference engine* and *situation rules* to construct a service instance, then the *service transformation engine* to transform or format the service based on the access device and then the *service routing engine* to forward the required service to the access device. It is possible that many similar service instances can be created during service request. For this, the SOM will maintain a 'service catalogue' to keep information about 'live' instances and destroy those that their life has expired.

To create a service instance, the SOM performs three fundamental processes. First, it will check the service catalogue to determine that no "live" service instances match the requested service. If not, it finds the relevant COTS components using component abstract profiles. Secondly, it dynamically constructs the required application based on the application concrete profile. Then finally, it executes the constructed application to create service instances that are delivered to the access device. Sometimes, existing instances can be reused or reconfigured and connected together with the newly created ones.

6. CONCLUSION

The proposed research aims to provide e-government systems developers with a smart systems engineering process. The systems development process envisioned in this research aims to provide a flexible, mixed system development environment with process instances that allow some unique parts of the EPSD system to be custom developed; some parts of the system to be configured and integrated from multiple COTS software components; some parts of the system to be rented; some parts of the system that rely on the state-of-the-art technology to be leased competitively from the market while some parts of the system or technologies to be tailored from systems or technologies that are developed for other business domains. The advantages of building systems by integrating COTS component have been extensively reported in Ncube (2000). This chapter therefore concludes by identifying some key future research initiatives that are necessary to realize the proposed process:

- **e-Personal Agent (e-Pa) :** this research envisions e-government services that are so inter-connected that everyone – private citizen, business and government people – would have computerized individual agents (e-Pa) updating them on almost anything or topic of their choice. This would allow private citizens and businesses the ability to have their entire government on-demand on their screens anywhere, anytime, whenever they want it. The e-Pa will enable users to subscribe to be updated about any service they require from the government.
- **A Service Oriented Middleware (SOM)**: this is needed to handle all the service requests by subscribers and for transforming, adapting and routing service instances to the access device used. When the SOM receives a service request, it uses the request to search for service instances that are still 'alive' which can provide the requested service. If there are no 'live' service instances it then searches the COTS component repositories for relevant candidate components using abstract profiles, reconfigure the components to create a concrete application that is then executed to create service instances that are then adapted and routed to the access devices based on the user context. To achieve this,

the SOM will need 'reasoning' capabilities, therefore a *rules inference engine* with its *situation rules*, a *service transformation engine* with its *transformation rules* and an *intelligent service routing engine* with its *routing rules* need to be developed.

- **Profile Description Language** – there is a need to develop or specify a complete Profile Description Language (PDL) in order to understand and describe the capabilities and characteristics of both the COTS-components and access devices. The PDL will provide formal semantics for modeling abstract, concrete and instance profiles for COTS-software components, access devices and services that will be used by the SOM to match and adapt service instances to the right context and the right access device. XML-based formats can be used for the development of the PDL meta-language.
- **e-Authentication Management Framework**: the e-AMF is needed to build and enable mutual trust that is required to support wide spread use of electronic interactions between the public and the government. The e-AMF will be an integral part of the proposed COTS-Oriented Process and will be tightly integrated with the SOM. The e-AMF will be responsible for establishing 'a method for satisfactorily establishing' unique user identities. It will provide a secure consistent method for proving identity whenever a user requests or accesses government service. Latest COTS technologies such as electronic signatures, e-IDs or universal identification can be used.

REFERENCES

Capra L, Emmerich W and Mascolo C., (2000), Reflective Middleware Solutions for Context-Aware Applications

Forman M., (2002), E-Government Strategy, USA Federal Strategic Report, February 2002

Gouscos D, Mentzas G. and Georgiadis P.,(2001), Planning and Implementing e-Government Service Delivery: Achievements and Learning from On-lone Taxation in Greece, 8th Panhellenic Conference on Informatics, Nicosia, Cyprus, November 8-10, 2001

Marvie R, Pellegrini M-C and Potonniee O., (1998), Smart Cards: A System Support for Service Accessibility from Heterogeneous Devices, LIFL, UPRESA 8002, Bat M3, UFR d'IEEA, France

Ncube C. (2000), *A Requirements Engineering Method for COTS-Based Systems* Development, PhD Thesis, City University London, May 20

Voas J, (2000), Deriving Accurate Operational Profiles for Mass-Marketed Software, Reliable Software Technologies

von Hoffman C. (1999), The Making of e-Government, CIO Magazine, November 15, 1999 Issue

Chapter 32

ANALYSIS OF THE RELATION BETWEEN SERVICE PARAMETERS FOR SERVICE LEVEL MANAGEMENT AND SYSTEM UTILIZATION

Masaharu Akatsu, Shoji Konno, Norihisa Komoda

Abstract: System utilization was investigated with the aim of selecting appropriate service parameters for defining the quality of information systems such as SLA (service level agreement). This investigation is based on the hypothesis that parameters having a strong influence on user satisfaction depend on how the system is used. Accordingly, the parameters for availability and responsiveness, which are known as the two major factors in SLA, were investigated. First, it is explained that parameters having a strong impact on user satisfaction are not common among information systems. Next, three parameters for availability (mean time between halt, mean time to restore, and both of them) and ones for responsiveness (mean time, maximum time, and variance) were introduced. To analyze the relation between these service parameters and system utilization, a questionnaire about system utilization was designed and administered to several experienced system engineers. Quantification theory type II was applied to the results of this questionnaire, and the validity of the hypothesis was demonstrated. The results also classified information systems into six groups in terms of availability and responsiveness. The characteristics of each system group were also clarified.

Key words: service level management, user satisfaction, availability, responsiveness, quantification theory type II

1. INTRODUCTION

Accompanying the rise of IT service providers such as ASPs (application service providers) and iDCs (Internet data centers), it has become popular to define the quality of information systems as SLA (service level agreement) (Muller, 1999; Hiles, 1999; Lewis, 1999). SLA is based on the concept that information systems do not only provide functions but also provide services to users. Information systems are the key factors to make or break a business's success; therefore, they are managed from the viewpoint of business managers and system users. This tide affects systems management IT professionals perform. Traditional systems management is concerned primarily with the health of the information systems and networks. On the other hand, it is becoming common to manage systems from the users' viewpoint in consideration of the impact of system quality on business, such a management style is called SLM (service level management).

The concept of "service level" is widely pervasive; however, its practice is still on the road to full realization. For example, there are no explicit criteria for the selection of service parameters. Some industry groups such as the ASP Industry Consortium and the Information Technology Association of America have presented SLA templates and guidelines for SLA. They provide a variety of service parameters, but they are little use for selecting parameters when SLA is individually defined. As a result, most service providers present almost the same SLAs. On the other hand, the true goal of SLA is to guarantee a valuable service to users. It is therefore desirable to identify service parameters that are strongly related to user satisfaction. "Cookie-cutter" parameters do not always satisfy users. The service parameters concerning each system should be different. This would lead to a way for service providers to differentiate themselves.

Over the years, numerous studies on user satisfaction have been conducted. Doll and Torkzadeh (1988) developed a standard measure of end-user satisfaction. Adelman et al. (1985) tried clarifying factors users consider most or least important when evaluating information systems. Jiang and Klein (1999) demonstrated that important factors for users change with system types. The assumption that evaluation factors can change according to system types is also applied to classification of software (Glass and Vessey, 1995).

The factors these studies have considered for system evaluation are almost the same, for example, reliability, performance, usability, accuracy. They are also considered from the viewpoint of the service level management (Strum et al., 2000; Park et al., 2001). In addition, the findings of one survey revealed that availability and performance are indispensable factors in SLA (Nextslm.org, 2000).

In spite of this result, application of these earlier works to SLA/SLM is not sufficient. Most of the research addresses user satisfaction, concentrating on evaluating important factors rather than on parameters by which the factors are measured. In the case of SLA/SLM, we need to define not only evaluation factors, but also parameters that enable us to evaluate service quality quantitatively. For example, it is not sufficient to say that availability strongly affects on user satisfaction: we must consider user satisfaction including parameters for availability.

This paper deals with this issue on availability and responsiveness, which are two major factors in SLA. First, we explain that the parameter for each factor which strongly affects on user satisfaction is not necessarily the same. It is different for each system. So, we provide three parameters for availability and responsiveness, respectively. Next, we frame a hypothesis that parameters having a strong influence on user satisfaction depend on how the system used. To validate the hypothesis, a questionnaire on system utilization is constructed. The questionnaire was administered to several experienced system engineers. The answers are analyzed by Quantification theory type II, and the validity of the hypothesis is demonstrated. We also classify information systems into three groups to determine the parameters referred in SLA. Finally, we show how our results are used in constructing SLA/SLM.

2. SERVICE PARAMETERS FOR SERVICE LEVEL MANAGEMENT

2.1 Service Parameters for Availability

Availability is generally defined as the percentage of the time that a service is available for use. In other words, availability is calculated from Equation (1) as follows.

$$\text{availability} = \frac{\text{mean time between halt}}{\text{mean time between halt} + \text{mean time to restore}} \quad (1)$$

This equation shows that availability is determined by two parameters; mean time between halt and mean time to restore. Figure 1 expresses the relation between these two parameters in detail. Services with the same availability are located on a straight line which passes through the origin, so when availability becomes high, the inclination is large. Suppose that a service is plotted at point *a* in Figure 1. All parameters are improved as long as point *a* moves in the direction of domain A. This change is always preferable. On the contrary, moves in other directions cannot be concluded simply whether the change is desirable from the viewpoint of user satisfaction. One typical example is that the case where one of two mean times improves but another gets worse. It depends on the service type whether this change is good or bad.

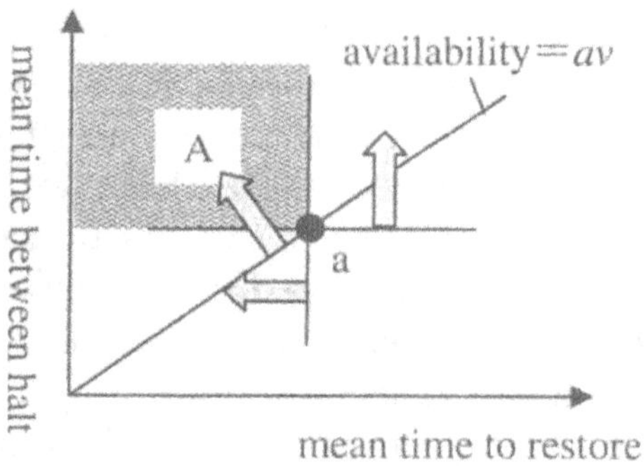

Figure 1. Relation among parameters for availability

Let us consider an example of this subtle situation. Suppose there are two service types. Service type α does not stop for 30 days, and then stops all day long on the 31st. Service type β repeats 7.5 hours work with halts of 15 minutes in between. Although the availability of both service types is equal, that is, about 97%, user satisfaction with each service is not necessarily the same. Consider a communications service via an in-house web portal, which employees use once a day. If this service belongs to type α, user satisfaction may be low because there is a day when employees cannot access the service. On the other hand, if the communications service is type β, user dissatisfaction is relatively low because employees can find a time when the service is available on any day. A movie distribution service is an opposite example. This service is continuously used for a couple of hours, so user satisfaction with type β is lower than that with type α. This is because the probability of service halts during use is higher in the case of type β.

The two above scenarios lead to the following conclusion. Availability defined by Equation (1) is not sufficient for an index for evaluation of user satisfaction. Certain parameters for calculating availability must also be considered. Accordingly, we have studied three possibilities for the parameters expressing the availability factor: mean time between halt, mean time to restore, and both of them.

2.2 Service Parameters for Responsiveness

Mean response time is a typical parameter for responsiveness. However, many studies have shown that satisfaction with responsiveness is not a simple matter of speed of delivery (Butler, 1983; Culwin and Faulkner, 2001). For example, low variance and predicted response can be a key to satisfaction (Ritchie and Roast, 2001). Moreover, an extremely slow response has a bad influence on users (Endo and Seltzer, 2000). Hence,

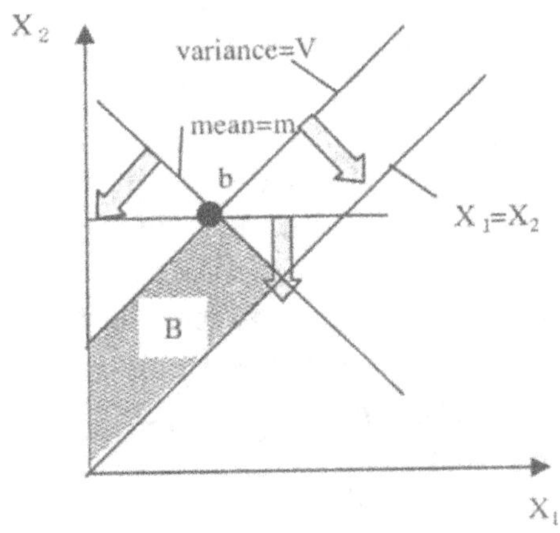

Figure 2. Relation among parameters for responsiveness

lowering maximum response time contributes to the improvement in user satisfaction. Even if mean response time gets worse, the variance and the maximum response time can be improved. Figure 2 shows the situation briefly. For simplicity, the figure assumes that there are only two response times (x_1, x_2; x_1=<x_2) and that both of them occur with the same frequency. Services with the same variance are plotted in a straight line with the slope 1. Services with the same mean response time are plotted in a straight line with the slope -1. Suppose that a service is plotted at point *b* in Figure 2. Move in the direction of domain B is always preferable, but it is not necessarily appropriate to suggest that moves in other directions are good or bad. Hence, it is not simple to select a parameter to be improved, that is, the most important parameter. The importance of each parameter is independent of user personalities (Guynes, 1988), but depends on system utilization. Accordingly, we prepare three parameters for the response factor: mean time, maximum time, and variance.

3. INVESTIGATION ON SYSTEM UTILIZATION

It is desirable to select a parameter that is the most influential to user satisfaction from the parameters mentioned in the previous section. One of the simplest ways to select a parameter is to make users choose a parameter that they regard as the most important. However, this is not appropriate because users tend to think all parameters are important.

Then, we pay our attention to system utilization. This is based on a hypothesis that parameters having a strong influence on user satisfaction depend on how the system is used. If the hypothesis is correct, investigating the utilization of many systems and analyzing them statistically reveals significant relations between system utilization and important parameters. By using such statistical analysis, the parameters that should be selected as a suitable measure for user satisfaction can be identified.

To validate the hypothesis, a questionnaire on system utilization is constructed. First, utilization time, frequency of use by the typical users, and available time are considered as items that can distinguish the services mentioned in the previous section. Next, people capable of using systems, number of users, and function the system provides are added in a questionnaire. These items are based on "reach" (the locations and people the system is capable of connecting) and "range" (functionality in terms of business activities) that Weill and Broadbent (1998) introduced with an aim to evaluate the value of infrastructure. Moreover, impact on business incurred by the loss of the system and peak load time is added. They are expected to have a strong relation with availability and responsiveness, respectively. Table 1 shows a questionnaire about system utilization. The answer to each question item is split into three or four categories in advance, and respondents select one of the prepared answers.

4. ANALYSIS BY QUANTIFICATION THEORY TYPE II

To analyze the relation between service parameters and system utilization, we gave the questionnaire survey to several experienced system engineers. They answered the questions in Table 1 for each system they had developed. They also chose parameters that they thought are the most influential on the user satisfaction in terms of availability and responsiveness. All data were given as categorical variables. Thus, Quantification theory type II can handle the analysis of the relation between important parameters and system utilization (Komazawa, 1982). In fact, the answers to the eight questions in Table 1 were taken as predictor variables, and the replies to parameters for availability and responsiveness were taken as criterion variables.

Table 1. Questionnaires on system ulitization

Item		Category
1	people capable of using systems	1. designated members 2. rank-and-file employees 3. anyone
2	number of users	1. less than 10 2. from 10 to 100 3. from 100 to 1000 4. over 1000
3	available time	1. 24 hours a day, everyday 2. predetermined period of time (ex. business hours)
4	peak load type	1. peak load at a specific day or time 2. peak load at an unpredictable time 3. no peak load
5	frequency of use by the typical users	1. everyday 2. a couple of times a week 3. once a week or less
6	utilization time	1. a few minutes 2. a few minutes per access and repeating it 3. over several tens of minutes without a break
7	function the system provides	1. send messages 2. access to information 3. perform transactions
8	impact on business incurred by the loss of the system	1. direct financial loss such as decline of sales 2. indirect financial loss such as operating inefficiency 3. no financial impact

The results of the questionnaire provided data on 42 systems. There were a variety of systems, such as workflow, CRM, financial application. Quantification Theory type II was applied to 37 data sets selected at random. The five remaining data sets were used for validation. Table 2 and Table 3 list the results of this analysis and Figure 3 and Figure 4 plot 37 sample scores as scatter plots for discrimination. The discriminant ratios (percentages correctly classified) are 92% for availability and 95% for responsiveness. Table 4 lists the correlation ratios for each axis. The test statistics F_0 is calculated as :

$$F_o = \frac{R^2 / p}{(1-R^2) / (n-p-1)} \tag{2}$$

where R^2 is correlation ratio, n is sample size, and p is parameter size.

All of the ratios are significant at the 1% significance level. Furthermore, the discriminant function correctly classified the five remaining systems.

The validity of our hypothesis is thus demonstrated. Namely, there is a close relation between parameters that reflect user satisfaction and system utilization. It is therefore concluded that the questionnaires shown in Table 1 are applicable for selecting the parameters for service level management.

.*Table 2.* Analysis by Quantification theory type II on availability

		Axis 1			axis 2		
item	category	category score	range	partial correlation coefficient	category score	range	partial correlatio n coefficient
1	1	-0.97	1.84	0.54	-0.62	0.90	0.31
	2	0.09			0.28		
	3	0.87			0.00		
2	1	1.02	1.21	0.40	1.21	1.69	0.44
	2	0.53			0.83		
	3	-0.17			0.03		
	4	-0.19			-0.48		
3	1	-0.93	1.32	0.61	0.01	0.01	0.01
	2	0.39			0.00		
4	1	-0.37	0.65	0.34	0.13	1.73	0.49
	2	0.28			0.56		
	3	-0.01			-1.17		
5	1	-0.08	0.89	0.33	0.33	1.37	0.47
	2	-0.27			-1.04		
	3	0.62			-0.09		
6	1	-0.03	0.30	0.10	0.06	1.75	0.36
	2	0.14			0.18		
	3	-0.16			-1.57		
7	1	1.25	2.11	0.74	0.57	0.98	0.33
	2	0.44			0.26		
	3	-0.86			-0.41		
8	1	-2.16	2.90	0.76	-0.21	0.78	0.23
	2	0.74			-0.10		
	3	0.14			0.57		

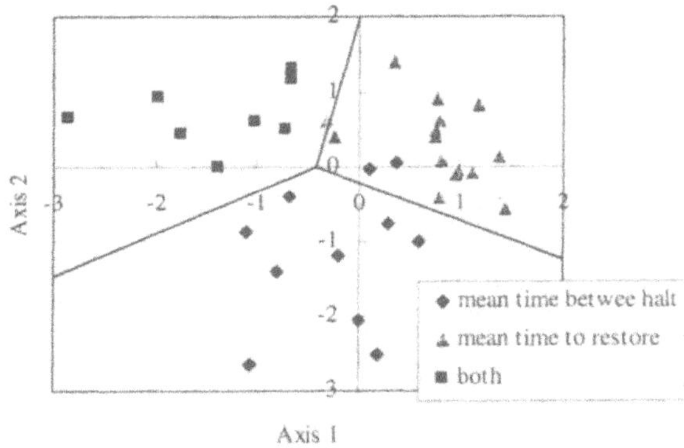

Figure 3. Scatter plots for discrimination on availability

Table 3. Analysis by Quantification theory type II on responsiveness

item	category	Axis 1			axis 2		
		category score	range	partial correlation coefficient	category score	range	Partial correlatio n coefficient
1	1	-0.73	0.99	0.46	0.26	1.50	0.56
	2	0.26			-0.51		
	3	0.16			0.99		
2	1	1.45	2.21	0.70	-0.45	1.75	0.57
	2	1.08			1.30		
	3	0.22			-0.09		
	4	-0.76			-0.36		
3	1	0.60	0.85	0.50	0.78	1.11	0.47
	2	-0.25			-0.33		
4	1	-0.26	0.45	0.28	0.25	1.07	0.40
	2	0.19			0.27		
	3	0.02			-0.80		
5	1	0.08	0.28	0.16	-0.02	0.33	0.12
	2	-0.11			-0.12		
	3	-0.20			0.21		
6	1	-0.01	0.69	0.22	-0.21	3.25	0.60
	2	-0.10			1.08		
	3	0.59			-2.17		
7	1	0.30	0.37	0.15	-0.11	1.23	0.56
	2	0.05			-0.67		
	3	-0.07			0.56		
8	1	-0.23	1.47	0.61	-1.02	2.05	0.51
	2	-0.29			0.04		
	3	1.18			1.03		

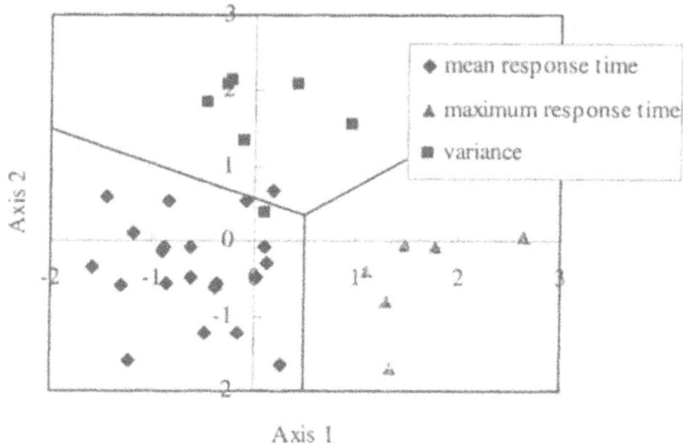

Figure 4. Scatter plots for discrimination on responsiveness

Table 4. Correlation ratio

	first axis	second axis
Availability	0.6888*	0.596*
Responsiveness	0.734*	0.625*

* indicates significance at $p < 0.01$ lelel

5. SYSTEM CLASSIFICATION

In Section 4, we divided systems into three groups to determine the service parameters in terms of availability and responsiveness. In this section, we clarify characteristics of each system group by using the analysis results listed in Table 2 and Table 3. The item (i.e., predictor variable) that has the larger value of a range and a partial correlation coefficient highly contributes to classify systems. After paying attention to such items, we sort out categories that have a larger absolute score. The rest of this section explains the characteristics on each system group.

5.1 System Classification in terms of Availability

(1) Group A1: mean time between halt as key parameter

A system in this group has a small absolute score for the first axis and a negative score for the second axis. Thus, a system that belongs to Category 3 of Item 6 (utilization time) tends to be classified to this group. In other words, a system used for a relatively long time without a break fits this group. Longer utilization leads to a higher risk of a system down when users are connected to the system. Therefore, mean time between halt is more important than mean time to restore.

Furthermore, a system belonging to Category 3 of Item 4 (peak load type) and Category 2 of Item 5 (frequency of use) also tends to go into group A1. Such systems have a characteristic that users can use them at any time and that the date and hour of use are not fixed. This means that users are likely to get the impression that a system whose halt interval is long is more user-friendly than one whose stop time is short when there are two systems of this group with the same availability.

(2) Group A2: mean time to restore as key parameter

A system in this group has positive values for both the first and the second axis. Hence, a system falling into Category 1 of Item 2 (number of users) and Category 1 or 2 of Item 7 (impact on business) is identified as a member of this group. Such a system has a simple function such as information access for relatively fewer users. A typical case of system utilization is that users find out the time when the system is available on any day. It does not bother users very much even if they cannot use the system at a particular time. In this situation, if the service-halt time is not so long, effects on users are small. The mean time to restore is thus the critical parameter.

(3) Group A3: both parameters are important

The gravity of this group is located in the first quadrant in Figure 3. Category 1 of Item 8 (business impact) is the most influential to on this location. This means that the systems in this group have the characteristic that they incur large financial losses when they are down. Since it is necessary to avoid stopping the service as much as possible, the halt interval must be long. In addition, quick restoration is also required even if the service stops. Both parameters should thus be considered.

5.2 System Classification from a Viewpoint of Responsiveness

(1) Group R1: mean response time as key parameter

A system in this group has negative values for both the first and the second axis. Category 4 of Item 2 (number of users) and Category 1 of Item 8 (business impact) indicate this characteristic. The former category shows that the system has many users. This means that mean response time is a common parameter for evaluating responsiveness and is universally accepted. The latter category shows that the system supports mission-critical business operations. Slow response is fatal in the case of a mission-critical system; namely, the quicker is better. Mean response time is thus an indispensable index because overall speedup is required.

(2) Group R2: maximum response time as key parameter

A system in this group has a large positive value for the first axis and a negative value for the second axis. Hence, a system belonging to Category 1 of Item 2 (number of users), that is, a system for a small user group, is classified into this group. This can be interpreted as follow: If the response happens to be extremely slow, such a bad experience has a strongly negative effect on satisfaction. The ratio of users who had such a bad experience is high in the case of this type of system. In view of this fact, shortening the guaranteed maximum response time is more important than improving mean time for this group.

(3) Group R3: variance as key parameter

This group has a small absolute value for the first axis and a positive value for the second axis. Category 2 of Item 6 (utilization time) represents this group most strongly. A system in this category is used for day-to-day operations. So stable operation is critical in the case that designated people perform the same type of transaction repeatedly. Namely, the lower the variance of response time is, the more users are satisfied. This is because they can predict the response time and establish their pace of work easily. In this case variance is therefore the key parameter.

6. APPLICATION TO SLA/SLM METHODOLOGY

Our SLA/SLM methodology is enhanced by our analysis on the relation between service parameters and system utilization. Figure 5 shows our SLA/SLM methodology. This is the flow until service level is stabilized for the first time. It goes without saying that SLM process does not be terminated at step 6 and that constant improvement activity goes on.

Our analysis improves the second step: mutual understanding between service providers and users. This step is broken down into three works. First, based on the analysis of step 1, important factors for service users are clarified. Second, in terms of the each factor, the key parameters are discussed. Third, both parties recognize their responsibility each other. The mutual understandings of step 2 are followed by negotiation over the service parameters and their targeted values in step 3.

In general, the above second work is executed through interview with service users. However, users tend to insist that everything is important when they are asked a vague question on what is important to their business.

Thus, it is difficult to find out valuable information. We improved this situation by using our analysis. First, we analyze the system utilization from the data gathered in step 1. Second, we estimate the critical service parameters. Then, we confirm with users on our estimation. This approach leads us to effective consensus-building on the important parameters.

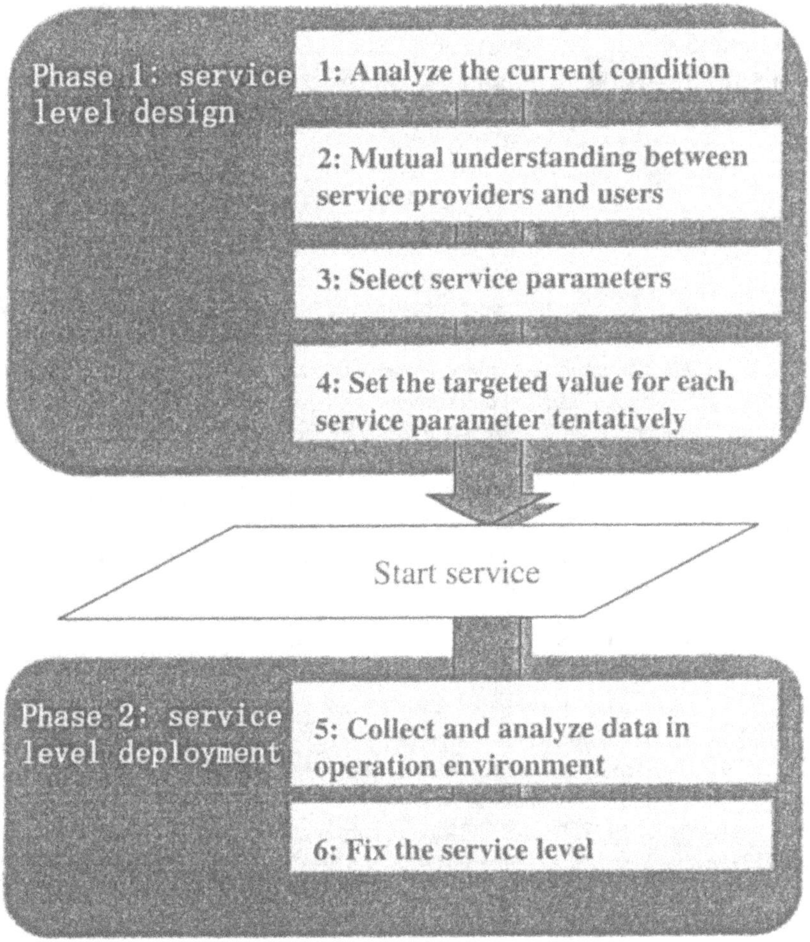

Figure 5. SLA/SLM methodology

7. CONCLUSION

The main purpose of the service level management is to increase the user satisfaction. The study outlined in this paper focused attention on system utilization to select the service parameters that reflect user satisfaction. To investigate parameters for availability and responsiveness, which are known as two major factors in SLA, the relation between service parameters and system utilization was analyzed by Quantification theory type II. The results demonstrated that this relation is strong and enabled us to classify information systems. Therefore, investigation on system utilization can identify the appropriate service parameters for service level management.

A similar approach to that described here will be applied to factors other than availability and responsiveness, such as workload and accuracy. In addition, future works will aim to construct a guideline for determining which factors should be considered for defining SLA/SLM.

REFERENCES

Adelman, L. et al., 1985. User and R&D Specialist Evaluation of Decision-Support Systems. In *IEEE Trans. on Syst., Man, Cybern.*, vol. SMC-15, No.3, pp.334-342.

Butler, T.W., 1983. Computer Response Time and User Performance. In *Proc. of ACM CHI'83 Conf. on Human Factors and Computing Systems*, pp.58-62.

Culwin, F. and Faulkner X., 2001. Brewsing the Web: Delay, Determination and Satisfaction. In *Proc. of 34th Hawaii Int'l Conf. System Science*, pp.1802-1806□

Doll, W. J. and Torkzadeh, G., 1988. The Measurement of End-User Computing Satisfaction. In *MIS Quarterly*, 12, pp. 259-274.

Endo, Y. and Seltzer, M., 2000. Improving Interactive Performance Using TIPME. In *ACM SIGMETRICS Performance Evaluation Review, Proceedings of the Int'l Conf. on Measurements and modeling of computer systems*, vol.28, Issue 1, pp. 240-251, June.

Glass, R. L. and Vessey I., 1995. Contemporary Application-Domain Taxonomies. In *IEEE Softw.*, vol.12, pp. 63-76, July.

Guynes, J. L., 1988. Impact of System Response Time on State Anxiety In *Communications of ACM*, Vol. 31, No. 3, pp.342-347, March.

Hiles, A., 1999. *Complete Guide to IT Service Level Agreements: Matching Service Quality to Business Needs*, Rothstein Associates Inc.

Jiang, J.J. and Klein G., 1999. User Evaluation of Information Systems: By System Typology. *In IEEE Trans. on Syst., Man, Cybern. Part A*, vol. 29, No. 1, pp.111-116, Jan.

Komazawa, T., 1982. *Quantification theory and data processing*, Asakura Books (in Japanese).

Lewis, L., 1999. *Service Level Management for Enterprise Networks*, Artech House Inc.

Muller, N. J., 1999. Managing Service Level Agreements. In *Intl. J. Network Management*, 9, pp.155-166.

Nextslm.org, 2000. Service Level Management – North America Survey 2000. On *http://www.nextslm.org/SLMSurvey2000.pdf*

Park, J. et al., 2001. Management of Service Level Agreements for Multimedia Internet Service Using Utility Model. In *IEEE Communications Magazine*, pp.100-106, May.

Ritchie, I. and Roast, C., 2001. Performance, Usability and the Web. In *Proc. of 34th Hawaii Int'l Conf. on System Science*, pp.1797-1801.

Chapter 33

USE OF MODELS AND MODELLING TECHNIQUES FOR SERVICE DEVELOPMENT

Luís Ferreira Pires, Marten van Sinderen, Cléver Ricardo Guareis de Farias, João Paulo Andrade Almeida*

Abstract: E-applications are increasingly being composed from individual services that can be realized with different technologies, such as, e.g., Web Services and standard component technologies. A current trend in the development of these services is to describe their technology-independent and technology-specific aspects in separate models. A prominent development that leads this trend is the Model-Driven Architecture (MDA). An important feature of the MDA approach is the explicit identification of Platform-Independent Models (PIMs) and the flexibility to implement them on different platforms via Platform-Specific Models (PSMs), possibly through (automated) model transformations. A platform can be any technology that supports the execution of these models, either directly or after translation to code in a programming language. This paper aims at identifying the benefits of the MDA approach in the development of services for e-applications. The paper presents a short introduction to MDA, in the context of service development, and an overview of the modelling capabilities of the Unified Modelling Language (UML), one of MDA's main modelling languages.

Keywords: Service-oriented development; Model Driven Architecture; Unified Modeling Language.

1. INTRODUCTION

There is a growing need to compose e-applications from individual services that can be provided by both proprietary components and third-party service providers. This need arises from requirements with respect to, e.g., shorter time-to-market, reduced development costs, and reuse of proven technological solutions. The ideal of a service-oriented development or service-oriented architecture is also fuelled by the industrial uptake of technologies such as Web Services and standard component technologies.

A current trend in the development of services is to separate their technology-independent and technology-specific aspects, by describing them in separate models. The most prominent development in this trend is the Model-Driven Architecture (MDA) [10, 13] approach, which is being fostered by the Object Management Group (OMG). The MDA approach is not a design methodology, but rather a collection of guidelines to be applied in combination with a design methodology in order to develop distributed applications. The core of the MDA consists of a number of OMG standards, including: the Unified Modelling Language (UML) [18], the Meta Object Facility (MOF) [12], and the XML Metadata Interchange (XMI) [15].

The most important aspect of the MDA approach is the explicit identification of Platform-Independent Models (PIMs) and the flexibility to implement them on different platforms via Platform-Specific Models (PSMs). A platform can be any technology that supports the execution of these models, either directly or after translation to code. In the case of distributed applications, MDA can be applied to develop PIMs that are middleware technology-independent, and develop PSMs for specific middleware platforms like CORBA/CCM, EJB or Web Services. MDA also aims at facilitating the translation from PIMs to PSMs, by introducing profiles for defining PIMs and PSMs, and by standardising transformations between them, which can then be automated by tools. Some research is being carried out in order to define these transformations and automate them.

This paper presents a short introduction to MDA and an overview of the modelling capabilities of UML, which is one of MDA's main modelling languages. The paper is further structured as follows: section 2 introduces some basic modelling concepts and principles; section 3 introduces the MDA approach; section 4 discusses the modelling of services using UML; finally, section 5 presents some final remarks.

2. MODELLING PRINCIPLES

A *model* is a representation of structural or behavioural aspects of a system in a language that has a well-defined syntax, semantics, and possibly rules for analysis, inference, or proof [13].

Models can be used in different ways in the course of a development project. A model used to prescribe properties of a system or system part to be built is called a *prescriptive model.* In contrast, a model used to describe an existing system or system part is called a *descriptive model.* In the case of prescriptive models, designers produce models of a system introducing information that constrain the intended characteristics of the system being specified. The information required for modelling is obtained along the development trajectory, and documented in several ways.

In order to understand any non-trivial system, one has to cope with a large amount of interrelated aspects. Attempting to capture all aspects of the design in a single model yields too complex and useless models [6]. Therefore, models should be derived using specific sets of abstraction criteria, which allow one to focus on particular aspects of the system at a time.

2.1 Viewpoints and abstraction levels

A model is often characterized by the set of abstraction criteria used to determine what should be included in the model. *Viewpoints* and *abstraction levels* are examples of abstraction criteria.

A viewpoint defines a set of related concerns that play a distinctive role in the design of a system. A model defined from a particular viewpoint focuses on the particular concerns defined by the viewpoint. Viewpoints should be chosen with respect to requirements that are of concern to some particular group involved in the design process.

Examples of viewpoints are the five RM-ODP viewpoints [7]: enterprise, information, computational, engineering and technology. The use of different viewpoints in order to describe a system raises the issue of consistency. Descriptions of the same or related entities appear in different viewpoints. Therefore, one must assure that these multiple models are not in conflict with each other.

Abstraction is the process of suppressing irrelevant detail to establish a simplified model, or the result of that process [6]. A model M_1 is at a higher level of abstraction than a model M_2 if M_1 suppresses details of the system that are revealed by M_2. Specifically, the pair of models $\{M_1, M_2\}$ is in a refinement relationship, in which M_1 (the abstraction) is more abstract than M_2 (the realization).

Refinement and *abstraction* are opposite and complementary types of relationships or design activities. Through refinement, an abstraction is made more concrete through the introduction of details, entailing design or implementation decisions, while through abstraction, details of a more concrete abstraction are omitted. An important property of refinement is that the resulting model should conform to the original one [1].

Design methodologies normally define different abstraction levels to be used for particular viewpoints. In these methodologies, abstraction levels are usually related to milestones in the design trajectory, or are related with particular design goals. Several design methodologies also define refinement (and abstraction) relations in order to guide the development of related abstraction levels.

2.2 Metamodelling

Metamodels can be used to define the syntax and semantics of models. When instances of the elements of a model *B* are used to produce a model *A*, *B* is said to be the *metamodel* of *A*. In this case, one can say that the *abstract syntax* of the model *A* is defined in the metamodel *B* [4]. Furthermore, model *A* can be considered as an instance of metamodel *B*.

The abstract syntax of a metamodel *B* can also be described in yet another metamodel *C*, thus constituting a metametamodel. Although the number of metalevels is arbitrary, metamodelling frameworks should define a limited number of useful metalevels.

Whenever a metamodel is accompanied by natural language descriptions of concepts that correspond to its elements, we say that the *semantics* of the modelling elements are informally defined. This approach has been adopted by OMG in the Meta-Object Facility (MOF) [11] and in the UML proposed standards [16, 17]. More rigorous approaches define the semantics of modelling elements in terms of a mathematical domain (e.g., the formal semantics of the Specification and Description Language (SDL) in [8]), or in terms of concrete, formal and explicit representations of domain conceptualisations (e.g., an ontology [6]).

3. MODEL DRIVEN ARCHITECTURE

The MDA approach [13] to *system (application) specification, portability* and *interoperability* is based on the use of *formal* and *semi-formal models*. From the perspective of systems development, a significant quality of the MDA approach is the *independence* of system specifications (i.e., sets of models) from potential target implementation platforms. A system specification exists independently of any implementation platform and has formal or semi-formal transformation rules to many possible target platforms. The application development effort is consolidated in the platform-independent models, such that the investments necessary to move to another platform can be reduced. Furthermore, model transformation rules may be implemented in model-driven tools to (partially) automate the transformation of platform-independent models into platform-specific models, increasing the level of automation of the development trajectory.

From the perspective of systems interoperability, the use of platform-independent models facilitates the creation of different platform-specific models corresponding to the same set of platform-independent models, which results ultimately in implementations that can be easily (if not automatically) integrated.

Platform-independent models also play an important role in the re-use of legacy applications. In this case, integration is done at a platform-independent level, using platform-independent models that represent the legacy application. These platform-independent models are derived by reverse engineering.

3.1 MDA viewpoints

The MDA generally defines a platform as a set of subsystems or technologies that provide coherent functionality through interfaces and specified usage patterns. Any subsystem that depends on the platform can use this functionality without concern for the details of how it is implemented [13].

Three different viewpoints are considered [13]: computation-independent viewpoint, platform-independent viewpoint and platform specific viewpoint. The computation-independent viewpoint focuses on the system environment and its requirements. However, there is no concern for the details of the structure and processing of the system. The platform-independent viewpoint focuses on the system operation, but hides the details necessary for a particular platform. The platform-specific viewpoint combines the platform independent viewpoint with the details of the use of a specific platform by a system.

A computation independent model (CIM) is a model developed according to the computation independent viewpoint. Similarly, a platform independent model (PIM) and a platform specific model (PSM) are models developed according to the platform independent and platform specific viewpoints, respectively.

Platform independence is a relative term that depends on the potential target platforms. For example, if the set of technologies that define a platform comprehends middleware platforms, such as, e.g., CORBA and Web Services, a CORBA Interface Definition Language (IDL) specification is a platform-specific model, because it is bound to CORBA. In contrast, if the set of technologies that define a platform comprehends programming languages and CORBA ORB implementations, such as, e.g., the C++ language and the C++ ORB implementation, a CORBA IDL specification is a platform-independent model, because it can be mapped onto several programming languages.

3.2 Model transformation

Model transformation is basically seen as a mapping of elements of one model onto elements of another model. Consider, for example, the creation of software systems by code generation. Each generated artefact, either some code in a programming language or some textual deployment artefact can be manipulated as a model. These models are based on a defined structure, which itself forms a metamodel. This metamodel can be expressed in terms of the UML and/or MOF standards.

Model transformation is useful if formally or systematically defined. As depicted in *Figure 19*a, a transformation may be defined at the level of metamodels. When transformation is applied, a source model is transformed into a target model according to the defined transformation (rules).

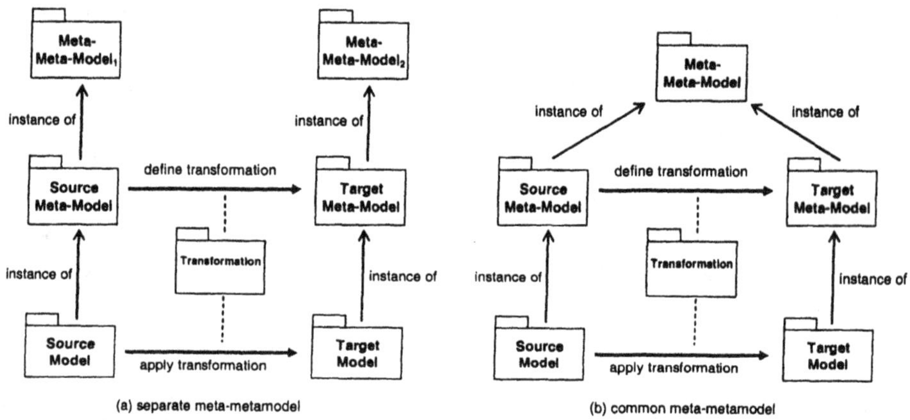

Figure 19: Model transformation

According to OMG definitions, a metamodel is based and constructed from elements of an underlying meta-metamodel (the MOF) and a model is constructed from elements of the metamodel. The use of a common meta-metamodel for the target and source metamodels, as illustrated in *Figure 19*b may facilitate the definition of transformations.

The model transformation pattern can be applied successively. In this case the notions of source and target models are relative. An intermediary model is considered a target model from the perspective of the transformation from the source model, and the same intermediary model is considered a source model from the perspective of the transformation to the final target model.

In order to allow a developer to guide the transformation of a source model when necessary, transformations may be parameterised. An annotation model may be used to hold the parameters for a transformation.

The application of the transformation may include a step that transforms the source model into an annotated source model and then proceeds with the transformation.

4. UNIFIED MODELING LANGUAGE (UML)

This section presents an overview of the modelling capabilities in the Unified Modeling Language (UML), which has been standardized under the auspices of the Object Management Group (OMG). Our discussion is primarily based on UML 1.4 and 1.5 [14, 18] specifications. UML 1.5 is the currently adopted UML specification by OMG. However, most of the currently available UML tools provide support only to UML 1.4 specification. The UML 1.5 specification extends UML 1.4 with the so-called *action semantics*, which mainly adds more preciseness to the definition of actions and procedures. Since the first documents of UML 2.0 [16, 17] have just been publicly released, we do provide a highlight of the main changes in this specification with respect to the previous one.

4.1 Structure modelling

UML defines a collection of diagrams for structure modelling, namely class diagrams, component diagrams and deployment diagrams. UML does not prescribe how these diagrams should be used in a development trajectory, but only their abstract syntax and intended semantics (to a certain extent and informally). A development methodology should be applied in an actual development project to define the UML diagrams that have to be produced to represent a certain model at the different phases, steps or workflows devised for the development trajectory. Since component diagrams are the most relevant type of diagram for the development of service architecture, we exempt ourselves from discussing class and deployment diagrams in this paper.

A component diagram captures dependencies among different kinds of software components, such as implementation classes, source code files, binary code files, executable files, and scripts. A component diagram has only a type form, not an instance form.

UML 1.5 defines a component as a modular, deployable, and replaceable part of a system that encapsulates implementation and exposes a set of interfaces. A component diagram is a graph of components connected by dependency relationships. Interfaces and calling dependencies among components can also be captured using a component diagram. A calling dependency occurs when a component uses a given interface. In such case, dependency arrows from components to the interface on other component must be employed.

A component diagram is used to model the static implementation view of a service. Thus, the architecture of a given service is captured as a collection of components and their dependency relationships. Since UML 1.5 does not consider a component as a unit of design, but a unit of deployment, the role of component diagrams in service development is rather limited. There is no support to component-based development, since UML 1.5 does not allow the representation of abstract components and does not support recursive decomposition of internal structures.

In UML 2.0, a component is a modular unit with well-defined interfaces, which allows it to be reusable and autonomous. The component concept has been introduced to support component-based development, in which components are modelled throughout the development trajectory, from abstract business components to concrete software components.

A component has one or more provided and required interfaces and its environment can only interact with it through these interfaces. The interfaces of a component shield the component's internal structure from its environment. Components can be composed together to form bigger components. This can be done by 'wiring' required interfaces to provided interfaces under the condition that these interfaces are compatible. This implies that a component *C* at some aggregation level is related to a collection of composed components that together realise component *C*.

A component can have required and provided interfaces: required interfaces are used by the component in order to perform its operation, while provided interfaces are those through which the component provides its capabilities. Required and provided interfaces are directly related to the direction of operation invocations, i.e., operations at the required interfaces are invoked by the component itself, while operations at the provided interfaces are invoked by the component's environment.

Alternatively, designers may group interfaces in a port, which defines an interaction point between a component and its environment, or between a component and some elements of its internal structure. Ports allow an even stronger decoupling of a component from its environment than what is already possible using only interfaces. A component can be defined separately from its ports, making it reusable in any environment that complies with the constraints imposed by these ports. Ports also group interfaces, so that the possibly different aspects of the interactions with a component can be properly separated.

UML 2.0 allows the specification of the internal details of a component in some different alternative ways. Because a component is also a class, one can define inside a component all the other classifiers (e.g., components and classes) that are non-shareable parts the component. A more detailed representation of the internal structure of a component can be defined by showing instances of the classes owned by the component and how they relate to the component's ports.

4.2 Behaviour modelling

UML 1.5 defines a collection of diagrams for behaviour modelling, namely use case diagrams, sequence diagrams, collaboration diagrams, statechart diagrams and activity diagrams. Similarly to the structural diagrams, UML 1.5 does not prescribe how these diagrams should be used in a development trajectory. Although, use cases diagrams are considered behavioural diagrams, these diagrams can only capture static behaviour. Thus, we exempt ourselves from discussing it in the scope of this work.

4.2.1 Sequence diagrams

A sequence diagram shows how roles interact with each other in time, by showing the messages they exchange; alternatively, it may represent this by means of instances of the roles and the stimuli they produce. A sequence diagram can be seen as a set of messages and their temporal ordering. It relates to the system structure in that the roles and messages defined in a sequence diagram should correspond to classifiers and operations defined in structural diagrams.

A sequence diagram shows classifier roles (or instances), and the messages (or stimuli) they exchange. A sequence diagram represents either: (1) an interaction, consisting of message exchange between classifier roles and possibly the consequences of these messages (the actions), or (2) an interaction instance set, consisting of stimuli exchanged between instances of classifier roles and possibly the consequences of these stimuli (the actions). Sequence diagrams are also capable of representing other aspects of the systems dynamics, like object creation and destruction, conditional stimuli and focus of control.

Sequence diagrams define behaviour 'by example'. They are normally used to show how the system performs some specific parts of its functionality. They are also related to some scenario of execution or operation phase. Sequence diagrams can be useful at a high abstraction level, when the designer wants to understand the global pattern of interaction between system parts, and at a low abstraction level, when details of the interaction between (more concrete) parts have to be described. Because they represent the partial behaviour of multiple system parts, they are suited as requirements for testing and verification, but are less suited for automatic code generation.

UML 2.0 introduces capabilities to define interaction fragments (smaller sequence diagrams) and to combine them together to form more complex sequence diagrams. These capabilities include (conditional) branching of interaction fragments, and references to interaction occurrences that can be defined separately. The gates concept has been introduced to connect different interaction fragments. The most important benefit of these new capabilities is the possibility of structuring sequence diagrams in terms of smaller fragments, increasing in this way the readability of sequence diagrams, mostly of importance in the case of complex diagrams. These capabilities make it possible to define behaviours more concisely and completely, approaching in this way the purpose and expressiveness of state charts.

Timing diagrams have been introduced in UML 2.0 to represent state changes and conditions on a timeline. Timing diagrams are expected to be useful for systems that have stringent timing constraints. UML 2.0 also defines the so-called interaction overview diagram, which allows sequence diagrams to be combined in the scope using the operator of activity diagrams. This implies that more alternative scenarios can be represented in a single diagram. Interaction overview diagrams bring interaction diagrams closer to activity diagrams (see section 4.2.3), by allowing them to represent behaviours in a more complete way.

4.2.2 Collaboration diagrams

The purpose of collaboration diagrams is similar to the purpose of the sequence diagrams (define behaviour through scenarios), but collaboration diagrams put more stress on the collaboration itself, i.e., on the roles participating in the collaborations and the associations between these roles. A collaboration diagram shows these roles and associations and plots an ordered set of directed message exchanges on the associations, in order to denote a specific interaction sequence.

A collaboration diagram shows classifier roles (or instances), their associations (or links), and the messages (or stimuli) they exchange. It contains the same information as sequence diagrams, but it represents the associations (links) explicitly.

Collaboration diagrams can play the same roles in the development trajectory as sequence diagrams. They only differ in that collaboration diagrams are not really suitable to represent complex interaction sequences, since the reader is forced to follow numbered messages throughout the diagrams to understand the sequences being described. The main benefit of collaboration diagrams is the combined representation of structural and behavioural aspects in a single diagram.

In UML 2.0 collaboration diagrams have been renamed to communication diagrams. They correspond to simple sequence diagrams, i.e., sequence diagrams that do not use the structuring capabilities that have been introduced in UML 2.0.

4.2.3 Activity diagrams

Activity diagrams in UML 1.5 are special cases of statechart diagrams. In activity diagrams one represents activities and their relationships, which allows one to define a behaviour that describes processes or workflows. The activities themselves may consist of actions, which are (smaller) tasks that are internal to the activity. Activity diagrams also include some capabilities to define decisions and merging of execution flows, and synchronisation states. Swim lanes can be used to partition an activity diagram, e.g., in terms of the roles that are responsible for the different activities.

An activity diagram represents an activity graph, which is a variant of a state machine. In activity graphs one defines activities (action states) and transitions triggered by the completion of these activities.

Most (software) development methodologies prescribe that business processes should be explicitly specified in the initial development steps. Particularly in the case of the Unified Process [9], these business processes can be used to specify the (behaviour of) use cases.

At the level of business models it is often necessary to model the processes that have to be performed by the business without necessarily assigning parts of these processes to some specific people, departments or software applications. These models allow business architects to reason about the procedural steps of these business processes, abstracting from how these steps are supposed to be performed. In the course of the implementation trajectory, choices have to be made concerning the allocation of these procedural steps to physical or logical entities.

Models of business processes normally consist of related activities that have to be performed in these processes. There are many alternative techniques that are suited for modelling activities and their relationships. In UML 1.5, activities can be modelled using activity diagrams.

In UML 2.0 activity diagrams are no longer a special case of statechart diagrams. Activity graphs in UML 2.0 have a semantics that is closer to Petri Nets, making them more suitable for the representation of business processes.

4.2.4 Statechart diagrams

Statecharts can be used to represent the behaviour of an object instance or other entities such as use cases, actors, subsystems, etc. Statecharts are used to define behaviour in terms of reactions to stimuli (discrete events). A statechart defines a collection of states and state moves, such that whenever a stimulus occurs, the behaviour performs some actions and transitions (state moves). Since behaviours defined using statechart diagrams can get rather complex in the case of complex behaviours, UML has some additional capabilities to structure these diagrams, like sub-states and sub-machines.

A statechart diagram represents a state machine. In essence a state machine is a graph that consists of states and state transitions triggered by events. A state may contain a list of internal transitions, which consist of internal actions or activities to be performed by the state machine while in this state. An event is some occurrence that may trigger a state transition. Events can be either the change of some Boolean value, the expiration of a timeout, an operation call or a signal. A transition is triggered by an event. A (simple) transition is a relation between two states (*state1* and *state2*), and defines that whenever the state machine is in *state1* and the event that triggers the transition is processed, the state machine moves to *state2*. Only one event is evaluated at a time and it is either discarded if it does not trigger any transition, or it triggers only one transition (interleaving semantics). A transition is said to be *fired* whenever it is performed (terminology derived from Petri Nets).

During the development of a (software) system one has to define the logical parts (objects) of the system and specify their behaviour. Statechart diagrams allow one to specify completely the behaviour of the logical parts of a system, as opposed to the partial specification through interactions supported by sequence diagrams and collaboration diagrams. A criticism on statechart diagrams is that although it is suitable for the specification of behaviours that are relatively close to the implementation code, its interleaving semantics makes it less suitable to specify the behaviour of logical parts that may be decomposed and distributed, i.e., behaviours at higher abstraction levels. These more abstract behaviours can be useful for early analysis (e.g., through simulation) and since they constitute a statement on functional requirements, they can also be used for conformance assessment (verification).

In UML 2.0 interfaces can own a (protocol) state machine. Entry and exit points and terminate pseudo-states have been introduced to facilitate the reference to state machines and to improve structuring, allowing reusability of state machines. UML 2.0 allows a state list to be represented by a single state symbol. State machine extension allows one to reuse the definition of a state machine and extend it with additional states and transitions. A state machine may own other state machines, which can be referenced from its internal states.

4.2.5 Action semantics

In UML, an action is a fundamental unit of executable functionality. Until UML 1.4, actions in an activity could only be defined as strings, typically an action-expression added to a transition definition. This implies that no standard semantics existed for actions, which has been a major obstruction, amongst other, for the interchange of information between simulation tools [21].

The Action Semantics initiative has been started to define more precisely the meaning of actions in UML; results of this initiative have been incorporated already in the UML 1.5 specification [18]. In the UML 1.5 specification, a package Action has been added to define action semantics, with minor needs for readjustment in the rest of the language. UML 2.0 is built upon UML 1.5 for the behaviour part, i.e., the action semantics work done in UML 1.5 has been reused in UML 2.0. In the remaining of this section we refer to UML 2.0 and how it handles action semantics.

From an abstract point of view, an action is a fundamental unit of executable functionality in an activity that contains the action. The execution of an action implies a transformation or processing in the modelled system. An action may have sets of incoming and outgoing activity edges, through which it gets its input and delivers its output values, respectively. Incoming and outgoing edges define the control and data flows that determine whether an action is allowed to be performed. The completion of an action may enable the execution of other actions that depend on this action. Actions may have pre- and post-conditions. Streaming parameters allow actions to start generating outputs while consuming inputs.

The different action types are used to define the semantics of individual actions. The following types of actions have been identified:

- Invocation actions: operation calls and the sending of signals, either to specific targets or broadcasting to potential targets;
- Read write actions: creation and destruction of objects, reading and modifying the values of variables and structural features (e.g., attributes) and creation and destruction of links;
- Computation actions: transformation of a set of input values to a set of output values by invoking a function.

Some additional read-write actions defined in UML 2.0 concern reading instances of a classifier, reading links of an association, determining the classifier of an instance and determining the association of a link. These specific actions allow one to specify system with introspective or reflexive capabilities, since they allow the behaviour to reason about the structure of the system itself. In order to completely define the semantics of events in a state machine, the acceptance of an event is defined as a specific form of action. The raising of an exception is also defined as a specific form of action.

The definition of action semantics makes it possible for tool vendors to develop tools for simulation and verification of behaviour specifications based on standard semantics. Action semantics has also enabled initiatives that aim at defining executable UML specifications, for simulation or even for prototyping. Executable UML [20] is an example of such an initiative.

4.3 Language extensibility

In the context of the MDA, an important characteristic of UML is its extensibility capabilities. UML is currently being positioned as a general purpose language that is expected to be customized for a variety of domains, platforms and methods [16].

A mechanism called profiling is used to enable lightweight extensions of the language. UML profiles extend the UML metamodel by specializing elements of the metamodel. The use of UML profiles enables the reuse of UML's notation and tools.

In case customization requirements exceed the capabilities offered by profiles, new languages may be defined via MOF metamodels. MOF 2.0 metamodels are being designed as instances of a subset of the UML 2.0, so that it will be possible to represent them using UML class diagrams.

5. FINAL REMARKS

Technologies such as Web Services and component models have been advocated as silver bullets for the service-oriented development of e-applications. However, if we are to draw any lessons from the past, then the most important would be: there are no (lasting) one-fits-all technology solutions, and therefore technologies will always differ and evolve. Direct mappings onto whatever specific technology, independently of how popular this technology may be at a particular point in time, will lead to inflexible systems: it results in technology lock-ins and hinders interoperability, portability and integration when, inevitably, the technology changes or new technologies enter the scene. MDA has been introduced against this background. The MDA approach separates business (computation-independent) models, computation (platform-independent) models and platform models, enabling reuse of design artefacts at any of these levels, thereby providing possibilities for better return on investment. In addition, by defining transformations between these models, development projects can be done in a shorter time span and with higher quality. Not surprisingly, MDA led to a massive attention for models, modelling languages and transformations, primarily concentrated around UML.

UML has been developed for the specification of object-oriented systems and, as such, the concepts and abstractions used in UML facilitate the development of software using object-oriented implementation technologies. This yields a language that can potentially be translated into executable code or even executed directly, supporting the last stages of a model-driven development trajectory.

Despite that UML behaviour specification capabilities may be suitable to support final stages of a model-driven development trajectory, the use of UML for behaviour specification in earlier stages of a model-driven development trajectory should be further investigated. Recently, constructs for component-based development have been incorporated into UML 2.0, allowing the concept of component to be recursively applied through the development trajectory. Nevertheless, the refinement of behaviour specifications is a deficiency of UML's behaviour representation capabilities. Still lacking is a notion of behaviour conformance in order to relate behaviours defined at a high-level of abstraction and the refined realizations of these behaviours [4]. Consequently, we cannot formally assess the correctness of component compositions. Activity diagrams used to express behaviour in an integrated perspective, e.g., for the purpose of business modelling, are not related by refinement to statecharts that distribute responsibilities of a business process to specific services and components that support this business process.

In the context of MDA, much effort has been invested in language definition and extension mechanisms, metamodelling, model transformation specification and tool support. The study of platform-independence, however, has been somewhat overlooked. We have been striving to address this in our research [1, 2], by providing guidelines for the selection of abstraction criteria and modeling concepts for platform-independent modeling. Further research is necessary in order to define criteria to ensure the beneficial exploitation of the PIM-PSM separation of concerns adopted by MDA.

REFERENCES

1. Almeida, J.P.A., Sinderen, M. van, Ferreira Pires, L. and Quartel, D.: A systematic approach to platform-independent design based on the service concept. In: *7th IEEE Intl. Enterprise Distributed Object Computing (EDOC) Conference*, Sept. 2003, pp. 112-123.
2. Almeida, J.P.A., Sinderen, M. van, Ferreira Pires, L. and Wegdam, M.: Handling QoS in MDA: a discussion on availability and dynamic reconfiguration. In: *Workshop on Model Driven Architecture: Foundations and Application (MDAFA) 2003*, CTIT Technical Report TR–CTIT–03–27, University of Twente, The Netherlands, June 2003, 91-96.
3. De Farias, C.R.G.: *Architectural Design of Groupware Systems: a Component-Based Approach.* PhD thesis, University of Twente, Enschede, the Netherlands, 2002.
4. Dijkman, R.M., Quartel, D., Ferreira Pires, L. and Sinderen, M. van: An Approach to Relate Viewpoints and Modeling Languages. In: *7th IEEE Enterprise Distributed Object Computing (EDOC) Conference*, Sept. 2003, pp. 14-27.
5. Harel, D. and Rumpe, B.: *Modelling Languages: Syntax, Semantics and All That Stuff.* Technical Report, The Weizmann Institute of Science, Rehovot, Israel, MCS00-16, 2000.
6. Guizzardi, G., Ferreira Pires, L. and van Sinderen, M.: On the role of Domain Ontologies in the Design of Domain-Specific Visual Languages. In: *2nd Workshop on Domain-Specific Visual Languages, ACM OOPSLA*, 2002.
7. International Telecommunications Union (ITU): *Open Distributed Processing Reference Model. Part 1 - Overview.* ITU Rec. X.901 | ISO/IEC 10746-1, Geneva, 1997.
8. International Telecommunications Union (ITU): *SDL Formal Semantics Definition.* ITU Rec. Z.100, Annex F, Geneva, 2000.
9. Jacobson, I., Booch, G. and Rumbaugh, J.: *The unified software development process.* Addison Wesley, USA, 1999.
10. Object Management Group: *MDA Guide Version 1.0.* May 2003.
11. Object Management Group: *Meta Object Facility (MOF) 2.0 Core Proposal.* April 2003.
12. Object Management Group: *Meta Object Facility (MOF) Specification 1.4.* April 2002.
13. Object Management Group: *Model Driven Architecture (MDA).* July 2001.
14. Object Management Group: *OMG Unified Modelling Language Specification, version 1.4.* September 2001.
15. Object Management Group: *OMG XML Metadata Interchange (XMI) Specification*, Version 1.2, January 2002.
16. Object Management Group: *UML 2.0 Infrastructure Specification.* September 2003.
17. Object Management Group: *UML 2.0 Superstructure Specification.* August 2003.
18. Object Management Group: *Unified Modeling Language Specification, version 1.5.* March 2003.
19. Quartel, D.A.C.: *Action relations. Basic design concepts for behaviour modelling and refinement.* CTIT Ph.D-thesis series, no. 98-18, University of Twente, Enschede, The Netherlands, 1998.
20. Mellor, S.J. and Balcer, M.J.: *Executable UML. A foundation for the Model-Driven Architecture.* Addison-Wesley, 2002.
21. Mellor, S.J., Tockey, S., Arthaud, R. and Leblanc, P.: *Software-platform-independent precise action specification for UML.* White paper. http://www.projtech.com

GPSR Compliance
The European Union's (EU) General Product Safety Regulation (GPSR) is a set of rules that requires consumer products to be safe and our obligations to ensure this.

If you have any concerns about our products, you can contact us on

ProductSafety@springernature.com

In case Publisher is established outside the EU, the EU authorized representative is:

Springer Nature Customer Service Center GmbH
Europaplatz 3
69115 Heidelberg, Germany

www.ingramcontent.com/pod-product-compliance
Lightning Source LLC
LaVergne TN
LVHW011900200926
848484LV00002B/493
* 9 7 8 1 4 0 2 0 7 7 9 5 1 *